D0205740

ECOLOGICAL CONCEPTS

THE CONTRIBUTION OF ECOLOGY
TO AN UNDERSTANDING OF
THE NATURAL WORLD

ECOLOGICAL CONCEPTS

THE CONTRIBUTION OF ECOLOGY TO AN UNDERSTANDING OF THE NATURAL WORLD

The First Jubilee Symposium to Celebrate
the 75th Anniversary of the British Ecological Society,
University College, London, 12–13 April 1988
Published as the 29th Symposium of the Society

EDITED BY

J. M. CHERRETT
School of Biological Sciences,
University College of North Wales,
Bangor

WITH THE ASSISTANCE OF

A. D. BRADSHAW
Department of Botany
University of Liverpool

F. B. GOLDSMITH
Department of Botany,
University College London

P. J. GRUBB
School of Botany
University of Cambridge

J. R. KREBS
Edward Grey Institute
of Field Ornithology,
Oxford

BLACKWELL SCIENTIFIC PUBLICATIONS
OXFORD LONDON EDINBURGH
BOSTON MELBOURNE

and published for them by
Blackwell Scientific Publications
Editorial offices:
Osney Mead, Oxford OX2 0EL
(*Orders:* Tel. 0865 240201)
8 John Street, London WC1N 2ES
23 Ainslie Place, Edinburgh EH3 6AJ
3 Cambridge Center, Suite 208
Cambridge, Massachusetts 02142, USA
107 Barry Street, Carlton
Victoria 3053, Australia

First published 1989

Set by Times Graphics, Singapore;
Printed and bound in Great Britain
at the University Press, Cambridge

DISTRIBUTORS

USA
Publishers' Business Services
PO Box 447
Brookline Village
Massachusetts 02147
(*Orders:* Tel. (617) 524-7678)

Canada
Oxford University Press
70 Wynford Drive
Don Mills
Ontario M3C 1J9
(*Orders:* Tel. (416) 441-2941)

Australia
Blackwell Scientific Publications
(Australia) Pty Ltd
107 Barry Street
Carlton, Victoria 3053
(*Orders:* Tel. (03) 347-0300)

British Library
Cataloguing in Publication Data

British Ecological Society, *Symposium.*
(*29th: 1988: University College, London*)
Ecological concepts.
1. Ecology
574.5

ISBN 0-632-02569-7
ISBN 0-632-02571-9 pbk

Library of Congress
Cataloging-in-Publication Data

British Ecological Society. Symposium
(29th: 1988: University College,
London, England)
Ecological concepts: the
contribution of ecology to an
understanding of the natural
world/edited by J.M. Cherrett with the
assistance of A.D. Bradshaw . . .
[*et al.*].
p. cm.
'The first jubilee symposium to
celebrate the 75th anniversary of the
British Ecological Society, University
College, London, 12–13 April 1988,
published as the 29th symposium of
the Society.'
Includes index.
ISBN 0-632-02569-7. —
ISBN 0-632-02571-9 (pbk.)
1. Ecology—Congresses.
I. Cherrett, J. M.
II. Bradshaw, A. D. (Anthony
David) III. Title.
QH540.B75 1988 89-34749
.574.5—dc20 CIP

CONTENTS

FOREWORD

The British Ecological Society was founded in 1913, the first national ecological society in the world. For our jubilee 75 years later, we decided that the time was ripe for taking stock of the achievements of our subject, and looking forward toward new developments. Accordingly, the Society held two complementary symposia in 1988. The first was planned as an assessment of the contribution of ecology to understanding the natural world under the title *Ecological Concepts*; it was held at University College London, the venue of the inaugural meeting of the society, on its anniversary date, 12–13 April. The second was focused on the future, and called *Toward a More Exact Ecology*, aimed at considering the most fruitful current approaches and technologies, and determining the major obstacles and likely profitable lines of advance; it was held at St Catherine's College, Oxford, on 13–15 September. The two symposium volumes thus represent a vision of the ecological community's understanding of the past, present and future of the subject; it is hoped that both will be of value in the planning of future research.

R. J. BERRY
President 1988–89

PREFACE

This jubilee symposium had its origins during preparations for a workshop on the teaching of ecology organized by the British Ecological Society in July 1985. Professor A. D. Bradshaw, then President of the Society, was persuaded to give the opening lecture entitled 'What is Ecology?' When pressed for more details, he tentatively volunteered to deal with the main principles of the discipline, and immediately provoked somewhat cynical enquiries as to what he thought these might be.

Thus, the idea was born that the 75th anniversary of the society's foundation might be an appropriate time to examine this question seriously, and to review the intellectual contribution ecology has made to our understanding of the natural world. The way in which the opinions of our members was sought is the subject of the first, introductory chapter, and will not be repeated here.

Once the key concepts had been identified, expert authors were chosen by the organizing committee (A. D. Bradshaw, J. M. Cherrett, F. B. Goldsmith, P. J. Grubb and J. R. Krebs) to deal with each of the ten topics in this volume (originally twelve in the symposium). Each author was asked to contribute an essay, so this volume must be seen as a collection of personal views. It is hoped these will stimulate readers to reconsider, expand or modify this ecological inventory, so producing deeper insights for our 150th anniversary.

I am most grateful to the members of the organizing committee for their guidance in planning the meeting, choosing the speakers and organizing referees. An especial debt of gratitude is owed to Dr F. B. Goldsmith for his work as local organiser of the Symposium, which was held at University College London, the site of the inaugural meeting of the society, on its exact anniversary. The task which he and his coworkers undertook was particularly onerous, as the numbers attending exceeded our wildest expectations.

Finally, I would like to thank the contributors for accepting their briefs with good grace, and for the thought-provoking contributions they have produced.

J. M. CHERRETT
Bangor

1 KEY CONCEPTS: THE RESULTS OF A SURVEY OF OUR MEMBERS' OPINIONS

J. M. CHERRETT
School of Biological Sciences, University College of North Wales, Bangor, Gwynedd LL57 2UW, UK

INTRODUCTION

In celebrating the 75th Anniversary of the founding of the British Ecological Society, we are also close to celebrating 100 years of British ecology, as Sheail (1987) has pointed out, and now seems an appropriate time to try to assess the contribution which ecology has made to our understanding of the natural world. A century of research has produced a formidable volume of factual data, but some ecologists suspect, as did Miall (1897) (reported in Elton 1927) about natural history, that it may be '... encumbered by multitudes of facts which are recorded only because they are easy to record' and there is unease that we still do not have an equivalent to the Newtonian Laws of Physics, or even a generally accepted classificatory framework. Southwood (1977) put this clearly when he wrote:

> In some ways I think we may see ourselves at a similar point to the inorganic chemist before the development of the periodic table; then he could not predict, for example, how soluble a particular sulphate would be, or what was the likelihood of a particular reaction occurring. Each fact had to be discovered for itself and each must be remembered in isolation.

More recently, McIntosh (1985), in an excellent review of the development of ecological ideas, has written: 'Although theoretical ecology is one of the more highly touted aspects of recent ecology, it is not easy to find consensus among ecologists about established theories, their basic postulates, sources or even their names or pseudonyms.'

In our attempt to assess the contribution of ecology, it seemed sensible to invite our entire membership to give us their views. Wishing to avoid the confusion which surrounds the definitions of such terms as generalization, hypothesis, law, postulate, principle or theory, we decided simply to ask our members the broader question, what they thought were the most important concepts (or ideas, or general notions) in ecology. In conducting such a survey, we had three objectives:

(i) To provide a framework for a symposium, in which distinguished speakers would be invited to discuss the origins, status and validity of these key ideas.

(ii) To provide a check list of the ecological concepts which a broad, international cross-section of ecologists currently believe to be of the greatest significance in our discipline. This might at least help educationalists concerned with curriculum development in ecology, who feel that the main principles should be covered, but who are uncertain what they are.

(iii) To establish a 1987 benchmark of ecological opinion, against which similar surveys in the future might be compared, to establish growth points and changed perceptions in the subject.

We anticipated that different ecologists would hold different views, and it seemed a good opportunity to see if such differences were purely individual, or were measurably influenced by such factors as nationality, age, training or employment. Whilst not a primary objective of the survey, it was thought that such research might be illuminating.

METHODOLOGY

We considered two contrasting ways of assessing members' views. The first was to ask each member to list their ten key concepts in order of importance. An open, unguided list of this sort has the advantage of eliciting uninfluenced opinions. It has the disadvantages of presenting each member with a time-consuming and somewhat daunting task which might produce a low rate of response, and of making the analysis of the replies very difficult. Someone has to classify a near continuum of concepts into discrete categories, a subjective and a time-consuming task, especially with a membership of over 4300. The second way was to produce a prepared questionnaire containing a large number of concepts from which members would be invited to select their top ten. A closed, guided list of this sort has the advantages of being less time-consuming both to fill in and to analyse, but the disadvantage of limiting the field to the prejudiced choice of its compiler.

A compromise was eventually adopted, which it was hoped would maximize advantages and minimize disadvantages in the two approaches. In a pilot survey, 148 ecologists selected by the symposium steering committee from twenty-four countries were invited to submit an unguided list in rank order of their ten most important concepts. From the seventy usable replies obtained from fifteen countries, a grand list of the fifty most frequently mentioned was complied. It was used to produced the structured questionnaire sent to all our members, with the November 1986 issue of the Bulletin, in which was a short explanatory article. A deadline for replies was set for 3 months later.

Inevitably, editorial bias was involved in preparing the list of fifty concepts included in the questionnaire, from the many, differently worded concepts listed in the replies; 'The problem of sorting out linearly independent ideas from the continuum . . . received', as one correspondent put it. Very general concepts such as the 'interdependence of organisms' were omitted, as they subsumed many other concepts and would not advance the three objectives we set ourselves at the outset. However, we felt confident that the list which eventually appeared in the questionnaire was based on a broader range of opinions than any individual could have produced. Because of time constraints, the twelve concepts selected as topics to be considered at the symposium were based on those most frequently referred to in this pilot survey.

Members were asked to score their top ten concepts in order of importance from ten, the most important, to one the least. They were also given an opportunity to add any concepts not covered by the list, and to rank them within their choice. The questionnaire included questions on the subject of the member's initial degree, the length of time since graduation that he or she had been professionally concerned with ecology, and the current area of employment. The country in which the member worked was noted from the address or postmark, and all respondents were given the opportunity to remain anonymous.

RESULTS

Sample of members replying

There were 645 questionnaires returned, a response rate of 14.7%. The geographical distribution of respondents was classified into the nine areas used by FAO in their Production Year Books, and the data are summarized in Table 1.1 Not surprisingly 71% were from the UK. The 644 respondents giving information on the subject of their first degree were classified under nine headings (Table 1.2), zoology, botany and biology representing the predominant kinds of training. The years of postgraduate experience possessed by the 641 respondents who supplied information were classified in 10-year intervals (Table 1.3). The mode lay in the 10–20-year band. Finally, the 643 respondents supplying information on their current employment were divided into eight categories (Table 1.4). The overwhelming majority were employed in further and higher education, scientific research coming second. Many respondents pointed out that they were employed to undertake both activities, but whenever advanced level teaching was mentioned, we placed them in the

TABLE 1.1. Geographical location of respondents to the questionnaire

Region	Number of respondents (out of 645)	Percentage
United Kingdom	458	71
North America	79	12
Europe	49	8
Oceania	41	6
Africa	9	1
Asia	5	1
Central America	4	1
South America and USSR	0	0

TABLE 1.2. Initial training of respondents to the questionnaire

Subject of first degree	Number of respondents (out of 644)	Percentage
Zoology	184	29
Botany	139	22
Biology	119	18
Ecology	71	11
Geography	28	4
Applied biological sciences	25	4
Environmental science	20	3
Agriculture	18	3
Miscellaneous	40	6

TABLE 1.3. Length of postgraduate experience of respondents to the questionnaire

Years since graduation	Number of respondents (out of 641)	Percentage
< 10	213	33
10–20	254	40
21–30	118	18
31–40	41	6
> 40	15	2

TABLE 1.4. Current employment of respondents to the questionnaire

Area of employment	Number of respondents (out of 643)	Percentage
Further and higher education	257	40
Scientific research (excluding above)	186	29
Conservation	65	10
Agriculture and forestry	26	4
Secondary education	16	2
Planning	14	2
Consultancies	13	2
Miscellaneous	66	10

further and higher education category. Although the most characteristic respondent was a UK zoologist with between 10 and 20 years of postgraduate experience, currently employed in further or higher education, the profiles show that ecologists from all over the world, with a wide range of experience, backgrounds and jobs took the trouble to answer.

Most popular concepts

Only thirty-six of the 645 respondents (5.6%) added topics of their own to the fifty topics included in the questionnaire. In all, 236 recognizably different concepts were suggested, 210 from the people originally approached in the pilot survey, and an additional twenty-six from members answering the questionnaire. Table 1.5 shows the number of respondents choosing each of the fifty concepts, the mean score it was given, which is used for ranking, and its standard deviation. As the standard deviations show, each mean score had considerable variability, all fifty concepts obtaining a minimum score of zero, and forty-six attaining the maximum score of ten. Of the four which did not obtain ten, seven was the lowest maximum recorded. The ecosystem, mentioned by 69% of respondents, emerges as the most popular concept by a wide margin.

Effects of background on choice of concepts

The choice of ten concepts, ranked in order of importance, can be taken as some sort of indication of what each respondent thinks ecology should be all about. This raises interesting questions concerning the degree of consensus shown, and whether differences in view are solely individual, or reflect schools of thought. Also, to what extent can choices be

TABLE 1.5. The most important fifty concepts in ecology

Rank	Concept	Score* from 645 respondents $\bar{X}$	SD	No. choosing topic (out of 645)
1	The ecosystem	5.18	4.50	447
2	Succession	2.98	3.38	347
3	Energy flow	2.56	3.48	274
4	Conservation of resources	2.29	3.41	257
5	Competition	2.23	3.10	268
6	Niche	2.14	3.20	245
7	Materials cycling	2.13	3.13	243
8	The community	1.88	3.32	189
9	Life-history strategies	1.88	2.94	238
10	Ecosystem fragility	1.76	3.13	194
11	Food webs	1.73	2.98	194
12	Ecological adaptation	1.71	3.03	199
13	Environmental heterogeneity	1.57	2.94	170
14	Species diversity	1.47	2.75	176
15	Density–dependent regulation	1.43	2.78	172
16	Limiting factors	1.34	2.67	166
17	Carrying capacity	1.27	2.52	162
18	Maximum sustainable yield	1.24	2.55	157
19	Population cycles	1.17	2.50	141
20	Predator–prey interactions	1.14	2.40	148
21	Plant–herbivore interactions	1.10	2.37	148
22	Island biogeography theory	0.99	2.27	132
23	Bioaccumulation in food chains	0.84	2.11	113
24	Coevolution	0.82	2.12	115
25	Stochastic processes	0.78	2.12	99
26	Natural disturbance	0.77	2.18	83
27	Habitat restoration	0.75	2.22	89
28	The managed nature reserve	0.69	2.06	92
29	Indicator organisms	0.68	1.83	108
30	Competition & the conditions for species exclusion (Gause)	0.65	1.99	77
31	Trophic level	0.61	1.89	73
32	Pattern	0.57	1.93	65
33	*r* and *K* selection	0.53	1.69	79
34	Plant animal/coevolution	0.51	1.79	64
35	The diversity/stability hypothesis	0.49	1.70	65
36	Socioecology	0.47	1.74	57
37	Optimal foraging	0.45	1.60	62

TABLE 1.5. The most important fifty concepts in ecology (*Cont.*)

Rank	Concept	Score* from 645 respondents		No. choosing topic (out of 645)
		$\bar{X}$	SD	
38	Parasite–host interactions	0.45	1.54	61
39	Species–area relationships	0.42	1.46	69
40	The ecotype	0.40	1.57	51
41	Climax	0.36	1.52	45
42	Territorality	0.33	1.40	46
43	Allocation theory	0.32	1.50	34
44	Intrinsic regulation	0.29	1.35	36
45	Pyramid of numbers	0.27	1.29	30
46	Keystone species	0.25	1.21	33
47	The biome	0.23	1.32	23
48	Species packing	0.10	0.86	10
49	The 3/2 thinning law	0.05	0.51	10
50	The guild	0.05	0.44	9

*Score runs from 10, the most important concept, to 1 the least important. Concepts not ranked by the respondent were given a score of zero.

associated with any of the four background factors we assessed in the questionnaire? It was decided to investigate these questions using correspondence analysis as adapted for use in ecological problems. In essence, each response to the questionnaire was treated as a stand (or quadrat), each concept chosen was treated as a species, and the score it was given was treated like a measure of cover or abundance. Each of the four background factors (geographical location, first degree, years of postgraduate experience and current employment) were then treated as environmental variables (like pH or moisture content). Techniques of community analysis familiar to ecologists such as classification and ordination (Greig-Smith 1983) were then employed.

Figure 1.1 is a diagram of the divisions obtained in the first two levels of a TWINSPAN (Hill 1979a) classification of the fifty concepts. Only the indicator concepts are listed, together with the number of respondents falling into each category. All the default options were taken in the programme, except that the order of importance, recorded on a scale from 0 to 10, was reduced to six 'pseudospecies' cut levels of score 0, 1+2, 3+4, 5+6, 7+8, and 9+10. Figure 1.1 suggests a level one division into 432 ecologists characterized by *the ecosystem, conservation of resources, materials cycling, ecosystem fragility* and *energy flow* as

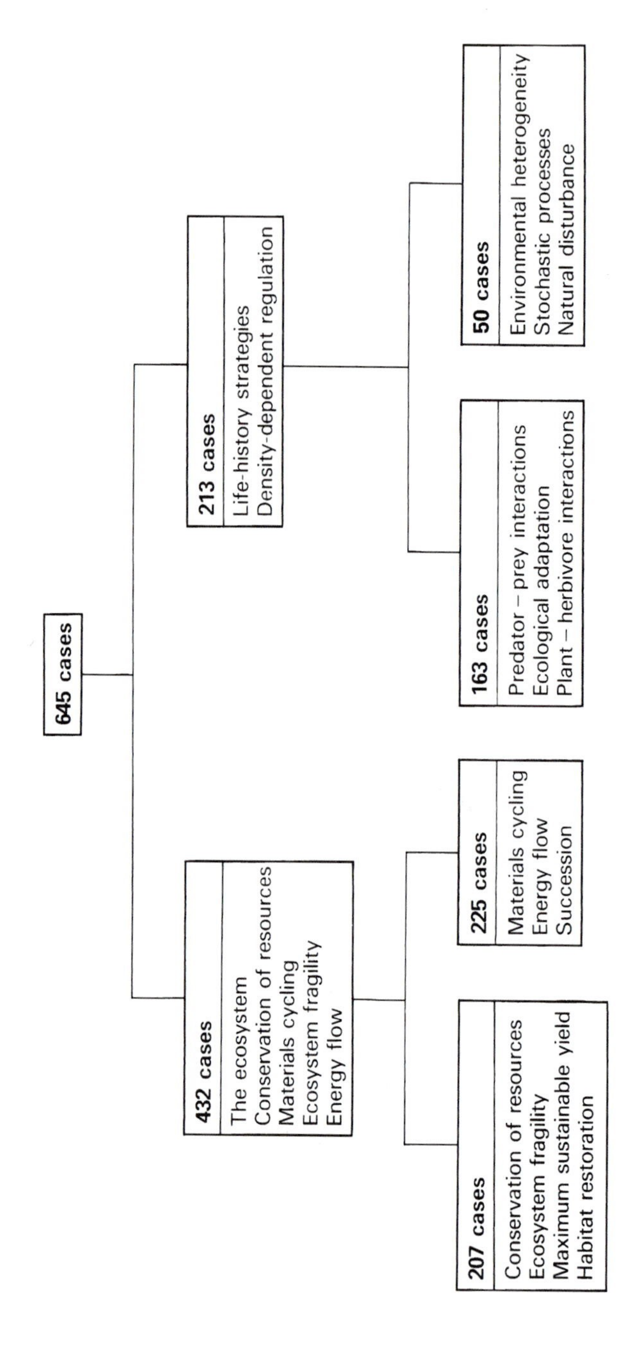

FIG. 1.1 Classification divisions obtained in the first two levels of a TWINSPAN analysis of the fifty concepts, based on 645 cases. Indicator concepts only are listed.

indicator concepts — we may perhaps call them 'practical holists', and a further 213 ecologists characterized by *life-history strategies* and *density-dependent regulation* — we may perhaps call these 'theoretical reductionists'. At level two, the 'practical holists' appear to be divided into 207 'practitioners' (characterized by *conservation of resources, ecosystem fragility, maximum sustainable yield* and *habitat restoration*) and 225 'more theoretical holists' (characterized by *materials cycling, energy flow* and *succession*). The 'theoretical reductionists' on the other hand seem to be divided into 163 advocates of 'biotic influence' (characterized by *predator–prey interactions, ecological adaptation* and *plant–herbivore interactions* and fifty 'external influence — environmentalists' (charac-

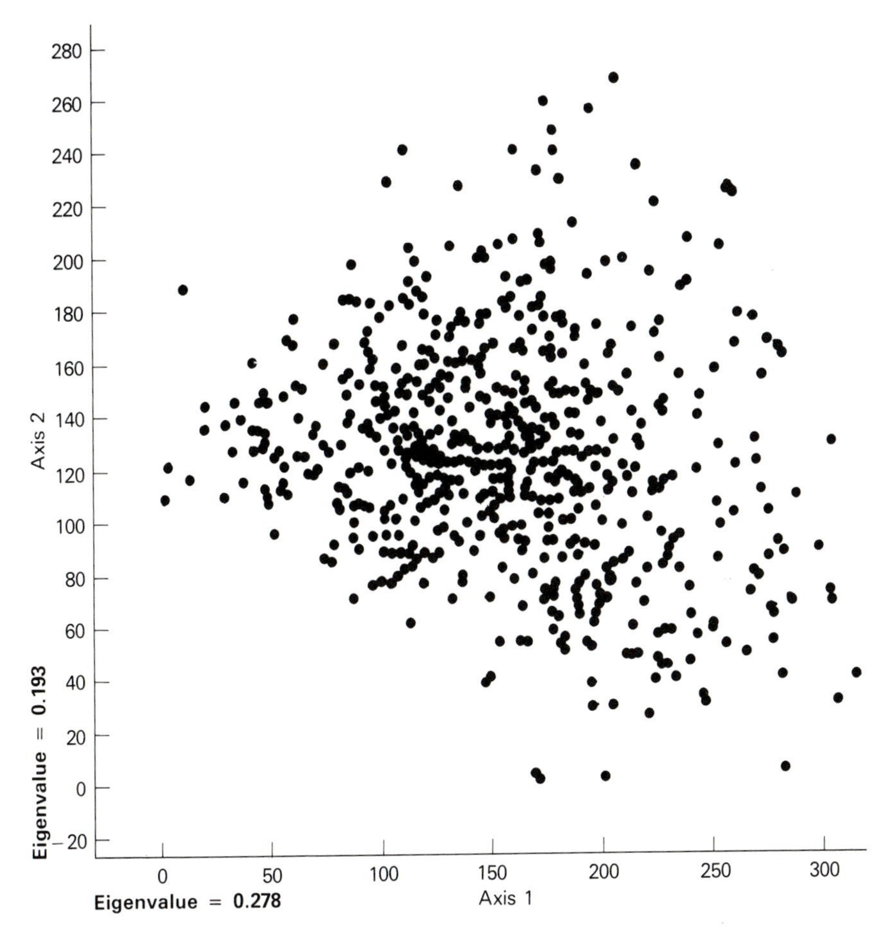

FIG. 1.2. Ordination of 645 responses (stands) to the questionnaire, based on the top ten concepts chosen. DECORANA analysis used.

terized by *environmental heterogeneity, stochastic processes* and *natural disturbance*).

The data were then subjected to a detrended correspondence analysis (Hill & Gauch 1980) using DECORANA (Hill 1979b), and the scatter of stand (individual respondent) points on the first two axes of the ordination (Fig. 1.2) suggests a continuum of variation, with no obvious discrete clustering. The scatter of species points (the fifty concepts) permits some interpretation of the nature of the two primary axes (Fig. 1.3). Axis 1 seems to reflect the 'practical holist'/'theoretical reductionist' gradient, with such concepts as *habitat restoration, biome, conservation of resources, the managed nature reserve* and *ecosystem fragility* at one end, and at the other, *species packing, allocation theory, the guild, 3/2 thinning law* and *optimal foraging*. Axis 2 although less clear appears to reflect the 'biotic influences'/'external influences (environmentalism)' gradient, with *territoriality, optimal foraging, pyramid of numbers, 3/2 thinning law, diversity/stability hypothesis* and *predator/prey* and *parasite/host interactions* at one end, whilst at the other are

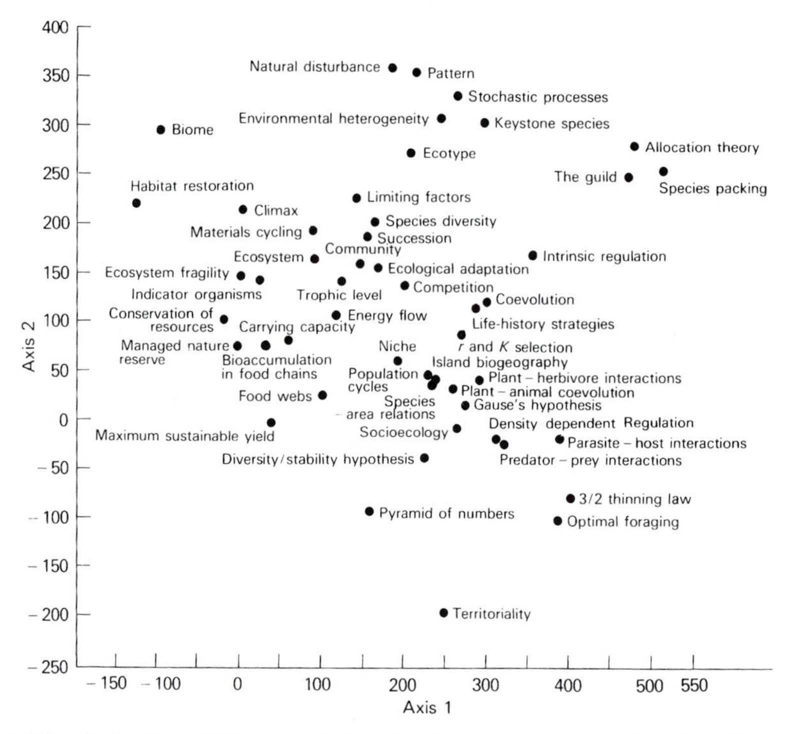

FIG. 1.3. Ordination of fifty concepts (species) in the questionnaire, based on 645 responses. DECORANA analysis used.

natural disturbance, pattern, stochastic processes, environmental heterogeneity, keystone species and *biome.*

With 645 stand points, investigating the effects of background factors, such as first degree, by labelling each stand point with the appropriate degree, and then looking for clusters of particular degrees along the two axes was not practical, and a method of simplifying the data was sought. Accordingly, the mean x and y axes positions were calculated for each category within a background factor (e.g. respondents with zoology degrees). To give some idea of the uncertainty of the mean positions, one standard error was plotted on either side, and the four interval points obtained (two on the x, and two on the y axes) were jointed by a curve to give a central position, surrounded by an area of uncertainty. Fig. 1.4 shows the effect of first degree. As expected of mean values, both axes are greatly truncated compared with the raw stand data in Fig. 1.2. Using our interpretation of the axes, those trained in geography tend to adopt a holist/environmental position in contrast to those trained in zoology who are reductionist/biotic. Botanists and ecologists adopt environmentalist perspectives, but are progressively more theoretical and reductionist, whilst biologists and applied biologists are reductionist, and increasingly concerned with biotic interactions.

Current employment (Fig. 1.5) suggests great similarity between respondents engaged in further and higher education and in research, reflecting perhaps the rather artificial distinction between the two noted

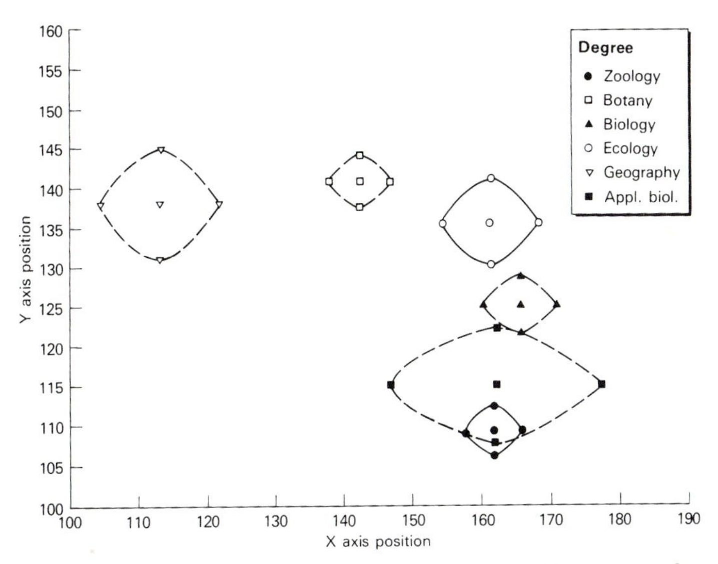

FIG. 1.4. Effect of first degree on the mean axis positions of 645 responses. One standard error is plotted around each mean.

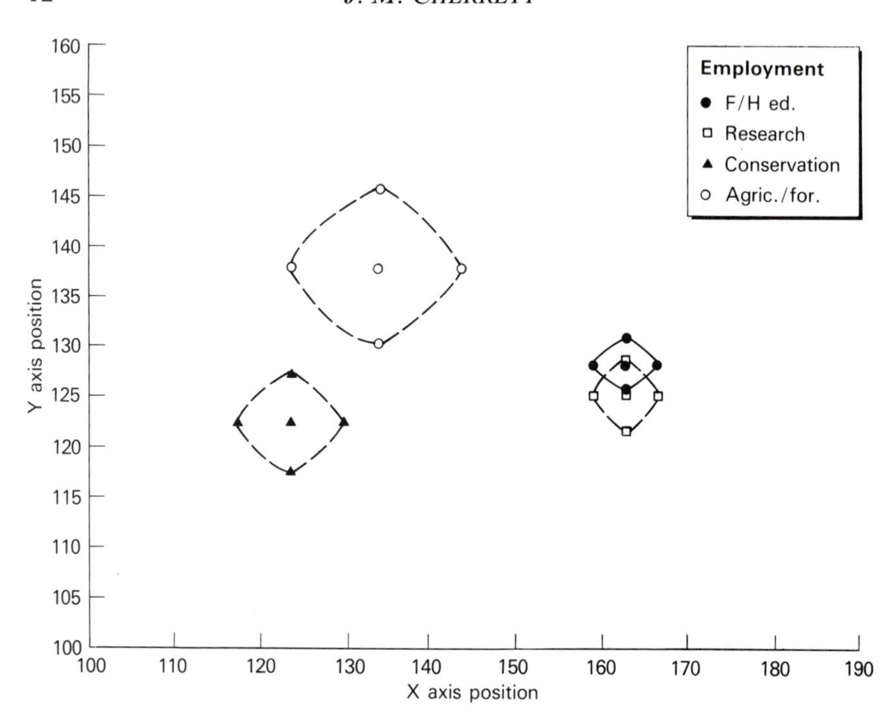

FIG. 1.5. Effect of current employment on the mean axis postions of 645 responses. One standard error is plotted around each mean.

earlier. Both are reductionist in outlook. By contrast, conservationists, and agriculturalists and foresters are holist, the conservationists being somewhat more concerned with biotic interactions. Years of post-graduate experience are clustered in the centre of the axes, and have no obviously interpretable effects (Fig. 1.6). The country of origin (Fig. 1.7) has a curious effect which is not easy to interpret. UK respondents (the narrow uncertainty band reflecting the large sample size) appear to be markedly more holist and biotic than everyone else. However, no attempt has been made in this anaylsis to look at the interaction effects of the four background measures. Such effects clearly exist, and a 2 × 2 contingency analysis shows that the UK membership contains a significantly greater proportion of geographers, zoologists, and conservationists, and a significiantly smaller proportion of further and higher educationalists and researchers than does the membership from the rest of the world.

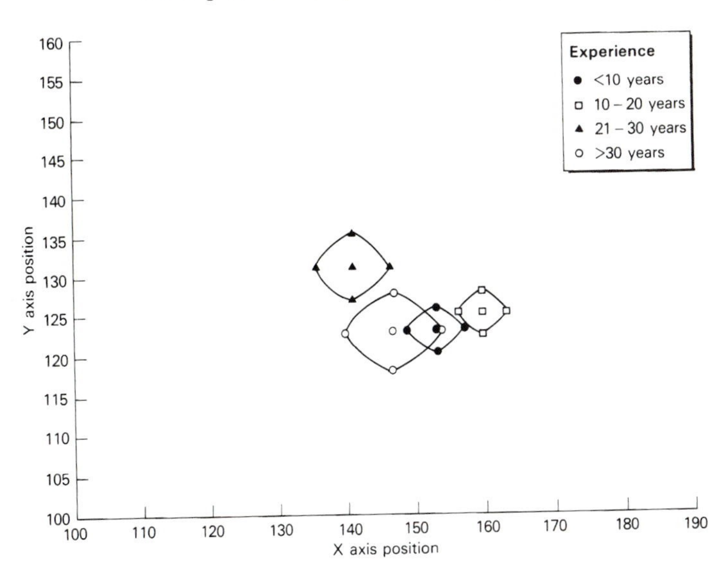

FIG. 1.6. Effect of years of postgraduate experience on the mean axis positions of 645 responses. One standard error is plotted around each mean.

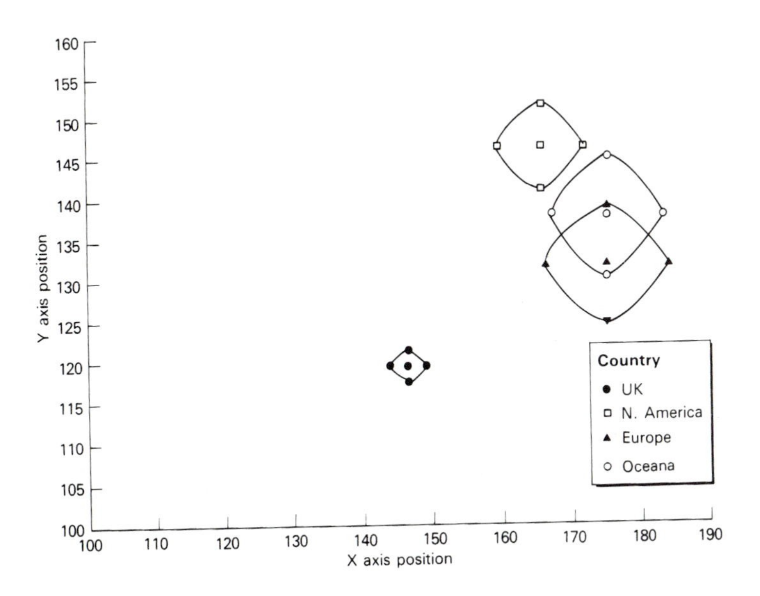

FIG. 1.7. Effect of country of origin on the mean axis positions of 645 responses. One standard error is plotted around each mean.

DISCUSSION

The response to the questionnaire (14.7% of the membership) was encouraging for a world-wide postal exercise, and although the sample drawn was undoubtedly biased (e.g. 17.4% of UK members replied compared with 10.7% of overseas members; $\chi^2(1) = 36.85, P < 0.001$), it was large and contained a wide spread of training, experience and employment (Tables 1.1 to 1.4). There is no reason therefore to doubt that the opinions expressed represent the views of a wide cross-section of professional ecologists.

As over 230 different concepts were mentioned in at least someone's top ten list, ecologists are not short of ideas, and as all fifty concepts in the questionnaire scored zero from some respondents, whilst forty-six were given maximum marks by at least one person, it is difficult to disagree with McIntosh (1985) who wrote: 'It is clear that ecologists are not unprincipled, but it is very difficult to find consensus among ecologists on what a principle is, or on specific principles.'

The most popular concepts to emerge are familiar, and perhaps reflect the conservatism which the 'lowest common denominator' effect of mass opinion polls generates. Thus few if any of the top ten concepts do not appear in early texts such as Allee *et al.* (1949). Nevertheless, it is the broad ideas such as *ecosystem*, *succession*, *energy flow*, *conservation* and *competition* which are regarded by our members as key ideas, lending some credence to Odum (1977) who wrote 'Among academic subjects, ecology stands out as one of the few dedicated to holism.'

Although there were wide individual differences in the concepts members thought important, the correspondence analysis showed a continuum of variation, implying that exclusive schools of ecological thought were not an important phenomenon. Nevertheless, training, employment and possibly country do exert a modest influence on mean axis position, and this may be important. The majority of respondents to our questionnaire were employed in further and higher education, and in scientific research (Table 1.4), both being categories which particularly value a theoretical reductionist approach (Fig. 1.5). If serving academic research is perceived as the *raison d'etre* of the Society, then its priorities may dominate policy to a degree not justified by the interests of the membership as a whole. It seems possible that in selecting the 148 ecologists to take part in the pilot survey, the steering committee may unconsciously have committed this error, as the top twelve concepts obtained, which formed the basis for this Symposium, were somewhat more theoretical and reductionist than those which ultimately emerged from the full-scale survey of the membership (Table 1.5). It is also interesting to note that the Society's publication *Ecology in the Curriculum* (Hale 1988) lists nineteen 'concepts and themes' in Appendix 3 as components of a school ecology course. If we assume *niche* is subsumed in *community structure*, only two of the top ten concepts in the present survey are omitted from the curriculum suggestions. However, these are *conservation of resources* and *ecosystem fragility*, ideas which perhaps are relevant to the education of children, the overwhelming majority of whom will not go on to become professional academic ecologists.

It will be interesting to see how our perceptions of key concepts change as the discipline develops — the subject perhaps for a study to mark the Society's 150th Anniversary! In the meantime, it is as well to keep in mind the warning issued by one of the respondents to the questionnaire who wrote: 'I am not willing to rank these "in order of importance" . . . surely if we have learned anything in ecology, it is to appreciate the multidimensional nature of our subject. It would seem to me to be pointless and even absurd to try now to force everything into one dimension of "importance".'

ACKNOWLEDGMENTS

I am grateful to the members of the Symposium Steering Committe (A.D. Bradshaw, F.B. Goldsmith, P.J. Grubb and J.R. Krebs) for advice on the formulation of the questionnaire; to S. Jones for advice on computing; to P.Greig-Smith for statistical advice and for reading the manuscript; to F.J. Cherrett for secretarial assistance; and to the members of the Society who took time to answer the questions.

REFERENCES

Allee, W. C., Emerson , A. E., Park, O., Park, T. & Schmidt, K. P. (1949). *Principles of Animal Ecology.* Saunders, Philadelphia.

Elton, C. (1927). *Animal Ecology.* Sidgwick & Jackson, London.

Greig-Smith, P. (1983). *Quantitative Plant Ecology*, 3rd edition. Blackwell Scientific Publications, Oxford.

Hale, M. (1988). *Ecology in the Curriculum*, 5–19. British Ecological Society, London.

Hill, M.O. (1979a). TWINSPAN: *A Fortran program for arranging multivariate data in an ordered two-way table by classification of the individuals and attributes.* Cornell University, New York.

Hill, M.O. (1979b). DECORANA: *A Fortran program for detrended correspondence analysis and reciprocal averaging.* Cornell University, New York.

Hill, M.O. & Gauch, H.G. (1980). Detrended correspondence analysis: an improved ordination technique. Vegetatio, **42,** 47–58.

McIntosh, R. P. (1985). *The Background of Ecology, Concept and Theory.* Cambridge University Press, Cambridge.

Odum, E. P. (1977). The emergence of ecology as a new integrative discipline. *Science,* **195,** 1289–1293.

Sheail, J. (1987). *Seventy-five Years in Ecology: the British Ecological Society.* Blackwell Scientific Publications, Oxford.

Southwood, T. R. E. (1977). Habitat, the template for ecological strategies? *Journal of Animal Ecology,* **46,** 337–365.

2 ECOSYSTEMS: FLUXES OF MATTER AND ENERGY

R. H. WARING
College of Forestry, Oregon State University, Corvallis, Oregon 97331, USA

INTRODUCTION

Members of the British Ecological Society, when asked to list ten important ecological concepts, ranked the 'ecosystem' concept as number 1 (see Chapter 1). The closely related concepts of *energy flow, conservation of resources, material cycling* and *ecosystem fragility* were also among the top ten. Were he here today, A.G. Tansley, would be pleased that his definition of an ecosystem as having both living and abiotic components has been so useful heuristically (Tansley 1935). The concept has been slightly refined, and perhaps best enunciated by Evans (1956), who suggested that those who study ecosystems are concerned with 'the circulation, transformation, and accumulation of energy and matter through the medium of living things and their activities'. Major (1969) provides a thorough review of the scientific development of the ecosystem concept.

The ecosystem concept is dimensionally undefined. An ecosystem may be a pond, a catchment basin, a biome, or the Earth's biosphere. Selecting the appropriate boundary is dependent on the problem and the time-scale. One lesson learned from evaluating nuclear fallout, pesticide residues, acid rain and global climatic change is that some of the most perplexing ecosystem problems operate at scales that cross political boundaries and extend beyond a single human generation.

Much of the educated public now accepts some implications of the ecosystem concept by assuming 'that everything really is connected to everything else, and someday, somewhere, somebody or something is going to pay for resource use'. The scientific literature backs the public's impression through documentation of how food chains accumulate toxins, materials are transferred through air and water, and life in its various forms, profits or pays for changes in the availability of critical resources.

Ecosystem ecologists have succeeded in identifying control points (energy gates) for fluxes of matter and energy that affect the stability and productive capacity of streams, fields, forests, and oceans. Systems theory (Von Bertalanffy 1950; Klir 1969), control theory (Ashby 1963), energy flow diagrams (Odum 1971) and simulation modelling (Patten 1971) have contributed to the identification. They provide today a general diagrammatic and computational framework for estimating difficult-to-measure fluxes and identifying the relative importance of processes (Gardner *et al.* 1981).

Ecosystem studies have changed from a descriptive to a predictive emphasis. This new predictive phase of the science is exciting but dangerous if model assumptions are inadequate. New combinations of environmental stresses and new combinations of organisms require that predictions be based on fundamentally sound principles.

In this paper I review four areas where fundamental insights now provide a more general basis for predicting ecosystem fluxes and for recognizing stability and instability. I will discuss: (i) hydrology, with emphasis on evapotranspiration; (ii) the carbon cycle, with emphasis on photosynthesis and net primary production; (iii) biogeochemistry, with emphasis on changing chemical equilibria; and (iv) ecosystem disturbance, with emphasis on resistance and resilience.

THE HYDROLOGICAL CYCLE

The hydrological cycle of the Earth is conceptually simple. The ocean is the major source of water, radiant energy is an evaporator, weather is a distributor, temporary storage occurs in ice fields, lakes and soils, extraction occurs through evapotranspiration, and liquid movement through surface and subsurface aquifers returns the water to the sea. Energy budgets and conservation of mass equations are principles borrowed from thermodynamics that help predict the expected forms and movement of water. Fundamental to these calculations is a gradient defining the potential rate of evaporation, condensation, or flow, and a series of resistances inhibiting the processes.

Meteorologists have defined the critical climatic variables necessary to predict the physical state of water and the potential gradient in evaporation. These variables are net solar radiation at the surface, temperature, humidity, precipitation, and wind speed. Whether the surface is a rock, a tree, bare ground or a pond affects the amount of water available and its resistance to movement (Fig. 2.1). Vegetation affects the resistance term in three significant ways: by the interception or absorption of water on its surfaces, by the geometry of the surfaces, and through direct biological control associated with deep rooting and leaf stomata.

The application of basic physics, championed by Monteith (1965, 1973) and Penman (1948), allows estimation of water flux from various

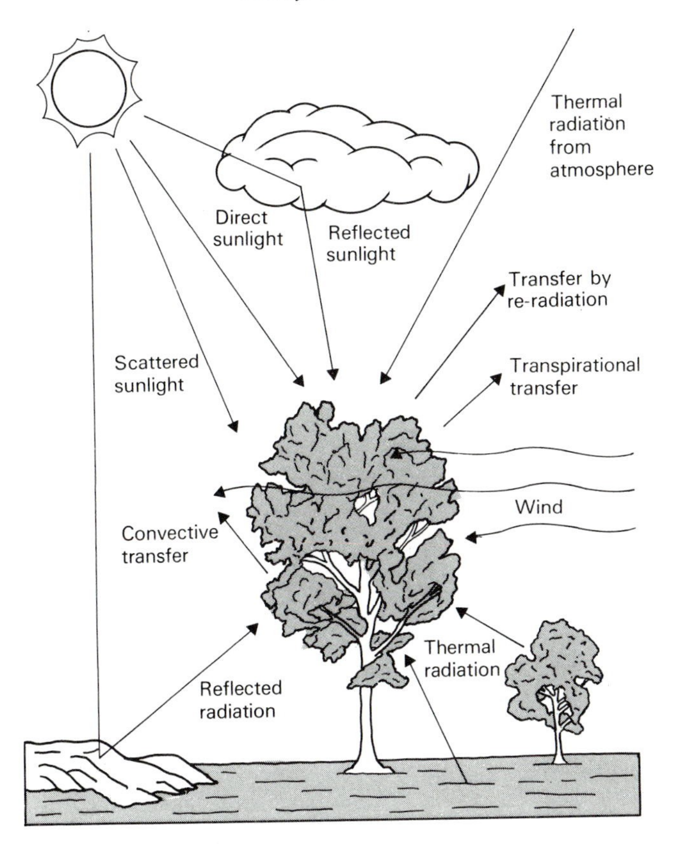

FIG. 2.1 Components of energy exchange. Solar radiation reaches plant canopies as direct, scattered and reflected sunlight, all of which contain some short-wave components important for photosynthesis. In addition, plant surfaces absorb thermal radiation emitted at long-wave, infrared-frequencies by the atmosphere, the ground and other plants. The bulk of the heat load on plants is re-radiated; evaporative cooling by transpiration (or evaporation) and heat transfer by convection remove the rest. Some heat is stored temporarily in the soil, rocks and plant tissue and is later re-radiated. [After Gates (1980).]

types of ecosystems and has brought new understanding and predictive power (Rutter, Morton & Robins 1975; Morton 1983).

Canopy reflectance properties and aerodynamic roughness (resistance) of the vegetation need to be characterized because they affect energy absorption and dissipation. These two variables explain why forests remain cooler than neighbouring fields and are nearly unaffected by large differences in the radiation on sunny and shaded slopes (McNaughton & Jarvis 1983). The geometry of leaves is important because forests composed of trees with needle-shaped leaves may catch and evaporate water on rainy, windy days at rates comparable to transpiration on sunny days (Rutter, Morton & Robins 1975; McNaughton & Jarvis 1983).

Physiological ecologists have shown across a variety of ecosystems the importance of biological control on maximum water use (Monteith 1976; Whitehead & Jarvis 1981; Roberts 1983). The maximum rates are quite similar because leaf stomata increasingly close and limit water loss as the humidity deficit increases (Fig. 2.2) or as light dims in the lower branches (Chazdon & Field 1987).

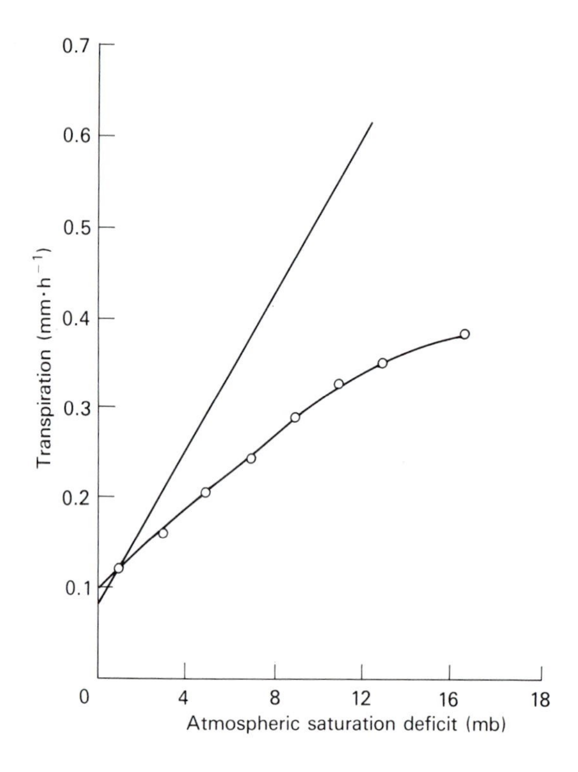

FIG. 2.2. Relationship between atmospheric saturation deficit and transpiration rate as observed shortly after sunrise in a dense Sitka spruce plantation in Scotland. Because the atmospheric saturation deficit tends to increase during clear days as temperature rises, partial stomatal closure is induced, reducing transpiration rates below those predicted if stomatal conductance remained near maximum (upper line). The actual transpiration was much reduced (lower curve). [After Jarvis & Stewart (1979), redrawn by Waring & Schlesinger (1985).]

When these general principles, and related ones controlling water movement through soil (Beven & Wood 1983) and from snow, (Waring, Rogers & Swank 1981) are applied to different ecosystems, the calculated water budgets predict, within errors of measurement, the observed streamflow in gauged catchment basins (Fig. 2.3).

As these approaches are scaled regionally, the local control of evapotranspiration becomes less important as advection moves energy from one area to another. At larger scales the net radiant energy and temperature are most affected by the presence or absence of tall or short vegetation (Priestley & Taylor 1972; Eagleson 1986; Jarvis & McNaughton 1986).

This predictive success in hydrology encourages ecosystem scientists to look harder for similarities in processess operating across diverse

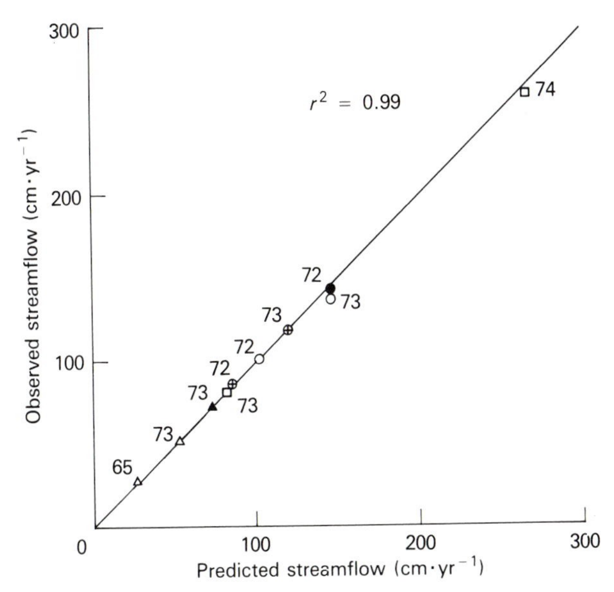

FIG. 2.3. Comparison of predicted annual streamflows from application of a hydrological model with measured values for a range of sites, climatic conditions and vegetative cover. The slope of the regression line is 1.0 indicating close agreement. Gauged catchment basins for which independent verification of streamflow could be made included those at Coweeta in the southeastern states of America with a cover of oak–hickory (O), white pine (⊕), and a recent clear felling (•); the Beaver Creek area in the southwestern states of America with ponderosa pine–oak–juniper (△) and conversion to shrub and grass (▲); and the Andrews Experimental Forest in the Pacific Northwest dominated by 450-year-old Douglas-fir (□). Examples are provided for extremely wet and dry years between 1965 and 1974. [After Waring, Rogers & Swank (1981).]

systems. For example, evapotranspiration also relates well to the rate at which plant litter decomposes because similar temperature and moisture conditions favour both (Meentemeyer, Box & Thompson 1982).

THE CARBON CYCLE

Carbon budgets were one of the first products of well-delineated aquatic ecosystem studies (Lindeman 1942; Odum 1957). Woodwell & Whittaker (1968) were the first to make direct estimates of gross primary production, autotrophic respiration, net primary production, heterotrophic respiration and net ecosystem production in forests. During the 1960s a major goal of the International Biological Programme was to compare carbon (energy) and material flow through representive types of natural ecosystems from the tundra to the tropics (Fig. 2.4). A large

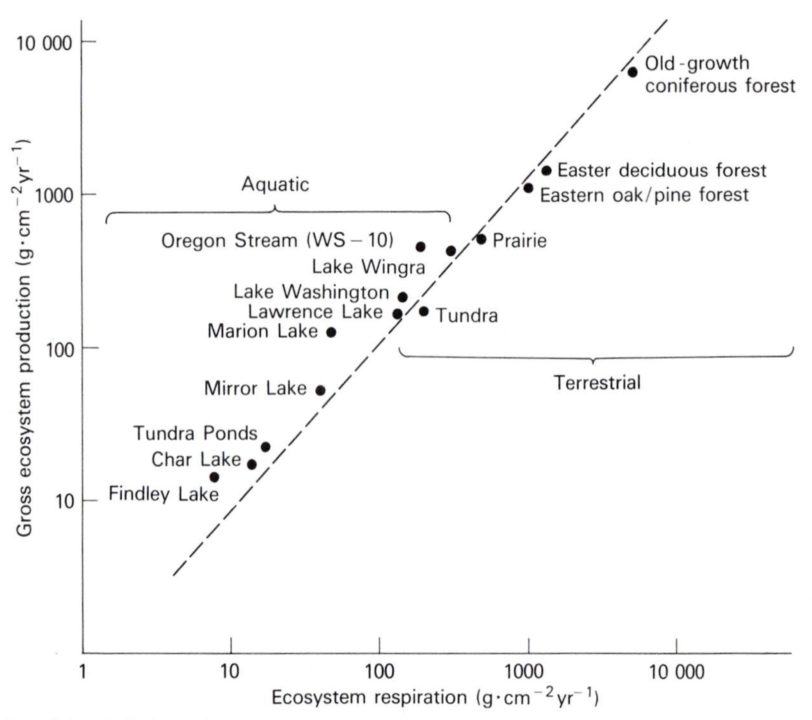

FIG. 2.4. Relationship between gross production and respiration for a series of aquatic and terrestrial ecosystems. Ecosystem respiration rises in parallel with gross ecosystem production (GEP) for a selection of aquatic and terrestrial ecosystems from sites throughout the United States. For aquatic ecosystems, GEP includes both gross primary production and carbon inputs from terrestrial environments. [After Wissmar *et al.* (1982).]

amount of data was accumulated describing the pools and annual transfers in terrestrial (Reichle 1981) and freshwater systems (Wetzel & Rich 1973; Odum & Prentki 1978; Rich & Devol 1978; Wetzel & Richey 1978; Wissmar & Wetzel 1978).

In spite of the immense effort, the International Biological Programme researchers could not explain the large differences observed in net primary production (O'Neil & DeAngelis 1981). The effort did show that small diameter roots can require an unexpectedly large amount of the total gross primary production, and it emphasized the importance of discovering what causes changes in the way photosynthate is used (Edwards *et al.* 1981).

Carbon allocation by plants is now better understood (Chung, Rowe & Field 1982; Bloom, Chapin & Mooney 1985; Landsberg 1986; Hameed, Reid & Rowe 1987; Vessey & Layzell 1987). An important generality is that the peak accumulation of nitrogen in photosynthetic tissue correlates well with above-ground net primary production in any environment (Ågren 1983; Ågren & Ingestad 1987). Variation in the slope of the relationship (Fig. 2.5) depends on the availability of other nutrients, the genetic capacity of the vegetation, and the metabolic cost required to support biomass as trees grow in stature (Gholz 1982; Bloom, Chapin & Mooney 1985; Waring 1987).

It has been discovered that woody biomass may contain enough parenchyma cells to require maintenance support in excess of that required by the foliage (Waring & Schlesinger 1985). In general, those organisms that occupy advanced stages in succession support more foliage with less structural maintenance cost than the organisms they replaced (Waring & Schlesinger 1985; Waring 1987). A distinction between maintenance and growth respiration is important because the first is temperature dependent whereas the latter has the same carbon demands regardless of the temperature of biosynthesis. By recognizing differences in carbon allocation and maintenance respiration, above-ground net primary production can be more closely related to photosynthesis.

Monteith (1981) related the light-capturing ability of plant canopies to crop yields by applying Beer's Law, which defines how light is filtered though any medium. The relationship is not linear; each additional layer of leaves absorbs a much smaller fraction of sunlight than the previous one.

Plant physiologists have shown that the photosynthetic capacity of leaves adjusts to the amount of light available within the range (400–700 nm), and that shaded leaves are more cheaply constructed and cost less to

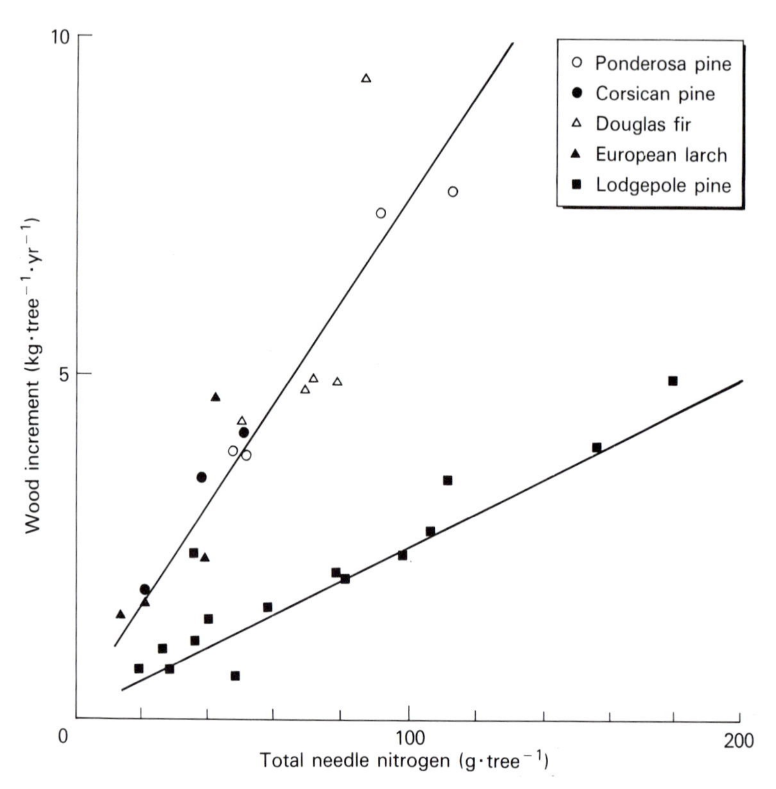

FIG. 2.5. Relationship between wood production and the nitrogen content of needles. Conifer trees produce wood in proportion to the peak seasonal nitrogen content of their needles. Wood increment is greater when nitrogen is in nutritional balance with other essential elements (upper line) than when it is not (lower line). [After Nordmeyer *et al.* (1987).] Lodge pole pine performs similarly to other coniferous species on nutritionally well-balance soils (Benecke & Nordmeyer 1982).

maintain than sun-grown leaves (Berry & Downton 1982; Chadzon & Field 1987; Choudhury 1987). Recently the influences of humidity and atmospheric carbon dioxide have been incorporated in models that accurately predict daily variation in the rates of photosynthesis and in the degree of stomatal control exerted on transpiration (Fig. 2.6), (Ball, Woodrow & Berry 1987). Thermal infrared radiation, which can be sensed remotely, may also provide indications of changes in canopy stomata conductance (Jackson 1983; Goward 1989).

The principle that light absorption by vegetation is related to photosynthesis and indirectly to net primary production is global. Net primary production in North and South America correlates with satellite-derived data on vegetation activity (Fig. 2.7). Moreover, the relative

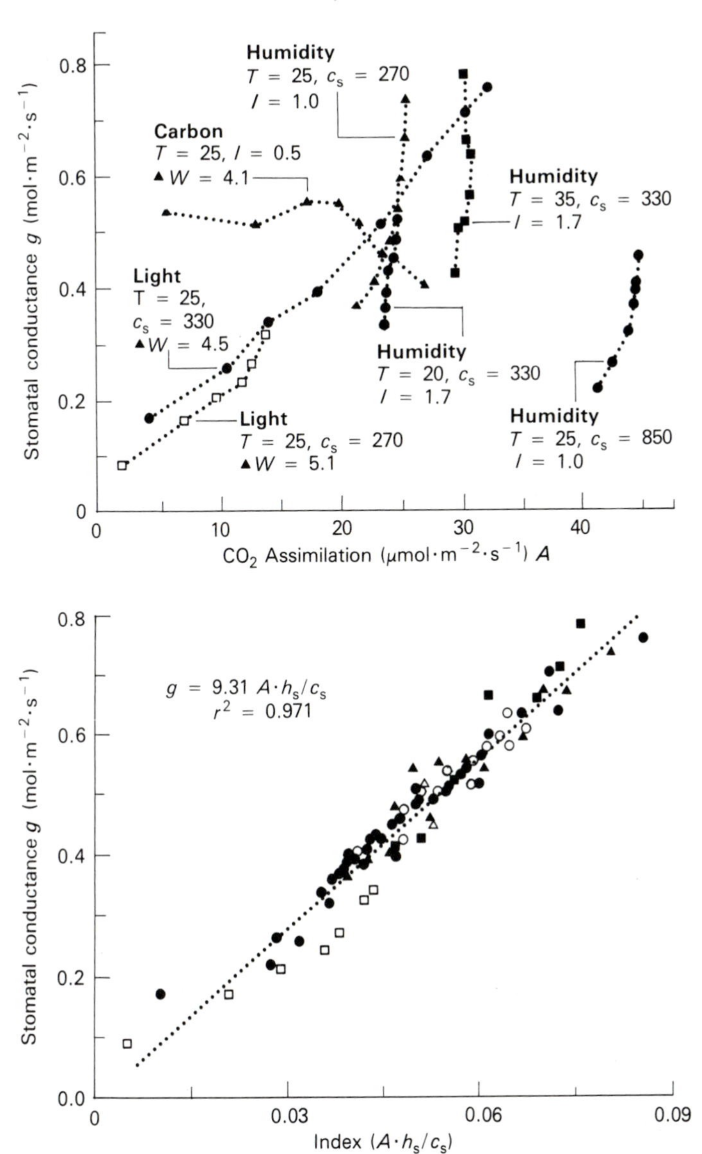

FIG. 2.6. The lower figure shows that in *Glycine max* stomatal conductance (g) is linearly related to a combined environmental–physiological index: the product of net CO_2 assimilation (A) times the relative humidity at the leaf surface (h_s) divided by the CO_2 concentration at the leaf surface (c_s). A is in units of μmol m^{-2} s^{-1}, h_s is a decimal fraction, and c_s is in μmol mol^{-1}. In the upper figure, the same data set is presented showing relationships varying with temperature (T); CO_2 (c_s), water vapour concentration gradient ($\triangle$ W in *mmol mol*$^{-1}$), and irradiance useful for photosynthesis (I in mmol m^{-2} s^{-1}). [After Ball, Woodrow & Berry (1987).]

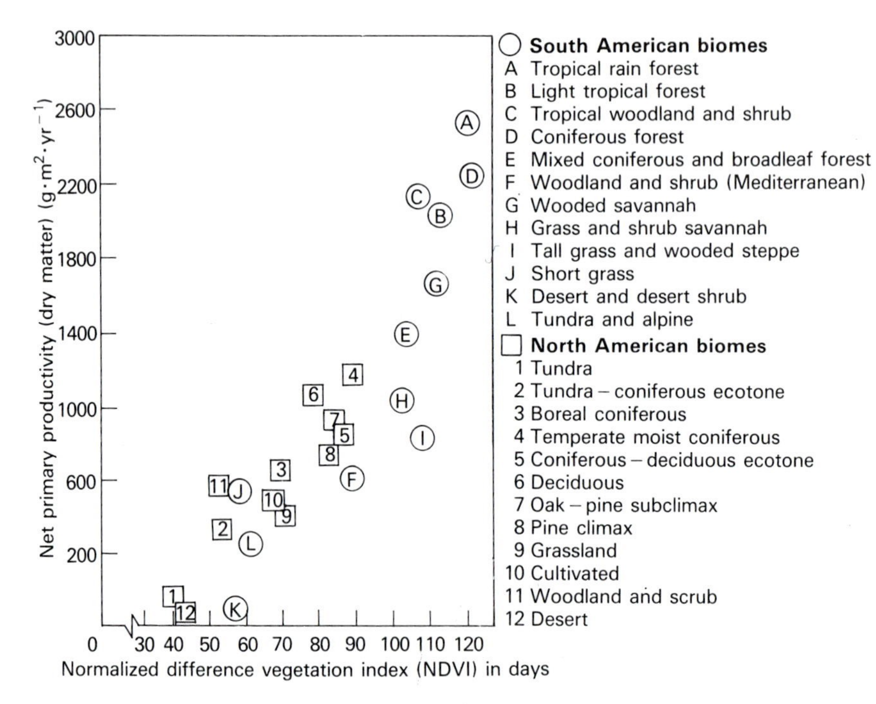

FIG. 2.7. Relationship between net primary production and the normalized difference vegetation index (NDVI) for twenty-four American ecosystems. From reflectance measurements in the visible (VIS) and near-infrared (NIR) spectra, satellite observations provided an index of the vegetation's ability to absorb photosynthetically useful radiation in all of North and South America. This normalized index (NDVI)=(NIR−VIS)/NIR+VIS in day equivalents, when averaged for the clearest day each week and integrated for the 12 months, April 1982–March 1983, compared well with conventional estimates in the literature of above-ground net primary production. [After Goward *et al.* (1987).]

seasonal flux in regional photosynthesis closely parallels the seasonal variation observed in the atmospheric levels of carbon dioxide measured at a series of ground stations (Fig. 2.8).

Explaining why canopy absorption of solar radiation correlated so well with photosynthesis and net primary production in such a wide variety of plant life required the expertise of plant physiologists. Their success came from a search for common features that both describe and explain carbon uptake, use and loss.

BIOGEOCHEMISTRY

The same mass balance approach used in the study of water and carbon cycles has been successfully applied to quantifying various elemental

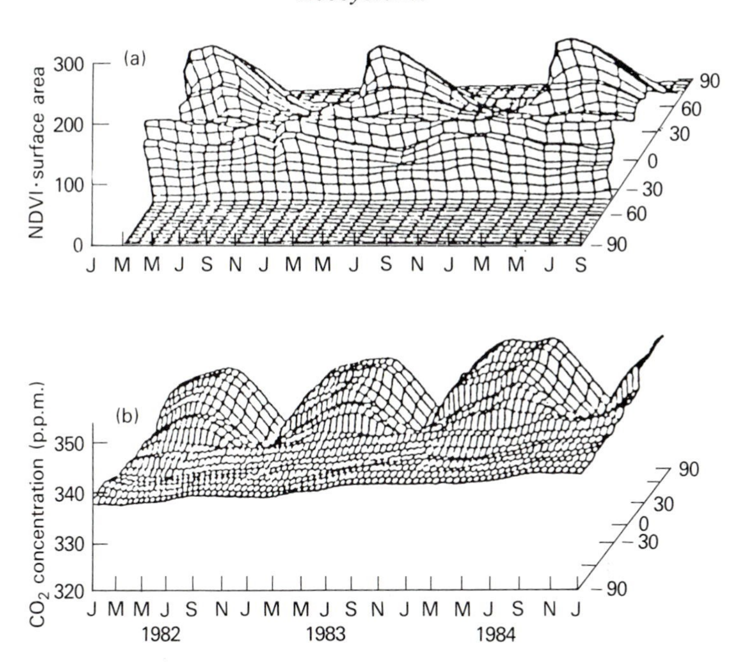

FIG. 2.8. Relationship between (a) the normalized difference vegetation index (NDVI) and (b) seasonal variation in atmospheric CO_2. The NDVI index derived from satellite measurements of reflectance in the visible and near-infrared spectra indicate large seasonal changes when weighted by the land surface area in the northern hemisphere (a), which correspond with the seasonal variation in atmospheric CO_2 recorded at twenty ground stations (b). [After Tucker *et al.* (1986).]

cycles. Large numbers of elemental cycles have been quantified for particular ecosystems, as in the pioneering studies by Bormann, Likens and their colleagues on the Hubbard Brook catchment basin (Bormann & Likens 1967, 1979). Such studies have been repeated at selected places around the world and provide explicit examples of how mineral cycling is affected by differences in parent materials, the type of vegetation, microbial activity and rates of transfer from one system to another (Viner & Smith 1973; Likens & Bormann 1974; Waring & Schlesinger 1985; Mooney, Vitousek & Matson 1987).

The controls on some processes such as nitrification, are known from the results of a large series of comparative studies. General rules have been formalized by ecosystems scientists that help recognize systems most likely to lose nitrogen, in the form of nitrate, following disturbance (Vitousek *et al.* 1982). The comparative approach has been extended to measuring and interpreting trace gas and aerosol emmissions over large regions (McNeal 1988). In these cases, empirical evidence and basic chemistry are yielding increased understanding and predictive capabilities.

Buffering reactions

Interest in the long-term effects of acid rain has prompted the construction of elemental budgets accounting for changes in hydrogen ion fluxes in forest ecosystems (Lindberg *et al.* 1986; Binkley & Richter 1987). These studies indicate that buffering mechanisms operate at many points but may be lacking in some critical steps, such as in nitrate uptake by roots. Acid rain has also been the impetus for studying the release of aluminium into soils and freshwater systems. These studies have highlighted the role of organic complexes, iron, and other elemental reactions that buffer some freshwater and soil systems (Carignan 1985; Rudd, Kelly & Furutani 1986; Schindler *et al.* 1986).

Redox analysis

Schlesinger (1988) developed a more general approach to understanding nutrient cycles by emphasizing oxidation and reduction reactions. He suggested that energy flow studies at all levels are best addressed through measurements of the biogeochemical flux of carbon in oxidized or reduced form. The reduced carbon is measured as an indicator of the energy available for biotic transformations of other elements. Changes in the global pool of reduced carbon are directly linked to changes in the form and amount of other elements of biochemical interest.

Schlesinger cited as an example the research of Garrels & Lerman (1981), who addressed the question of whether oxygen produced by photosynthesis of carbon dioxide has been balanced by the oxidation of sulphide in pyrite (FeS_2) to sulphate in gypsum ($CaSO_4$) in marine sediments. They developed this hypothesis after carefully studying the chemical stoichiometry involved in changing the gypsum reservoir. They considered possible transfers among ten reservoirs, including the atmosphere, ocean, eight mineral phases and organic matter, and concluded that gypsum, pyrite, carbonate and organic carbon were of major importance in the precipitation reactions. They could estimate changes in the gypsum, pyrite, and carbonate reservoirs by observing shifts in the stable isotopic composition of sulphur and carbon found in sediments back a half-billion years. They then calculated organic carbon and sulphate fluxes from the ocean to sedimentary deposits and found they were in parallel, so supporting their original hypothesis (Fig. 2.9).

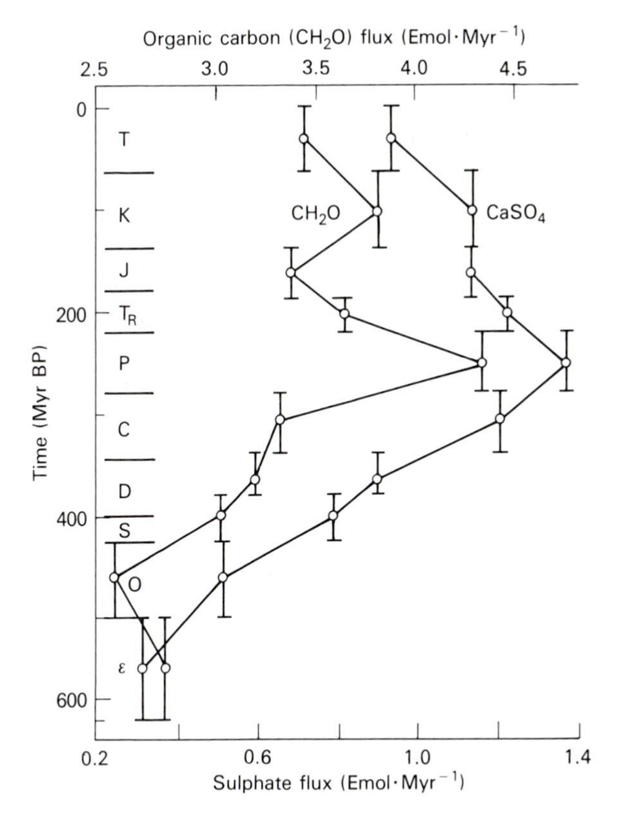

FIG. 2.9. Changes in organic carbon and sulphate fluxes during the last 500 million years. Using the oxidation state of carbon in combination with sulphur, iron, calcium and organic matter, Garrels & Lerman (1981) deduced from a stoichiometric model that oxygen produced by photosynthesis of CO_2 has been balanced by the oxidation of pyrite sulphide (FeS_2) to sulphate in gypsum ($CaSO_4$) in marine sediments. A parallel shift in estimated organic carbon and sulphate fluxes from the ocean to sedimentary reservoirs indicates variation in photosynthesis and related processes during the last 500 million years. Carbon fluxes were computed from Period mean shifts in the isotopic ratios of $^{13}C/^{12}C$ in carbonate rocks; sulphate fluxes were computed from the model but also could be confirmed by measured shifts in the isotopic ratios of $^{34}S/^{32}S$ in pyrite and gypsum.

Stoichiometric relationships

The internal chemistry of organisms can be linked to ecosystem operation. Redfield (1958) recognized that the stoichiometry of C : N : P in plankton approached a constant ratio. From this simple relationship and from observations of the geochemistry of oceans, he deduced patterns in the abundance of these elements and of oxygen in the sea and atmosphere. He further noted that the generally low phosphorus level in sea water must limit the growth rates of plankton (Schidlowski 1988).

Reiners (1986) recognized that the 'Redfield ratios' extend to the protoplasmic contents of all life. He stressed the need to recognize how evolution has affected global biogeochemistry. As life evolved, organisms incorporated more and more materials in supporting structures. These structures were made from readily available, but different resources, diatoms (rich in silica), corals (rich in $CaCo_3$ or $CaPO_4$), fungi and insects (rich in N), and woody plants (rich in C,H,O).

From the 'stoichiometry-of-life' principle, organic synthesis must be limited by the relative rates that essential elements are supplied. If changes in the cycling rates or deposition rates occur in a particular ecosystem, they should be recorded in some form in the composition of bones, shells, lignified material, and sediment.

Sediment changes have already been noted in the chemical equilibria of marine paleoenvironments (Fig. 2.9). Changes in toxic metals (McLaughlin 1985) and nitrogen concentrations (Cotrufo 1983) have been found in tree rings, suggesting the possibility of reconstructing growth efficiency estimates (Oren *et al.* 1987) (Fig. 2.5).

When investigating nutrient availability, the decomposition process is a logical focus. Biochemical assays might tell whether the decomposing substrate is becoming more enriched in tannins, lignin, or cellulose in comparison with key elements like nitrogen (Melillo, Aber & Muratore 1982). A more generally informative approach compares the rates at which carbon is metabolized with the mineralization rates of other elements (Staaf & Berg 1982; see special application by Schlesinger 1985). In most cases, N, P, and S are not released from organic substrates as rapidly as Ca, Mg or K (Fig. 2.10). In fact, additional nutrients may be added or lost through fungal links to the soil, or bacterial fixation, or denitrification at various stages in substrate decomposition. When an essential element such as P is added artificially, its concentration is decreased in the litter and it is mineralized at a rate equivalent to carbon loss (Bjorn Berg, personal communication, Swedish Agricultural University, Uppsala, Sweden).

There are many interpretations of the changing rates of ecosystem processes as well as of changes in past climates. All await more general application of stable isotope chemistry (Peterson & Fry 1987; Ehleringer & Rundel 1988). The technique often allows the identification of changing sources of water (Salati *et al.* 1979; White *et al. 1985;* Sternberg 1987); inputs of pollutants (Feyer 1978); the degree of change in climate (Edwards *et al.* 1985; Mix, Ruddiman & McIntyre 1986); the effects of

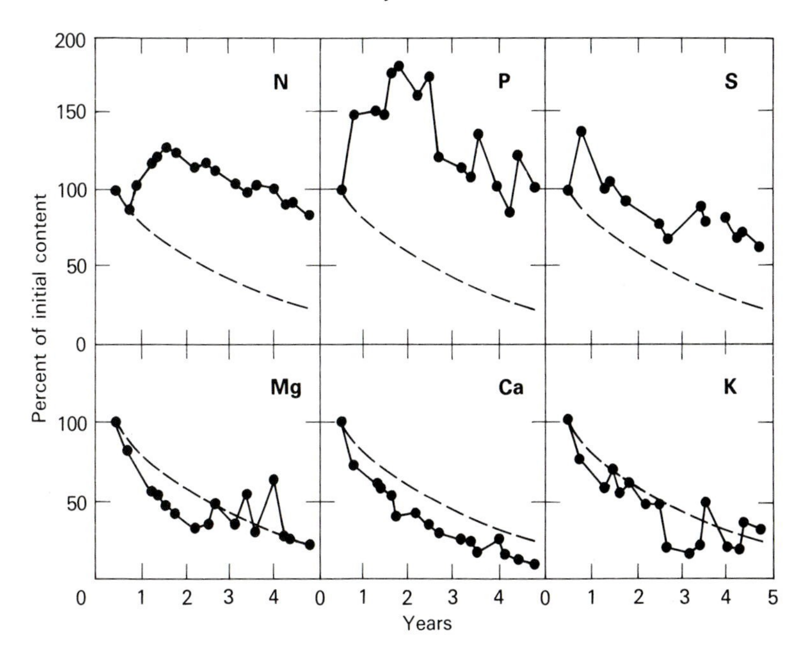

FIG. 2.10. Loss of nutrient elements from the litter of Scots pine in Sweden during the first 5 years of decomposition. For each nutrient, the solid line indicates the percentage of the initial content remaining at various intervals. Note that the loss of K, Ca and Mg is more rapid than is the disappearance of the organic mass of litter (dashed line), whereas N, P and S are retained during the period of litter decay. [After Staaf & Berg (1982).]

salt and drought on photosynthesis (Farquhar & Richards 1984; Downton, Grant & Robinson 1985), and changes in the substrates for food chains (Araujo-Lima *et al.* 1986; Peterson & Fry 1987).

Archaeologists, for example, determined that maize became a major part of the diet for American woodland Indians about 1000 years ago (Fig. 2.11). This discovery was made possible because maize, with a C_4 photosynthetic pathway, has a different stable isotope ratio of $^{13}C/^{12}C$ compared with other plants in the diet of the Indians or of the animals they ate.

Ecosystem studies in biogeochemistry have developed a number of important concepts. Buffer reactions, redox analysis, stoichiometric relationships and stable isotope analyses add powerful new insights to the more classical mass balance approach. Because the foundations of the principles are well understood, they may be applied safely at a variety of scales.

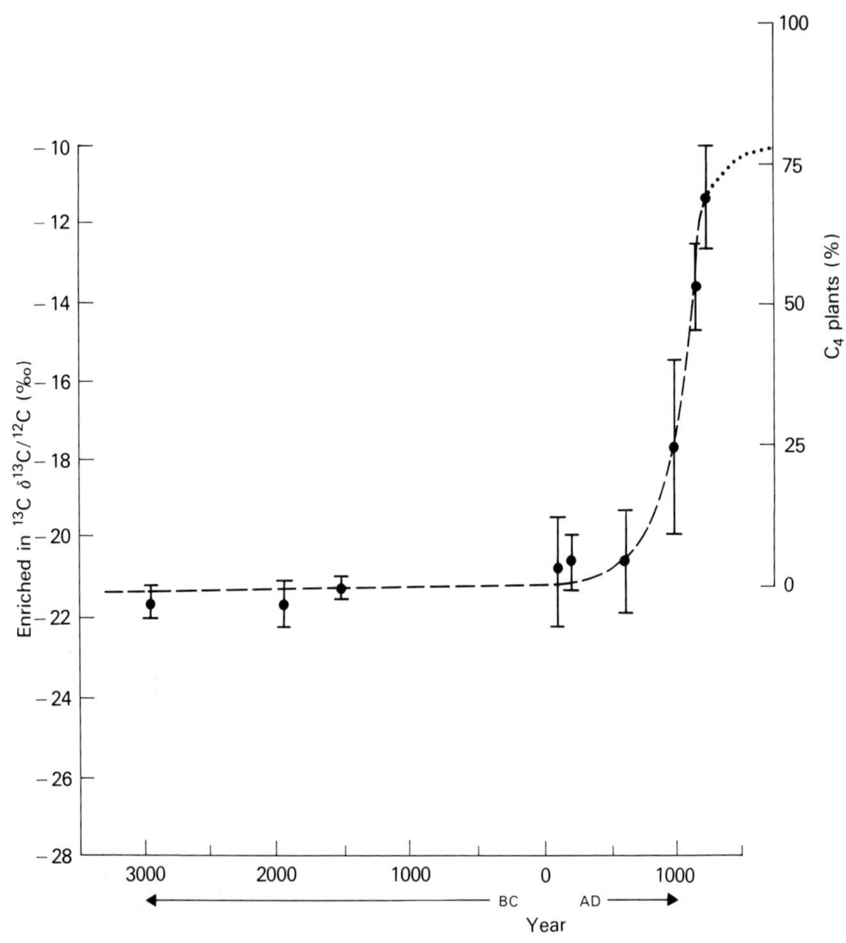

FIG. 2.11. Change in the diets of North American Indians as recorded by the carbon isotopic composition in the collagen bone protein. With the introduction of maize, an enrichment in the $^{13}C/^{12}C$ isotope composition, relative to a standard limestone fossil indicates a 70% reliance on this plant. Maize has a C_4 photosynthetic pathway that allows its carbon to be discriminated from that of other plants with the common C_3 pathway. [After van der Merwe (1982); drawing from Peterson & Fry (1987).]

ECOSYSTEM DISTURBANCE

What influence does a change in the environment or structure have on ecosystem function? This question has led to a number of related concepts. The concept of 'resistance' or susceptibility to change focuses attention on internal controls and redundancies that buffer against change. 'Stability' addresses a system's capacity to recover to an equilibrium state following disturbance. How rapidly the functional operation (or state) returns is a measure of the system's 'resilience'. The idea that ecosystems have internal controls allowing for maintenance of an equilibrium condition has great appeal and is still occasionally presented in some textbooks as a defining characteristic (Nebel & Kormondy 1981).

Resistance

The idea of homeostatic control, derived from cybernetic theory (Ashby 1963), and it was reinforced by early observations of trophic dynamics in a small lake (Lindeman 1942) and by studies of the flow of energy through a freshwater spring (Odum 1957). Margalef (1968) emphasized the importance of biochemical diversity in organisms, rather than species diversity, in searching for controls on various metabolic functions at different stages in succession. Lovelock (1979) extended the idea of ecosystem control to maintaining a global atmosphere conducive to supporting life within supposedly narrow limits of chemical composition.

The search for positive and negative feedbacks has been beneficial toward understanding ecosystem operation. Care must be taken, however, in assuming that a true cybernetic-like control operates in ecosystems. Even the Earth's atmospheric composition has wandered far from equilibrium, with variations in CO_2 and O_2 exceeding 40% of present levels over the last 100–150 million years (Barnola *et al.* 1987; Berner & Landis 1988).

Interactions and replacements do occur among organisms which damp the rate of change in major processes in ecosystems under chronic stress. In a continuing lake acidification experiment, biomass and production of phytoplankton and zooplankton remained stable as long as populations of organisms could be replaced. Likewise, the decomposition process exhibited no measurable change with continual modifications in composition (Schindler 1987). Similarly, in a forest subjected to ionizing radiation, the composition of organisms changed more rapidly than did organic production (Woodwell 1967).

Functional simplicity in an ecosystem would be expected to reduce resistance to change. However, a few highly buffered systems stand as apparent exceptions. Floodplain forests of redwood survived periodic floods, which deposited more than 10 m of silt, and fires which scarred the sides of 1000-year-old trees (Stone & Vasey 1968).

The most stable ecosystems in regard to nutrient loss are those with large internal storage capacities. These ecosystems are often complex in structure. Such systems show high rates of internal recycling and often have elaborate food webs. Nutrients, when available, are quickly incorporated. Continued stress simplifies the structure, reversing the normal succession (Woodwell 1967). Before there is notable change in structure, however, systems become less efficient in capturing resources. More energy is expended with less return and leakage from the system increases. Increased leaching of nutrients is thus a common precursor to structural changes in ecosystems (O'Neill *et al.* 1977; Swank *et al.* 1981; Hauhs & Wright 1986).

Physiologists recognize a number of biochemical changes in microbes (Dawes 1986), animals (Robbins 1983), and plants (Levitt 1980a, 1980b) that are indicative of stress. In higher plants the ratio of amino-acids to carbon-rich defensive compounds increases with stress, making plants more susceptible to attack from many pathogens and herbivores (Matson & Waring 1984; Bryant *et al.* 1987). The remote sensing community recognizes that changes in the fluorescence or reflecting properties of photosynthetic pigments in phytoplankton and plant leaves indicate stressed conditions (Carder & Steward 1985; Rock *et al.* 1986; Chappelle & Williams 1987).

Resilience

Disturbance to an ecosystem may lead to nearly irreversible change. This occurs when salt accumulates in the surface soil of irrigated crops, when a lake dries, or a smelter continually distributes toxic compounds across a basin. In many cases, however, the disturbance is temporary, as following a fire, an oil spill, or a flood. If the disturbance is sufficiently regular, then components of the ecosystem may adapt and eventually require disturbance to maintain a normal, resilient system. Tidal pools require daily exchange with the sea, pastures require mowing or grazing, and many forests are adapted to periodic fires which reduce competition, accelerate nutrient cycling, and limit disease and herbivore activity (Waring & Schlesinger 1985).

Ecosystem characteristics that buffer against disturbance, such as large storage reservoirs and high internal recycling may inhibit rapid recovery (Fig. 2.4). Once the large structures in the system are lost, nutrients are dispersed and important symbiotic relationships broken. Extreme examples are some impoverished rain forests and coral reefs (Pomeroy 1970).

Rapid recovery takes place where nutrients cycle easily through systems and the efficiency in energy and nutrient capture is high.

Agricultural systems, meadows and ponds are examples of ecosystems quick to recover following disturbance. These ideas on resilience have been formalized mathematically by DeAngelis (1980).

Ecosystem scientists have developed methods and experience in identifying unstable ecosystems and in recognizing those likely to recover slowly following a disturbance. The next challenge is to be able to predict the function of ecosystems when on new trajectories brought about by a permanent change in the environment, sometimes induced by introduced species (Vitousek 1986).

CONCLUSION

Over the last half century, Tansley's ecosystem concept has served ecology by focusing attention on how fluxes of material and energy influence and are modified by living organisms and their products. The empirical experience of constructing budgets of energy and matter for a wide range of small, well-bounded and relatively stable ecosystems has identified important components and processes. Concepts and approaches are now being tested on larger geographic scales and under changing environments. At these new levels of integration, complete energy and material budgets are extremely difficult to derive empirically.

The ability to extend principles across scales is a measure of their soundness. In estimating ecosystem fluxes the most robust principles appear to combine physics, physiology and chemistry.

From complex modelling exercises and some experiments, a number of generalities concerning ecosystem stability, rates of recovery, or change in trajectory have emerged. A challenge lies in predicting the functional activity and composition of ecosystems. To meet this challenge ecosystem studies can be designed to incorporate a longer historical perspective and be extended to a broader geographical scale. Regional-scale models, stable isotopes analyses and remote sensing technology are methods likely to play an increasing role in meeting these new challenges.

ACKNOWLEDGMENTS

I appreciate the opportunity provided by the British Ecological Society to develop this paper.

Extremely valuable comments were made on earlier drafts of the manuscript by W. H. Schlesinger and W. E. Winner. I am grateful also to B. Yoder, C. Rose and L. Waring for extensive editorial assistance.

REFERENCES

Ågren, G. I. (1983). Nitrogen productivity of some conifers. *Canadian Journal of Forest Research*, **13,** 494–500.

Ågren, G. I. & Ingestad, T. (1987). Root: shoot ratio as a balance between nitrogen productivity and photosynthesis. *Plant, Cell and Environment*, **10,** 579–586.

Araujo-Lima, C.A.R.M., Forsberg, B. R., Victoria, R. & Martinelli, L. (1986). Energy sources for detritivorous fishes in the Amazon. *Science*, **234**, 1256–1258.

Ashby, W. R. (1963). *An Introduction to Cybernetics.* Wiley, New York.

Ball, J. T., Woodrow, I. E. & Berry, J. A. (1987). A model for predicting stomatal conductance and its contribution to the control of photosynthesis under different environmental conditions. *Progress in Photosynthesis Research*, IV. **5** (Ed. by I.J. Bingins), pp. 221–224. Proceedings of the VII International Photosynthesis Congress.

Barnola, J. M., Raynaud, D., Korotkevich, Y. S. & Lorius, C. (1987). Vostok ice core provides 160 000 year record of atmospheric CO_2. *Nature*, **329,** 408–414.

Benecke, U. & Nordmeyer, A. H. (1982). Carbon uptake and allocation by *Nothofagus Solandri var. cliffortiodes* (Hook. f.) Poole and *Pinus contorta* Douglas ex Loudon ssp. *contorta* at montane and sub-alpine altitudes. *Carbon Uptake and Allocation in Subalpine Ecosystems as a key to Management* (Ed. by R. H. Waring), pp. 9–21. Oregon State University, Forest Research Laboratory, Corvallis.

Berner, R. A. & Landis, G. P. (1988). Gass bubbles in fossil amber as possible indicators of the major gas composition of ancient air. *Science*, **239,** 1406–1409.

Berry, J. A. & Downton, W. J. S. (1982). Environmental regulation of photosynthesis. *Photosynthesis: Development, Carbon Metabolism, and Plant Productivity*, Vol.**2** (Ed. by Govindjee), pp. 263–343. Academic Press, New York.

Beven, K. & Wood, E. F. (1983). Catchment geomorphology and the dynamics of runoff contributing areas. *Journal of Hydrology*, **65,** 139–158.

Binkley, D. & Richter, D. D. (1987). Nutrient cycles and H^+ budgets of forest ecosystems. *Advances in Ecological Research*, **16,** 1–51.

Bloom, A. J., Chapin, F. S., III & Mooney, H. A. (1985). Resource limitation in plants—an economic analogy. *Annual Review of Ecology and Systematics*, **16,** 363–392.

Bormann, F. H. & Likens, G. E. (1967). Nutrient cycling. *Science*, **155,** 424–429.

Bormann, F. H. & Likens, G. E. (1979). *Pattern and Process in a Forested Ecosystem.* Springer-Verlag, New York.

Bryant, J. P., Clausen, T. P., Reichardt, P. B., McCarthy, M. C. & Werner, R. A. (1987). Effect of nitrogen fertilization upon the secondary chemistry and nutritional value of quaking aspen (*Populus tremuloides* Michx.) leaves for the large aspen tortrix (*Choristoneura conflictana* (Walker)). *Oecologia*, **73,** 513–517.

Carder, K. L. & Steward, R. G. (1985). A remote-sensing reflectance model of red-tide dinoflagellate off west Florida. *Limnology and Oceanography*, **30,** 286–298.

Carignan, R. (1985). Quantitative importance of alkalinity flux from the sediments of acid lakes. *Nature*, **317,** 158–160.

Chappelle, E. W. & Williams, D. L. (1987). Laser induced fluorescence (LIF) from plant foliage. *IEEE Transactions of Geoscience and Remote Sensing*, **GE–25**, 726–736.

Chazdon, R. L. & Field, C. B. (1987). Determinants of photosynthetic capacity in six rainforest *Piper* species. *Oecologia*, **73,** 222–230.

Chung, G. C., Rowe, R. N. & Field, R. J. (1982). Relationship between shoot and roots of cucumber plants under nutritional stress. *Annals of Botany*, **50,** 859–861.

Choudhury, B. J. (1987). Relationships between vegetation indices, radiation absorption, and net photosynthesis evaluated by a sensitivity analysis. *Remote Sensing of Environment*, **22,** 209–233.

Cotrufo, C. (1983). Xylem nitrogen as a possible diagnostic nitrogen test for loblolly pine. *Canadian Journal of Forest Research*, **13**, 355–357.

Dawes, E. A. (1986). *Microbial Energetics*. Blackie & Son Ltd, Glasgow.

DeAngelis, D. L. (1980). Energy flow, nutrient cycling, and ecosystem resilience. *Ecology*, **61**, 764–771.

Downton, W. J. S., Grant, W. J. R. & Robinson, S. P. (1985). Photosynthesis and stomatal responses of spinach leaves to salt stress. *Plant Physiology*, **77**, 85–88.

Eagleson, P. S. (1986). The emergence of global-scale hydrology. *Water Resources Research*, **22**, 6S–14S.

Edwards, N. T., Shugart, H. H., Jr., McLaughlin, S. B., Harris, W. F. & Reichle, D. E. (1981). Carbon metabolism in terrestrial ecosystems. *Dynamic Properties of Forest Ecosystems*. (Ed. by D.E. Reichle),pp. 499-536. Cambridge University Press, Cambridge.

Edwards, T. W. D., Aravena, R. O., Fritz, P. & Morgan, A. V. (1985). Interpreting paleoclimate from ^{18}O and ^{2}H in plant cellulose: comparison with evidence from fossil insects and relict permafrost in southwestern Ontario. *Canadian Journal of Earth Sciences*, **22**, 1720–1726.

Ehleringer, J. & Rundel, P. (Eds) (1988). *Stable Isotopes in Ecological Research*. Springer-Verlag, Berlin.

Evans, F. C. (1956). Ecosystems as the basic unit in ecology. *Science*, **123**, 1127–1128.

Farquhar, G. D. & Richards, R. A. (1984). Isotopic composition of plant carbon correlates with water-use efficiency of wheat genotypes. *Australian Journal of Plant Physiology*, **11**, 539–552.

Feyer, H. D. (1978). Seasonal trends of NH_4^+ and NO_3^- nitrogen isotope composition in rain collected at Julich, Germany. *Tellus*, **30**, 83–92.

Gardner, R. H., O'Neill, R. V., Mankin, J. B. & Carney, J. H. (1981). A comparison of sensitivity analysis and error analysis based on a stream ecosystem model. *Ecosystem Modelling*, **12**, 173–190.

Garrels, R. M. & Lerman, A. (1981). Phanerozoic cycles of sedimentary carbon and sulfur. *Proceedings of the National Academy of Science, U.S.*, **78**, 4652–4656.

Gates, D. M. (1980). *Biophysical Ecology*. Springer-Verlag, Berlin.

Gholz, H. L. (1982). Environmental limits on above ground net primary production, leaf area, and biomass in vegetation zones of the Pacific North West. *Ecology*, **63**, 469–481.

Goward, S. N. (1989). Satellite bioclimatology. *Journal of Climate* (in press).

Goward, S. N., Dye, D., Kerber, A. & Kalb, V. (1987). Comparison of North and South American Biomes from AVHRR observations. *Geocarto International*, **2**, 27–39.

Hameed, M. A., Reid, J. B. & Rowe, R. N. (1987). Root confinement and its effects on the water relations, growth and assimilate partitioning of tomato (*Lycopersicum esculentum* Mill.) *Annals of Botany*, **59**, 685–692.

Hauhs, M. & Wright, R. F. (1986). Regional pattern of acid deposition and forest decline along a cross section through Europe. *Water, Air and Soil Pollution*, **31**, 463–474.

Jackson, R. D. (1983). Canopy temperature and crop water stress. *Advances in Irrigation*, **1**, 43–85.

Jarvis, P. G. & McNaughton, K. G. (1986). Stomatal control of transpiration: scaling up from leaf to region. *Advances in Ecological Research*, **15**, 1–49.

Jarvis, P. G. & Stewart, J. B. (1979). Evaporation of water from plantation forests. *The Ecology of Even-aged Forest Plantations* (Ed. by E.D. Ford, D.C. Malcolm & J. Atterson), pp. 327–350. Institute of Terrestrial Ecology, National Environmental Research Council, Cambridge.

Klir, G. J. (1969). *An Approach to General Systems Theory*. Van Nostrand Reinhold, New York.

Landsberg, J. (1986). *Physiological Ecology of Forest Production*. Academic Press, London.

Levitt, J. (1980a). *Responses of Plants to Environmental Stresses.* Vol. **I**. Academic Press, New York.

Levitt, J. (1980b). *Responses Of Plants to Environmental Stresses.* Vol. **II**. Academic Press, New York.

Lindberg, S. E., Lovett, G. M., Richter, D. D. & Johnson, D. W. (1986). Atmospheric deposition and canopy interactions of major ions in a forest. *Science*, **231,** 141–146.

Lindeman, R. L. (1942). The trophic-dynamic aspect of ecology. *Ecology*, **23,** 399–418.

Likens, G. E. & Bormann, F. H. (1974). Linkages between terrestrial and aquatic ecosystems. *Ecology*, **24,** 447–456.

Lovelock, J. E. (1979). *Gaia: A New Look at Life on Earth.* Oxford University Press, Oxford.

Major, J. (1969). Historical development of the ecosystem concept. *The Ecosystem Concept in Natural Resource Management* (Ed. by G.M. Van Dyne), pp. 9–22. Academic Press, New York.

Margalef, R. (1968). *Perspectives in Ecological Theory.* University of Chicago Press, Chicago.

Matson, P. A. & Waring, R. H. (1984). Effects of nutrient and light limitations on mountain hemlock: susceptibility to laminated root rot. *Ecology*, **65,** 1517–1524.

McLaughlin, S. B. (1985). Effects of air population on forests. *Journal of the Air Pollution Control Association*, **35,** 512–534.

McNaughton, K. G. & Jarvis, P. G. (1983). Predicting effects of vegetation changes on transpiration and evaporation. *Water Deficits and Plant Growth* Vol. **7,** (Ed. by T.T. Kozlowski), pp. 1–47. Academic Press, New York.

McNeal, R. J. (1988). Preface. *Journal of Geophysical Research*, **93,** (D2), 1349.

Meentemeyer, V., Box, E. O. & Thompson, R. (1982). World patterns and amounts of terrestrial plant litter production. *BioScience*, **32,** 125–128.

Melillo, J. M., Aber, J. D. & Muratore, J. F. (1982). Nitrogen and lignin control on hardwood leaf litter decomposition dynamics. *Ecology*, **63,** 621–626.

Mix, A. C., Ruddiman, W. F. & McIntyre, A. (1986). Late Quaternary paleoceanography of the tropical Atlantic. 1: spatial variability of annual mean sea-surface temperatures, 0–20000 years B.P. *Paleoceanography*, **1,** 13–66.

Monteith, J. L. (1965). Evaporation and environment. *Society for Experimental Biology Symposium*, **19,** 205–234.

Monteith, J. L. (1973). *Principles of Environmental Physics.* Edward Arnold, London.

Monteith, J. L. (ed) (1976). *Vegetation and the Atmosphere* Vol. **2**. Academic Press, London.

Monteith, J. L.(1981). Climatic variation and growth of crops. *Quarterly Journal of the Royal Meteorological Society*, **107,** 749–774.

Mooney, H. A., Vitousek, P. M. & Matson, P. A. (1987). Exchange of materials between terrestrial ecosystems and the atmosphere. *Science*, **238,** 926–932.

Morton, F. I. (1983). Operational estimates of lake evaporation. *Journal of Hydrology*, **66,** 77–100.

Nebel, B. J. & Kormondy, E. J. (1981). *Environmental Science: The Way the World Works.* Prentice-Hall, Englewood Cliffs, New Jersey.

Nordmeyer, A. H., Kelland, C. M., Evans, G. R. & Ledgard, H. J. (1987). Nutrients and the restoration of mountain forests in New Zealand. *Human Impacts and Management of Mountain Forests* (Ed. by T. Fujimori & M. Kimura), pp. 145–154. Forestry and Forest Products Research Institute, Ibaraki, Japan.

Odum, H. T. (1957). Trophic structure and productivity of Silver Springs, Florida. *Ecological Monographs*, **27,** 55–112.

Odum, H. T. (1971). *Environment, Power, and Society.* Wiley-Interscience, New York.

Odum, W. & Prentki, R. T. (1978). Analysis of five North American lake ecosystems, IV:

Allochthonous carbon inputs. *Verhandlungen Internationale Vereinigung fuer theoretische und angewandte Limnologie*, **20**, 574–580.

O'Neill, R. V., Ausmus, B. S., Jackson, D. R., Van Hook, R. I., Van Voris, P., Washburne, C. & Watson, A. P. (1977). Monitoring terrestrial ecosystems by analysis of nutrient export. *Water, Air, and Soil Pollution*, **8**, 271–277.

O'Neill, R. V. & DeAngelis, D. L. (1981). Comparative productivity and biomass relations of forest ecosystems. *Dynamic Properties of Forest Ecosystems* (Ed. by D.E. Reichle), pp. 411–449. Cambridge University Press, Cambridge.

Oren, R., Waring, R. H., Stafford, S. G. Barrett, J. W. (1987). Analysis of 24 years of ponderosa pine growth in relation to canopy leaf area and understory competition. *Forest Science*, **33**, 538–547.

Patten, B. C. (1971). *System Analysis and Simulation in Ecology*. Academic Press, New York.

Penman, H. L. (1948). Natural evaporation from open water, bare soil and grass. *Proceedings of the Royal Society London Series A*, **193**, 120–145.

Peterson, B. J. & Fry, B. (1987). Stable isotopes in ecosystem studies. *Annual Review of Ecology and Systematics*, **18**, 293–320.

Pomeroy, L. R. (1970). The strategy of mineral cycling. *Annual Review of Ecology and Systematics*, **1**, 171–190.

Priestley, C. H. B. & Taylor, R. J. (1972). On the assessment of surface heat flux and evaporation using large-scale parameters. *Monthly Weather Review*, **100**, 81–92.

Redfield, A. C. (1958). The biological control of chemical factors in the environment. *American Scientist*, **46**, 206–226.

Reichle, D. E. (ed.) (1981). *Dynamic Properties of Forest Ecosystems*. Cambridge University Press, Cambridge.

Reiners, W. A. (1986). Complementary models for ecosystems. *American Naturalist*, **127**, 59–73.

Rich, R. & Devol, A. H. (1978). Analysis of five North American lake ecosystems, VII: Sediment processing. *Verhandlungen Internationale Vereinigung fuer theoretische und angewandte Limnologie*, **20**, 598–604.

Robbins, C. T. (1983). *Wildlife Feeding and Nutrition*. Academic Press, Orlando, Florida.

Roberts, J. (1983). Forest transpiration: A conservative hydrologic process? *Journal of Hydrology*, **66**, 133–141.

Rock, B. N., Vogelmann, J. E., Williams, D. L., Vogelmann, A. F. & Hoshizake, T. (1986). Remote detection of forest damage. *BioScience*, **36**, 439–443.

Rudd, J., Kelly, C., & Furutani, A. (1986). The role of sulfate reduction in long term accumulation of organic and inorganic sulfur in lake sediments. *Limnology and Oceanography*, **31**, 1281–1289.

Rutter, A. J., Morton, A. J. & Robins, P. C. (1975). A predictive model of rainfall interception in forests. II. Generalization of the model and comparison with observations in some coniferous and hardwood stands. *Journal of Applied Ecology*. **51**, 191–203.

Salati, E., Dall'Olio, A., Matsui, E. & Gat, J. R. (1979). Recycling of water in the Amazon Basin: an isotopic study. *Water Resources Research*. **15**, 1250–1258.

Schidlowski, M. (1988). A 3800-million year isotopic record of life from carbon in sedimentary rocks. *Nature*, **333**, 313–318.

Schindler, D. W. (1987). Is the whole really more than the sum of the parts?. *Status and Future on Ecosystem Science* (Ed. by G.E. Likens, J.J. Cole, J. Kolasa, J.B. McAninch, M.J. McDonnell, G.G. Parker & D.L. Strayer). Occasional Publication of the Institute of Ecosystem Studies, No. **3**. The New York Botanical Garden, Millbrook, New York.

Schindler, D. W., Turner, M. A., Stainton, P. & Linsey, G. (1986). Natural sources of acid neutralizing capacity in low alkalinity lakes of the precambrian shield. *Science*, **232,** 844–847.

Schlesinger, W. H. (1985). Decomposition of chapavval shrub foliage. *Ecology*, **66,** 1353–1359.

Schlesinger, W. H. (1988). Contemporary theory for ecosystem structure and function. *Perspectives in Ecological Theory* (Ed. by J.R. Roughgarden, R.M. May, & S.A. Levin), pp. 268–274. Princeton University Press, Princeton.

Staaf, H. & Berg, B. (1982). Accumulation and release of plant nutrients in decomposing Scots pine needle litter. Long-term decomposition in a Scots Pine forest. II. *Canadian Journal of Botany*, **60,** 1561–1568.

Sternberg, L. D. S. L. (1987). Utilization of freshwater and ocean water by coastal plants of southern Florida. *Ecology*, **68,** 1898–1905.

Stone, E. C. & Vasey, R. B. (1968). Preservation of coast redwood on alluvial flats. *Science*, **159,** 157–161.

Swank, W. T., Waide, J. B. Crossley, D. A., Jr. & Todd, R.L. (1981). Insect defoliation enhances nitrate export from forest ecosystems. *Oecologia*, **51,** 297–299.

Transley, A. G. (1935). The use and abuse of vegetational concepts and terms. *Ecology*, **16,** 284–307.

Tucker, C. J., Fung, I. Y., Keeling, C. D. & Gammon, R. H. (1986). Relationship between atmospheric CO_2 variations and a satellite-derived vegetation index. *Nature*, **319,** 195–199.

van der Merwe, N. J. (1982). Carbon isotopes, photosynthesis, and archaeology. *American Scientists*, **70,** 596–606.

Vessey, J. K. & Layzell, D. B. (1987). Regulation of assimilate partitioning in soybean. *Plant Physiology*, **83,** 341–348.

Viner, A. B. & Smith, I. R. (1973). Geographical, historical and physical aspects of Lake George. *Proceedings of the Royal Society London*, Series B, **184,** 235–270.

Vitousek, P. M. (1986). Biological invasion and ecosystem properties: can species make a difference? *Biological Invasions of North America and Hawaii* (Ed. by H.A. Mooney & J. Drake), pp. 163–176. Springer-Verlag, New York.

Vitousek, P. M. Gosz, J. R., Grier, C. C., Melillo, J. M. & Reiners, W. A. (1982). A comparative analysis of potential nitrification and nitrate mobility in forest ecosystems. *Ecological Monographs*, **52,** 155–177.

Von Bertalanffy, L. (1950). An outline of general system theory. *British Journal for the Philosophy of Science*, **1,** 134–165.

Waring, R. H. (1987). Characteristics of trees predisposed to die. *BioScience*, **37,** 569–574.

Waring, R. H., Rogers, J. J. & Swank, W. T. (1981). Water relations and hydrologic cycles. *Dynamic Properties of Forest Ecosystems* (Ed. by D.E. Reichle), pp.205–264. Cambridge University Press, London.

Waring, R. H. & Schlesinger, W. H. (1985). *Forest Ecosystems: Concepts and Applications.* Academic Press, Orlando, Florida.

Wetzel, R. G. & Rich, P. H. (1973). Carbon in freshwater systems. *Carbon and the Biosphere* (Ed. by G.M. Woodwell & E.V. Pecan), pp. 241–263. U.S. Atomic Energy Commission Conference 720510, Washington, D.C.

Wetzel, R. G. & Richey, J. E. (1978). Analysis of five North American lake ecosystems, VIII: Control mechanisms and regulation. *Verhandlungen Internationale Vereinigung fuer theoretische und angewandte Limnologie*, **20,** 605–608.

White, J. W. C., Cook, E. R., Lawrence, J. R. & Broecker, W. B. (1985). The D/H ratios of sap in trees: implications for water sources and tree ring D/H ratios. *Geochimica et Cosmachimica Acta*, **49,** 237–246.

Whitehead, D. & Jarvis, P. G. (1981). Coniferous forests and plantations. *Water Deficits and Plant Growth* Vol. **6** (Ed. by T.T. Kozlowski), pp. 49–152. Academic Press, New York.

Wissmar, R. C., Richey, J. E., Devol, A. H. & Eggers, D. M. (1982). Lake ecosystems of the Lake Washington Drainage Basin. *Analysis of Coniferous Forest Ecosystems in the Western United States* (Ed. by R.L. Edmonds), pp. 333–385. Hutchinson Ross, Stroudsburg, Pennsylvania.

Wissmar, R. C. & Wetzel, R. G. (1978). Analysis of five North American lake ecosystems, VI: Consumer community structure and production. *Verhandlungen Internationale Vereinigung fuer theoretische und angewandte Limnologie*, **20**, 587–597.

Woodwell, G. M. (1967). Radiation and the patterns of nature. *Science*, **156**, 461–470.

Woodwell, G. M. & Whittaker, R. H. (1968). Primary production in terrestrial ecosystems. *American Zoologist*, **8**, 19–30.

3 FOOD WEBS

J. H. LAWTON
Department of Biology, University of York, now at:
Centre for Population Biology, Department of Pure and Applied Biology, Imperial College, Silwood Park, Ascot SL5 7PY, UK

THIRD FISHERMAN: Master, I marvel how the fishes live in the sea.
FIRST FISHERMAN: Why, as men do a-land — the great ones eat up the little ones.

Pericles, Prince of Tyre, Act 2

INTRODUCTION

The first ecologist to publish a diagram of a food web appears to have been V. E. Shelford in *Animal Communities in Temperate America*, published in 1913. The first to grasp the full theoretical significance and importance of food webs was Charles Elton in *Animal Ecology*, in 1927. The term 'food chain' is Elton's; his 'food cycle' we now refer to as a food web, that is a collection of food chains. It is worth quoting what Elton wrote over 60 years ago at some length, because much of it seems to have been forgotten:

. . . animals have to depend ultimately upon plants for their supplies of energy, since plants alone are able to turn raw sunlight and chemicals into a form edible to animals. [p. 56]

The herbivores are usually preyed upon by carnivores, which get the energy of the sunlight at third-hand, and these again may be preyed upon by other carnivores, and so on, until we reach an animal that has no enemies, and which forms, as it were, a terminus on this food cycle. There are, in fact, chains of animals linked together by food, and all dependent in the long run upon plants. We refer to these as 'food-chains', and to all the food-chains in a community as the 'food-cycle'. Starting from herbivorous animals of various sizes, there are as a rule a number of food-chains radiating outwards, in which the carnivores become larger and larger, while the parasites are smaller than their hosts. [pp. 56–57]

Size has a remarkably great influence on the organisation of animal communities. . . . A little consideration will show that size is the main reason underlying the existence of . . . food chains, and that it explains many of the phenomena connected with the food-cycle. There are very definite limits, both upper and lower, to the size of food which a carnivorous animal can eat. [p. 59]

Each stage in an ordinary food-chain has the effect of making a smaller food into a larger one, and so making it available to a larger animal. But since there are upper and lower limits to the size of animals, a progressive food-chain cannot contain more than a certain number of links, and usually has less than five. [p. 61]

If you make a list of the carnivorous enemies and of the parasites of any species of animal, you will see . . . certain curious facts about the sizes of the two classes of animals relative to their prey or host. . . . In fact, most animals have a set of carnivorous animals much larger than themselves, and a set of parasitic enemies much smaller than themselves . . . In all these cases we are, of course, speaking of the size relative to their prey or host. [pp. 71–72]

Apparently there are never very many stages in such food-chains of parasites. The reason for this is that the largest parasite is not very big, and any hyperparasite living on or in this must be very much smaller still, so that the fifth or sixth stage in the chain would be something about the size of a molecule of protein! [p. 78]

. . . the smaller an animal the commoner it is on the whole. . . . To put the matter more definitely, the animals at the base of a food-chain are relatively abundant, while those at the end are relatively few in numbers . . . Finally, a point is reached at which we find a carnivore . . . whose numbers are so small that it cannot support any further stage in the food-chain. [p. 69]

When we are dealing with a simple food-chain it is clear enough that each animal to some extent controls the numbers of the one below it. . . . Ultimately it may be possible to work out the dynamics of this system in terms of the amount of organic matter produced and consumed and wasted in a given time. . . . The effect of each stage in a food-chain on its successor is easy to understand, but when we try to estimate the effect of, say, the last species in the chain upon the first, or upon some other species several stages away, the matter becomes complicated. If A keeps down B, and B keeps down C, while A also preys on C, what is the exact effect of A upon C? [pp. 120–121]

These eight short paragraphs contain most of the essential ideas round which modern research on food webs is now focused, and some which have been lamentably ignored by Elton's successors. Food chains are energy transformers, supported ultimately by green plants. Body size offers a key to understanding structure in food webs, and so do dynamic interactions between species. Ecologists must study parasites as well as true predators when drawing up food webs.

Although not couched in the form of explicit hypotheses, there are also two explanations here for why food chains are short. The length of food chains is limited by constraints on species body sizes, and/or by the diminishing availability of food (energy) at successive steps in the chain. Elton obviously thought that both mechanisms were important.

We will have cause to return repeatedly to Elton's insights throughout this essay.

General remarks and organization

Food webs are diagrams depicting which species in a community interact. Obviously, the interactions are trophic, not competitive or mutualistic, although these distinctions are sometimes blurred. Published webs usually depict binary relationships — whether the species interact or not — and ignore a great deal of important biology, for example the frequency and intensity with which species A feeds on species B. In Pimm's words (1982), most published webs are therefore 'caricatures of nature'. My dictionary defines caricature as a representation exaggerated for comic effect, or a ludicrously inadequate or inaccurate imitation! This is a fair assessment of most of the published information on food webs.

Confronted with limited data of highly variable quality, hardly any of which is really good, food web studies face either hand-wringing paralysis, or cautious efforts to see what can be discovered in the existing information. If nothing else, the latter course of action should serve as a spur to gather more and better data, particularly if published webs reveal evidence of interesting regularities and patterns in nature. However, we have to accept that some of the patterns may eventually prove to be artefacts of poor information. Bearing in mind this considerable worry, I have nevertheless chosen to adopt an optimistic approach, and first summarize evidence for the existence of ten patterns in the structure of published webs. There then follows a major section on theoretical explanations put forward to explain them. Patterns and explanations are gathered together for ease of reference in Table 3.1. The final parts of the paper review problems in the published data, and point to significant gaps in our knowledge.

PATTERNS

Patterns define positional relationships between species in a food web that recur more often than expected by chance. Current interest in patterns in food webs stems primarily from the publication of two books; Cohen's *Food Webs and Niche Space* (1978) and Pimm's *Food Webs* (1982). Ecologists now have 113 webs compiled from the literature (Briand & Cohen 1987) from which to seek evidence of non-random, non-trivial patterns in binary links between components of the web. Raking and reraking these published entrails is a small growth industry. As explained in the Introduction, I want first to summarize the patterns that have emerged from these analyses, without questioning the validity of the data on which they are based. Later, I want to discuss the quality of the information. One feature of the data that we do need to note here is the fact that many of the components of published webs are not single species; they may be genera, groups of similar taxa (e.g. 'herbivorous gastropods' or 'algae'), or incredibly heterogeneous organisms that happen to feed roughly in the same way [one such group, by no means atypical, consists of dragonflies, spiders and passerine birds! (Teal 1962; web 24 in Cohen 1978)]. For simplicity, it is conventional to refer to all such groupings as 'species', even though we know it is an oversimplification.

The published patterns can be grouped under ten headings. This is not an exhaustive list; rather it contains those that seem to me to be the most interesting. They are:

(i) A miscellaneous group that distinguishes real food webs from models where interactions are placed entirely randomly. These include a restriction on loops of the kind: species A eats species B eats species C eats species A etc. [termed 'acyclicity' by Gallopin (1972)], and absurdities such as predators without prey (Pimm 1982; Pimm & Lawton 1983). In the analyses that follow I have largely ignored biological absurdities; most of them are trivially obvious. I have, however, commented on feeding loops, because despite assertions to the contrary, under some circumstances these can, and do, occur in real food webs.

(ii) On average each species feeds on a constant number of others, irrespective of the total number of species in the web (MacDonald 1979). Cohen & Newman (1985) call this property d the 'density of links per species'. As a consequence, as the number of species in the web increases, 'connectance', defined as the number of realized links, divided by the possible links, declines hyperbolically (Rejmanek & Stary 1979; Yodzis 1980; Pimm 1982; Cohen, Newman & Briand 1985).

(iii) 'Top' species have no predators: 'intermediate' species serve as predators and prey; 'basal' species are prey. Then, the average proportions of basal, intermediate and top species in published webs are constant and independent of the total number of species in the web (Briand & Cohen 1984; Cohen & Briand 1984). Based on an entirely different, taxonomically more reliable data set, but using a cruder trophic classification, Jeffries & Lawton (1985) found that the ratio of predator species to prey species in freshwater communities was roughly constant, and independent of the number of species in the sample. Similar observations have been made by others (e.g. Evans & Murdoch 1968; Arnold 1972; Cameron 1972; Moran & Southwood 1982).

(iv) Perhaps not surprisingly, given (ii) and (iii), the average proportion of links between basal, intermediate and top species is also independent of the number of species in the web, although variation round this average trend is large (Cohen & Briand 1984).

(v) In the food web literature, an 'omnivore' is a species that feeds on more than one level in the food chain. In published food webs, omnivores are rarer than expected (Pimm & Lawton 1978, 1983; Pimm 1982), which may be more or less the same as Cohen & Briand's (1984) observation that most published webs lack feeding links between basal and top species. There are, however, some apparent exceptions to this generalization, for example, webs of insects and parasitoids often show abundant, and very complex patterns of omnivory (Pimm & Lawton 1978,1983; Pimm 1982), as may assemblages of decomposers (Pimm 1982; Pimm & Lawton 1983). However, perhaps more than any other pattern, the real extent of omnivory in food webs may be seriously distorted by poor-quality and highly aggregated data (see, for example, the discussions in Sprules & Bowerman 1988). Patterns of omnivory therefore require particularly careful interpretation.

(vi) Webs from 'fluctuating' and 'constant' environments differ. Webs from constant environments have proportionately more links between basal and top species [i.e. more omnivory in the sense of pattern (v)] than webs from fluctuating environments, and show greater variance in many characteristics, for example in proportions of links (pattern iv) and species (pattern iii) in various trophic positions (Briand & Cohen 1984; Cohen & Briand 1984). Webs from fluctuating environments also appear to have lower connectance (Briand 1983; Cohen, Newman & Briand 1985).

(vii) Food chains are short (Elton 1927; Hutchinson 1959; Pimm & Lawton 1977). Typically the modal number of species between the base and top of published webs is three or four; six species or more is very rare (Pimm 1982). Two additional points are worth making at this juncture. First, here and throughout the chapter, I have tried to avoid the term 'trophic level'. It is practically impossible to assign many real species to fixed trophic levels, and the concept conceals more than it reveals (Darnell 1961; Cousins 1980, 1987; Platt 1985). Second, although it is a familiar idea that food chains are short, this particular pattern depends on the way in which steps are defined. Food chains are short only if detritus and bodies are regarded as the start of a new chain (Patten 1985).

(viii) Habitat structure influences the length of food chains, which appear to be shorter in 'two-dimensional' habitats (e.g. grassland) than in 'three-dimensional' (e.g. forest) or 'solid' (e.g. pelagic) habitats (Briand & Cohen 1987).

(ix) Theoretically, webs are said to be divided into 'compartments' when the majority of interactions between species take place within blocks or modules of species, and there are very few interactions between blocks. Published webs based on binary links between species are not divided into compartments within habitats, but evidence for compartments can be found at habitat boundaries (Pimm & Lawton 1980; Pimm 1982).

(x) Food webs are 'interval' more often than expected by chance (Cohen 1977, 1978, 1983). Basically, this property of webs describes patterns of overlap in prey use by predators. If overlaps can be expressed in one dimension, the web is interval; if it requires two dimensions to illustrate the pattern of overlap between predators, the web is non-interval. This somewhat abstract property of food webs is described further on p.62.

EXPLANATIONS

Energy flow

The standard, text-book explanation maintains that food chains are short because energy transfer between links is inefficient. Eventually, there is too little energy to support another link in the chain (Hutchinson 1959). As we have seen, Elton (1927) believed this was one reason why food chains are short, and in the limit, the proposition is obviously true. To maintain a single predator at a trophic level higher than twenty requires an area of land bigger than some continents (Slobodkin 1961). But the energy hypothesis conspicuously fails to explain why food chains are as short as they are. More specifically, it predicts that food chains should be longer in more productive ecosystems (Pimm & Lawton 1977). Oksanen *et al.* (1981) present some data that may be consistent with this prediction; other studies are not (Pimm 1982). However, both these tests are based on limited information, and neither is definitive. Nor is the observation that food chains differ in length definitive even though energy inputs are very similar (Kitching 1983; Kitching & Pimm 1985; Pimm & Kitching 1987), although such data suggest that the length of food chains is determined by something other than energy flow.

The most comprehensive test of the hypothesis that food chains should be longer in more productive ecosystems is by Briand & Cohen (1987). They found no difference in food chain length comparing twenty-two published webs from habitats with a mean productivity less than 100 g of carbon $m^{-2} \cdot year^{-1}$, with ten webs in which productivity exceeded 1000 g of carbon $m^{-2} \cdot year^{-1}$, and six webs from intermediate habitats. It is remarkable that other, similar tests have not been attempted based on good, new data on the structure of food webs along gradients of productivity. It is equally remarkable that manipulative, experimental tests have also been so slow to emerge. A pioneering attempt has now been made by Pimm & Kitching (1987). I admire their idea, but have reservations about what the results tell us.

Pimm and Kitching created small, artificial analogues of water-filled tree holes, sustained by a supply of leaf litter, in a Queensland rain forest where similar natural tree holes abound. They varied 'productivity' by setting litter inputs from half to twice the natural rates, a four-fold range, and argued that because 'the percentage of energy consumed that goes into new tissues is between 20 and 50% in short-lived ectotherms like insects' (typical tree-hole inhabitants), an extra trophic level is expected between the least and the most productive containers. I disagree. It is not production/consumption ratios for invertebrates that lie in the range 0.2–0.5, but ratios of production/assimilation (McNeill & Lawton 1970; Humphreys 1979). Hence to energy losses measured by the ratio of production to assimilation must also be added food that is consumed and not assimilated *plus* energy made available by production in the previous trophic level that is not consumed by the next. In other words, what matters is the overall 'ecological efficiency' of energy transfer between steps in the food chain, and as Slobodkin (1961) showed long ago for small aquatic invertebrates, this efficiency is unlikely to be much better than 10%. In consequence, I would not expect a four-fold range in productivity to extend the length of tree-hole food chains by more than an average of about 0.4 of a step. Pimm and Kitching did not find longer food chains developing in their more productive containers, and claim that this result is inconsistent with the energy hypothesis. That may be so, but it may also be that they did not vary energy inputs enough to see an effect. The result is therefore equivocal. These experiments are being repeated using a wider range of productivities (S.L. Pimm, personal communication); the results will be fascinating.

Consideration of ecological efficiencies permits an alternative test of the energy flow hypothesis. Endothermic animals have much lower ecological efficiencies than invertebrate ectotherms (about an order of magnitude less; Lawton 1981). Hence, for a given level of energy input, food chains of ectotherms should be longer than endotherm chains (Pimm 1982). Some clever accounting by Yodzis (1984a) based on thirty-four published food webs suggests that this is indeed the case. The number of additional links in food chains supported by vertebrate endotherms is significantly less than the number of additional links supported by invertebrate ectotherms; vertebrate ectotherms are intermediate both in ecological efficiency and in the number of additional links which they support. Yodzis' intriguing result stands as sole support for the hypothesis that food chains are short because of energetic constraints. It is not easily reconciled with Briand & Cohen's (1987) analysis, and at the moment serves to highlight how little we really know about so fundamental an ecological problem.

Dynamic constraints

Predictions derived from Lotka–Volterra models

Simple Lotka–Volterra models of food webs can be used to make a number of predictions about structure. The rules are straightforward. To mimic the real world, model food webs must be feasible (all populations with positive equilibria) and locally stable (returning to equilibrium after small perturbations), because (ignoring locally unstable, but persistent solutions, for example limit cycles) structures and parameter values that lead to non-feasible, unstable models will be transitory, rare or absent in the real world (Yodzis 1981; Pimm 1982). These simple dynamic constraints generate several patterns in model food webs that are also common in real webs (Table 3.1). They include a general lack of omnivory except in insect webs (Pimm & Lawton 1978; Pimm 1979); low connectance and an upper bound between connectance and species richness that declines hyperbolically as the number of species in the web increases (May 1972); reticulate webs that are more stable than compartmented webs (Pimm 1979, 1982), [although theoretical views on this point are contradictory, e.g. May (1972, 1979); O'Neill *et al.* (1986)]; and an approximately constant ratio of predator species to prey species (Mithen & Lawton 1986). Adding one more dynamic constraint to the modelling process, namely that populations with the ability to return rapidly to equilibrium after a disturbance are more likely to persist in the face of repeated environmental shocks, leads to a further prediction. Long food chains have long return times; hence food chains in the noisy real world must be short (Pimm & Lawton 1977), and should be shorter in more variable environments. Finally, elaborating Lotka–Volterra models to incorporate the effects of age and size/structure (Pimm & Rice 1987) reconfirms the destabilizing effects of omnivory, and finds that feeding loops (A eats B eats C eats A) are strongly destabilizing. Both omnivory and loops are more likely on natural history grounds if members of a food web change markedly in size during development, and feed on different things as they grow (e.g. Hardy 1924; Darnell 1961).

By inference, the more variable structure, greater connectance and more extensive omnivory shown by webs from constant environments may also be a product of more stringent dynamic constraints for persistence and stability in fluctuating environments (May 1981, 1986; Briand 1983), although there have been no formal tests of this possibility, for example by modelling Lotka–Volterra food webs in a stochastic environment. Last but not least, simple dynamic constraints do not

TABLE 3.1. Summary of patterns found in published food webs, together with a list of theoretical explanations advanced to explain the patterns. ● signifies that theory predicts the observed pattern; ○ signifies that it does not. A blank cell denotes that no attempt has been made to link that combination of pattern and theory. Parentheses signify tentative or uncertain predictions. (See text for further details)

	Observed patterns									
Theoretical explanations	Feeding loops absent i	Links per species constant. Connectance declines hyperbolically with increasing number of species ii	Constant proportion of basal, intermediate and top species (constant predator/prey ratio) iii	Constant proportion of links between basal, intermediate and top species iv	Omnivory rare, except in insect–parasitoid and donor controlled webs v	Webs from constant environments have more connectance, more variation and more omnivory vi	Food chains are short vii	Food chains shorter in 2-dimensional habitats viii	Webs are not compartmented except at habitat boundaries ix	Food webs are interval x
Energetic constraints							(●)			
Standard dynamic models in Lotka–Volterra form	●	●	●		●	(●)	●		●	○
Donor controlled dynamics		○			●		●			
Neutral interactions (no dynamics; links assigned randomly, subject to minimal biological constraints)					○		(○)			(○)
Cascade model generated by body-size constraints (no dynamics)	●	Assumed by model	●	●	(○)		●			(●)
Natural history, optimal foraging and other evolutionary constraints		●			●		(●)		●	

appear to explain why food webs are predominantly interval (Cohen 1978), although this problem has not been particularly well explored (see, for example, Pimm 1978, 1982 and Cohen 1983).

Drawing these arguments together, an impressive proportion of the patterns seen in real food webs enhance the stability of model webs, with the reasonable inference that real webs are strongly, though not entirely structured, by similar dynamic constraints (Pimm 1982; Pimm & Lawton 1983). Given that a central aim of science is to explain as much as possible about nature with the minimum number of assumptions, this is an exciting result. It has a level of elegance and generality that is all too rare in ecology. But elegant results are not necessarily true. There are two problems; first, the models may be making the right predictions for the wrong reasons, and second, there may be alternative ways of generating the same patterns. Next, I review a limited number of independent tests of the dynamic constraint hypothesis, before moving on to consider problems in the assumptions that underpin the Lotka–Volterra models. Alternative explanations for food web patterns follow in later sections.

Independent tests of model predictions

Schoener & Spiller (1987) show how dynamic constraints in real food webs may eliminate species and reduce omnivory, although this was not the purpose of the original experiment. On Caribbean islands, spiders prey on small insects and are themselves eaten by lizards. The lizards are omnivores in the sense of Pimm & Lawton (1978) because they too feed on small insects. Experimentally removing lizards greatly enhanced spider density and diversity; in the presence of lizards this middle trophic level was much reduced because not only were spiders preyed on by lizards, but they also competed with them for insect prey, exactly as the models predict (Pimm 1982). In other words, omnivory in real systems may be rare, because intermediate species are squeezed out of the food chain. In the case documented by Schoener and Spiller, only some species of spiders suffered this fate, but the general mechanism is well illustrated.

The only prediction arising from Lotka–Volterra models of food webs that has been subject to explicit, independent tests is the argument that food chains should be shorter in more variable environments. Although Briand & Cohen (1987) find that this is indeed the case, their sample of food webs confounds environmental variability and habitat structure, and they conclude that when the latter is factored out, constant environments do not support markedly longer food chains. My own view is that assigning webs to 'constant' or 'fluctuating' environments in

Briand and Cohen's compendium is so subjective (based on the impressions of the original authors who drew up the webs), and sample sizes are so small when habitat structure has been factored out, that no firm conclusions can be drawn. This is a classic case of a clever idea being frustrated by a shortage of suitable date; criticism is easy, but doing anything better is not!

The same water-filled tree holes that provided a test of energetic constraints on food chain length also provide a test of dynamic constraints. Although total energy inputs in the form of leaf litter are very similar in tree holes in England and Queensland, inputs are spread more constantly over the year in Queensland where climatic fluctuations are less. As predicted, food chains are shorter in English tree holes (Kitching 1983). Similar empirical observations on a wide range of 'phytotelmata' are made by Kitching & Pimm (1985), who conclude that variations in 'spatial and temporal uncertainty' are a major contributor to differences in food chain lengths in these particular systems. Moreover, species feeding higher in the food chain also recruited more slowly than lower levels after an experimental disturbance to the Queensland system (Pimm & Kitching 1987). A reasonable inference is that frequent disturbances make it impossible for high ranking carnivores to persist in the community. But it is not reasonable to conclude that these observations test and vindicate simple Lotka–Volterra models of food webs. The problem is that Pimm and Kitching's result is also consistent with an alternative dynamic model (see p. 57). It is to this, and other alternatives that we must now turn.

Problems with Lotka–Volterra models

Why use them at all?

Ecologists mistrustful of mathematics have frequently pilloried the use of simple models in ecology, I have no intention of rehearsing these arguments here. However, all models make assumptions, and in the present context some of these assumptions make a big difference to 'permitted' structures in model food webs, and need to be confronted.

This is not to say that I regard the existing models as a waste of time. Fifteen or 20 years ago, food web theory had progressed very little since Elton (1927), and interest in developing such theory was moribund; food webs were about energy flow, and that was that. There were a few fascinating exceptions to this otherwise lack-lustre field, for example MacArthur (1955), Elton (1958), Paine (1966, 1969) and May (1972,

1973). But the idea that there might be regular patterns in the structure of food webs was not part of the intellectual baggage of most ecologists. If they thought about the problem at all, they most probably subscribed to what Stuart Pimm and I privately referred to as the 'tangled knitting hypothesis' of food web structure; you can sometimes see interesting shapes, but unravelling them is impossible. It therefore seemed worth while to strike out in a new direction and discover whether population dynamic interactions between species could generate simple rules constraining the structure of food webs, and to seek evidence for dynamic constraints in the real world.

In reality, the idea was not that new. As we have already observed, Charles Elton clearly recognized the potential importance of dynamic interactions between species in food webs in 1927. Returning to the theme in *The Ecology of Invasions by Animals and Plants* (1958), he explicitly discusses the possibility of modelling interactions between species in a food chain using Lotka–Volterra equations (pp. 130–131) (cautioning, however that such models are 'oversimplified'). Of course they are simplified, but as in all science, it is often interesting to see what progress can be made with the minimum number of assumptions. The results of this adventure are summarized in Pimm (1982), who provides a bold, lucid and internally highly consistent review. Now the dust has settled, my own view is that simple Lotka–Volterra models get some of the answers right for more or less the right reasons; and some they get right for quite the wrong reasons. These arguments will become clearer as we proceed.

Interaction strength, closely coupled enemy–victim interactions, spatial heterogeneity and other assumptions

There are a number of important ways in which the predictions of simple Lotka–Volterra models of food webs can be altered by making different assumptions, or by using rather different models. For example, combining population dynamic and energy flow constraints in one model alters the prediction that high connectance necessarily creates instability (DeAngelis 1975; Kirkwood & Lawton 1981). Or consider the hypothesis, based on return-times, that long food chains are dynamically fragile. As May (1981) points out, this prediction hinges on some subtle mathematics; Pimm (1982, 1984) assembles arguments in its defence. The problem is that alternative models give contradictory signals. Difference equation models of insect host–parasitoid–hyperparasitoid interactions bolster the conclusion that long food chains are dynamically fragile, based on familiar local stability criteria rather than more contentious return-time arguments (Beddington & Hammond 1977). In contrast, DeAngelis *et al.* (1978) and DeAngelis (1980) show that long food chains may still be stable, providing the transit time for a molecule or unit of energy through the web is fast. An obvious corollary of this work is that species of small body size, with short generation times and high weight-specific metabolic rates (Peters 1983), should build up longer food chains than large-bodied species. The difficulty is that body size is implicated in food web structure in other, much simpler ways, making the argument a difficult one to test.

Over and above these specific examples are two much more general and serious problems. The first is the assumption that links in food webs all involve closely coupled enemy–victim interactions; the second is that the interactions take place in a spatially homogeneous world. It is inconceivable that both assumptions are always met in the sample of published food webs that ecologists have used to determine patterns in nature; they are probably approximately true for some of the links. Yet changing either assumption changes the predictions of food web models, making it hard to avoid the conclusion that for a significant, but unknown, proportion of real food webs, simple Lotka–Volterra models are making the right predictions for the wrong reasons.

Pimm (1982) argues forcefully that predators usually affect the dynamics of prey populations. Hence models in which predators affect prey populations and prey populations in turn influence predator dynamics (the classical, closely coupled Lotka–Volterra assumptions) are appropriate descriptions of the real world. Nobody denies that enemies can and do significantly influence the population dynamics of myriads of victim species (e.g. Huffaker 1971; Strong, Lawton & Southwood 1984; Sih *et al.* 1985, Carpenter *et al.* 1987), and that many food webs contain examples of strong interactions involving 'keystone' predators (Paine 1966, 1969, 1971, 1980; Krebs 1985; Menge & Sutherland 1987). But common sense and data both suggest a continuum of possibilities. At one extreme are closely coupled enemy–victim interactions; at the other are totally neutral feeding links of no consequence for the dynamics of either consumer or consumed. Between them are two possibilities, namely 'donor controlled' dynamics in which victim populations influence enemy dynamics but enemies have no significant impact on victim populations, and the alternative in which enemies influence victim populations, but not vice-versa. (This latter possibility may occur with very polyphagous predators, where individual species of prey have trivial effects on the predator's dynamics. The problem has received no attention in the food web literature, and will not be considered further here.) To some extent, Lotka–Volterra models that sample from a range of possible parameter values, embracing highly efficient as well as ineffective predators (e.g. Pimm & Lawton 1977, 1978; Pimm 1982; Mithen & Lawton 1986) may allow for all these possibilities. But my feeling is that in general, such models still give too much emphasis to tightly coupled enemy–victim interactions.

Consider donor control first. There are technical problems in designing field experiments that are both powerful enough and on the right spatial and temporal scales to detect significant predator impacts on some victim populations (e.g. Carpenter & Kitchell 1987; Johnson *et al.* 1987), but even with this caveat, donor controlled dynamics do not appear to be unusual. A familiar example is consumers in decomposer food chains that may have no influence on the rain of plant material and animal bodies that sustains them (Pimm 1982). Odum & Biever (1984) provide several other examples, including nectarivores and possibly many seed feeders. It seems extremely unlikely that many rare phytophagous insects, whose species are legion, have any impact upon the abundance of their host plants (Caughley & Lawton 1981; Strong, Lawton & Southwood 1984). Biological control agents frequently establish, but fail to have any measurable impact on the target pest (e.g. Kelleher & Hulme 1984; Hokkanen 1986). Parasitoids and predators of insects in more natural situations may display dynamics that appear to be predominantly donor controlled (Dempster & Pollard 1981). Evidence for significant top down control is also conspicuous by its absence in many freshwater food chains (Harris 1985; Thorp 1986). Even the classical food chain from trees to caterpillars, small birds and sparrowhawks appears to be donor controlled at the link between small birds and hawks; 'there is no good evidence of sparrowhawks having a widespread effect on the breeding numbers of their prey, (however) there is circumstantial evidence for the reverse, that sparrowhawk numbers are affected by prey numbers' (Newton 1986). The 'circumstantial' evidence, incidentally, is impressive.

In the limit, we must expect many documented feeding links in food webs to have trivial consequences for both enemy and victim. Paine (1980, 1983, 1988) provides some of the best examples (see also Menge & Sutherland 1987). One problem is that the population dynamics literature is undoubtedly biased towards 'significant' enemy–victim interactions. If newspapers were our sole source of information about society, we could imagine that people spend most of their time in highly unpleasant, exciting and strong interactions with one another. The reality is that for most of us, most of the time, life is humdrum; no news may be good news, but it doesn't sell papers. The extent of a similar bias in the ecological literature is hard to gauge, but it is undoubtedly there. However, in drawing up a food web, there is at least the opportunity to illustrate links that nobody seriously believes would be interesting to study in their own right. The problem is that there are no agreed criteria for what constitutes a significant link (May 1983, 1986; Paine 1988). Common sense suggests that only some of the links in published webs are best modelled by closely coupled enemy–victim interactions, or by donor control, and that many have trivial dynamic consequences.

The last, crucial assumption in the simple models is that encounters between enemy and victim populations take place at random in a spatially uniform world. In reality, most populations have clumped distributions, and many (though by no means all) predators and parasitoids search non-randomly and aggregate their attacks on the high density patches (e.g. Hassell 1978; Lessells 1985). Herbivores may behave in a similar manner (Kareiva 1983; Strong, Lawton & Southwood 1984).

Effect of changing the assumptions on model predictions

Compared with standard Lotka–Volterra models, donor controlled interactions have very different dynamics, and make very different predictions about web structure (Pimm 1982). There are no longer tight constraints on overall connectance (DeAngelis 1975), and rampant omnivory is not destabilizing. It is unclear how other food web patterns are affected. Intriguingly, preliminary evidence suggests that omnivory may be common in some decomposer food webs (Macfadyen 1979; Pimm 1982; Pimm & Lawton 1983; Walter 1987), where we expect to find donor control. This apparent pattern is crucial to the present arguments and deserves more attention.

Species feeding high in the food chain of a donor controlled system must still be vulnerable to stochastic extinction, simply because they will have smaller populations than more basal species (Leigh 1981; Fowler & MacMahon 1982). Hence experiments or observations showing that food chains are shorter in more unpredictable environments (see above) are consistent with both donor controlled models of food webs, and predictions based on standard Lotka–Volterra equations, and cannot be used to distinguish between them.

If feeding links are dynamically trivial, species can be assembled in food webs free of any dynamic constraints. Below, we shall explore models of food web assembly free of dynamic constraints, but with other limitations. Here I want to restrict attention to cases where assembly is virtually a free for all. The outcome is rather unsatisfactory; model food webs with links assigned at random, subject to minimal, biologically sensible constraints appear to have too many trophic levels and too much omnivory compared with real webs (Pimm 1980a, 1982), and may also be non-interval more often than real webs (Cohen 1978). Unfortunately, both studies are not without problems, and require cautious interpretation (Pimm 1978; Cohen, Briand & Newman 1986).

Yodzis (1984b) attempts a similar exercise, assembling food webs without dynamics, but subject to the rules of ecological energetics. He shows that it is possible to construct food webs that match many of the properties of forty real webs gathered together by Briand (1983), particularly for webs from fluctuating environments. However, I have serious reservations about this result. First, the assembly process terminates abruptly when energy availability at the top of assembled food chains falls below an arbitrary value; without such a rule, the simulated food chains would be too long. Second, concordance between real webs and their model counterparts is obtained by varying primary productivity and ecological efficiency. Although the ranges of parameter values used for both are defensible (though ecological efficiences tend to be rather low), it is unclear how they were selected to simulate individual webs. The primary productivities, in particular, bear no resemblance to actual productivities, where these are known for particular webs (compare values in table 1 in Briand & Cohen 1987, with those in table 1 of Yodzis 1984b). Accordingly, although the idea is interesting, I do not think that we learn very much that is useful from this particular analysis.

Worries about the effects of spatial heterogeneity on model predictions centre on cases where predators have a significant impact on prey populations. Spatially heterogeneous trophic interactions tend to be markedly stabilizing in a wide variety of population models (Hassell & May 1985; Chesson & Murdoch 1986; DeAngelis & Waterhouse 1987; Sih 1987). They have not been used to investigate properties of model food webs systematically, but some of their effects are not difficult to predict. For example, the standard Lotka–Volterra models used throughout Pimm's book incorporate 'self-damping' terms only at the base of food chains. Species higher in the trophic stack experience intraspecific competition via resource exploitation, but not via behavioural interactions, for example territoriality. Strong and effective self-damping is one way to produce donor control. The extent to which self-damping should be incorporated into model food webs has been debated by Saunders (1978), Lawton & Pimm (1978), Yodzis (1981) and Pimm (1982). These arguments focus on the significance of conventional intraspecific behaviours that lead to self damping, and by and large I believe they are correct to play down their importance in food web dynamics. What they conspicuously overlook is the fact that predator aggregation and the differential exploitation of prey in high density patches generates dynamic effects that are identical to more familiar forms of intraspecific behavioural interactions between predators (Murdoch & Reeve 1987). For this reason, Free, Beddington & Lawton (1977) coined the term 'pseudo-interference' to describe them. Most of the theoretical work on pseudo-interference has been confined to insect host–parasitoid systems, although there is no reason to believe that parasitoids have a monopoly on the problem. Permitting predators in general to contribute to stability by incorporating aggregative responses into food web models is likely to lead to significantly more robust systems, although it may not alter many of the qualitative predictions generated by simpler models (May & Hassell 1981; Pimm 1982). The problem therefore remains open.

CASCADE MODEL AND BODY SIZE

An alternative major hypothesis

It should now be clear that although simple Lotka–Volterra models correctly predict the existence of a number of patterns in published food webs, the worry must be that they are giving us the right answers for the wrong reasons, at least sometimes. As Pimm (1982) remarks (p. 125), science is most interesting when there are at least two rival hypotheses consistent with published data. A series of papers by Joel Cohen and his colleagues provide one such rival hypothesis (Cohen & Newman 1985; Cohen, Newman & Briand 1985; Cohen, Briand & Newman 1986; Newman & Cohen 1986).

The hypothesis is a static model of the assembly of species into food webs; no dynamics are involved. Links are assigned essentially at random. But it differs in one crucial way from the static models of food web assembly outlined above, by imposing a simple constraint. The model assumes that species can be arranged *a priori* into a cascade or hierarchy such that a given species can feed on only those species below it, and can itself be fed upon by only those species above it in the hierarchy. Cohen and his co-authors refer to it as the cascade model. Precise details of the model, and the way in which it is used to generate predictions about food web structure are in Cohen & Newman (1985). Over and above the apparently arbitrary assumption of a trophic hierarchy, the model also makes one other key assumption. In order to use it to assemble model food webs, the density of links per species (*d* on p. 46) must be specified and fixed in advance. Cohen *et al.* use the observed values in published webs, either an average over a sample of webs, or a single value from each web, depending upon their question. Hence the cascade model cannot say anything about pattern (ii) on p. 46, because it is already built into the model. We will return to this constraint on connectance later.

Food webs assembled according to this recipe have the correct proportions of basal, intermediate and top species, and the correct distribution of trophic links, except the cascade model tends to generate webs with too many links between basal and top species (Cohen & Newman 1985; Cohen, Newman & Briand 1985). It also gets close to predicting the lengths of most published food chains. However, there is a tendancy for the model to create food chains that are too long (S.L. Pimm, personal communication; and see Cohen, Briand & Newman 1986), and it also tends to be wide of the mark in predicting chain-lengths for 16 or 17 webs in a total of 113 which have either an unusually high (more than four) or an unusually low (less than two) mean number of actual links (Cohen, Briand & Newman 1986). The length of food chains grows very slowly as increasing numbers of species enter the assembly process under the cascade model; even with a million species, the median number of links in the longest chain is below seventeen (Newman & Cohen 1986).

An excess of predicted over observed links between basal and top species is not necessarily the same as too much omnivory. Top species that feed both on basal species and intermediate species are omnivores. But chains of two species (a basal species fed on by a top species) cannot have omnivores. Accordingly, it is not possible to tell whether the excess of basal to top links in Cohen *et al.*'s simulated webs is due to too much omnivory, or to too many two-species chains. The cascade model may also obscure patterns of omnivory generated by links within the category of intermediate species, and between intermediate and top and intermediate and bottom species. Hence, although it is plausible that the cascade model predicts too much omnivory, this is by no means certain.

Why the cascade model? The role of body size

Cohen *et al.* offer no explanation for why species in food webs feed only on individuals below them in the hierarchy, and are themselves fed on only by individuals higher in the hierarchy. What determines the order of species? How is the hierarchy created? One simple possibility, unequivocally recognized by Elton over 60 years ago is body size (Warren & Lawton 1987; and see Paine 1963). As Elton points out, although it is not universally true, very commonly, predators are larger than their prey, both in vertebrate (Gittleman 1985; Verezina 1985) and in invertebrate (Warren & Lawton 1987 and references therein) food chains. The result is a trophic hierarchy based on body size, sufficient to conform to all the requirements of the cascade model. Plants at the bottom of the cascade reinforce its structure, as do detritus particles of a wide range of sizes (Cousins 1980, 1987). Limnologists and marine biologists have long recognized the importance of size in trophic interactions (e.g. Kerr 1974; Platt & Denman 1977; Harris 1985; Platt 1985 and references therein). The concept of a 'trophic continuum' in food webs, in which body size plays a crucial role has been most clearly developed by Cousins (1980, 1987). The only additional argument advanced by Warren & Lawton (1987) is to link these familiar effects of body size to the cascade model.

There are, of course, exceptions to the generalization that predators are usually larger than their prey, but they are revealing. Again as Elton pointed out, insect parasitoids and parasites in general are *smaller* than their hosts. Imagine a matrix with species listed along the rows and again down the columns. Non-zero elements in the matrix define the presence of feeding links between species. Technically, feeding links in the cascade model are 'upper triangular'; that is all the links lie above the leading diagonal of the matrix (see Cohen & Newman 1985 for further details). A food web matrix with species arranged from the smallest to the largest along the rows and down the columns will be strictly upper triangular if predators can feed only on species smaller than themselves (Warren & Lawton 1987). Parasitoids and parasites that can attack only species larger than themselves create food web matrices that are 'lower triangular', and the cascade model still holds (flipping a lower triangular matrix over round the main diagonal makes it upper triangular). Mixed food webs of parasites and 'real' predators may contain entries in the matrix above and below the main diagonal, and violate the assumptions of the cascade model. But in the published food web literature, mixed parasite and predator food webs are very rare. People either study 'real' predators, or food webs of insects and their parasitoids; they rarely study both (but see Hawkins & Goeden 1984), and hardly anybody incorporates parasites or diseases into food web diagrams (see below). Hence published food webs conform to the assumptions of the cascade model, based on a hierarchy of body sizes, not because there are no exceptions, but because nobody has really studied them.

Much the same may be said about obvious exceptions to the rule that predators are usually bigger than their prey. Group hunters, employing what Elton called 'flock tactics', are capable of overpowering prey much larger than themselves (hunting dogs, or driver ants for example). However, such predators are again conspicuous by their absence in published food webs. In other words, a trophic cascade generated using body sizes will only describe a subset, albeit a major subset, of trophic links in real communities (Warren & Lawton 1987).

Intervality

Patterns of overlap in prey use by predators, sufficient to generate interval food webs are a possibility if feeding relationships are constrained by body size. If prey species increase in size from A to D, and predators increase in size from 1 to 3, the resulting food web is interval if predator 1 takes A and B, predator 2 takes B and C, and predator 3 takes C and D. Intervality is conserved with more overlap, providing, for example, that species 3 takes B, C and D, or indeed all four prey species. But the web is not interval if 3 feeds on A, C and D, but never on B. As Cohen (1978) points out, any food web in which at least one kind of prey is eaten by all predators is interval. In other words, a trophic cascade generated using body size as the ordering variable has the potential to produce interval food webs. However, a recent unpublished study (reported in Cohen & Newman 1988) finds that although food webs made up of small numbers of species, assembled according to the cascade model, have a high probability of being interval, this is not true of large webs. Nor is body size the only niche dimension capable of generating intervality (Cohen 1978; Sugihara 1983, 1984). The problem is therefore unresolved.

OTHER EXPLANATIONS

A rag bag of hypotheses and explanations for some of the observed patterns in published food webs may be brought together under the twin banners of natural history and optimal foraging. Natural history (a Jack of all trades is master of none) may constrain species to feed on only a limited number of others in the community, generating pattern (ii) (Pimm 1980b, 1982). If food webs are divided into compartments anywhere, it will be at obvious boundaries between habitats (Pimm & Lawton 1980; Pimm 1982). Omnivory may also be subject to simple biological constraints that are relaxed in detritus food chains; bodies don't bite back. Although patterns of omnivory do not seem to be entirely determined by simple natural history constraints (Pimm & Lawton 1978, 1983; Pimm 1982), a more recent and more thorough analysis by Yodzis (1984c) suggests that the difficulty of feeding simultaneously on both plants and animals is sufficient to account for the relative shortage of omnivory in published webs. Probably all these simple natural history constraints contribute to observed patterns.

Finally, it has been suggested that obvious advantages may accrue to species by feeding as low in the food chain as possible, where resources are more abundant, whilst avoidance of competitors may favour being higher in the chain; the length of food chains reflects these counterbalancing selective forces (Hastings & Conrad 1979; Pimm 1982; Pimm & Lawton 1983; Stenseth 1985). The length of food chains may also be constrained by 'engineering' problems; a predator large enough to feed on eagles may not be able to fly (see Pimm 1982 for further discussions).

Habitat structure and the length of food chains

None of the theoretical explanations gathered in Table 3.1 explains Briand & Cohen's (1987) observation of shorter food chains in 'two-dimensional' habitats. One possibility is that structurally complex 'three-dimensional' habitats create prey refuges that either stabilize predator–prey interactions (Hassell 1978) or generate donor-controlled dynamics; both favour the persistence of long food chains. The problem is that 'solid' habitats, open water for example, also have long food chains, but lack structural complexity. A more fundamental worry is the arbitrary nature of the habitat classification based on human scales and perceptions. A prairie may look two-dimensional to a buffalo or a biologist, but it is anything but two-dimensional for a beetle. Habitat surfaces are fractal (Morse *et al.* 1985; Lawton 1986a), so that 'living space' must be defined according to the size(s) of the organism(s) under investigation. *A priori* it is difficult to see why Briand & Cohen's anthropocentric habitat classification generates anything very interesting. But obviously it does. Why is a mystery.

DISTINGUISHING BETWEEN EXPLANATIONS

Table 3.1 summarizes the theoretical arguments from the last section, and shows that most of the empirical patterns teased out of published food webs could be generated by more than one mechanism. Despite a strong historical association between food chains, trophic levels and energy flow, it is clear from Table 3.1 that ecological energetics has little to contribute to an understanding of structure in food webs, aside from the important question of whether energy flow or something else limits the length of food chains. There are no less than five plausible explanations for why food chains are short.

Two major chunks of theory can each account for several, but not all, food web patterns. One focuses upon dynamic interactions between species, the other on simple, static assembly rules in which body size appears to play an important part. Both also have to compete with an untidy group of 'natural history' constraints that may also explain an important subset of patterns. On present evidence no one theory or group of theories can claim precedence over any other as the best or most general explanation for structure in food webs.

In trying to understand the likely relative contributions of these conflicting explanations, we need to remember one simple fact. The quality of the published data is so variable and idiosyncratic, and so highly skewed in the direction of 'poor' on a scale from awful to excellent (Cohen 1978; McCormick & Polis 1982; Pimm 1982; May 1983; Paine 1983, 1988; Cohen & Newman 1985; and many others) that several of the patterns may eventually prove to be artefacts. As we have already noted (p. 45), many 'species' are in fact aggregations of various taxa. Equally serious are major variations in the spatial and temporal scales over which data have been gathered. Figures 3.1 and 3.2 show how the perceived structure of two food webs changes with increasing spatial and/or temporal investigation (Kitching 1987; P. H. Warren, personal communication). These data speak for themselves.

Worries about artificial patterns aside, the poor quality of the data also implies that explanations for perceived patterns must be very simple. It defies logic to suggest that the motley collections of binary links identified by sampling food webs in nature with varying degrees of efficiency are all dynamically important, or that dynamics alone explain the structure of any particular web. However, providing only that real predators take prey smaller than themselves (or parasites attack hosts bigger than themselves), the cascade model should accurately predict

---

FIG. 3.1 Several versions of the same food web for animals inhabiting water-filled tree holes in south-east Queensland (from Kitching 1987). (a) shows the full regional web, based on all sampling stations and seasons. There are nine taxa in the regional web, indicated by circles 1–9. Feeding links between taxa are indicated by lines. In most cases these links are unambiguous (e.g. 9 feeds on 1, 6, 7, etc.), but it is not possible to identify enemies and victims with certainty for some of the links in Kitching's original diagram (e.g. it is unclear whether 6 feeds on 7, and/or 7 feeds on 6); fortunately these details are not important in the present context. The 'square' on species 9 indicates that it is cannibalistic. (b) and (c) show webs at particular sites on particular sampling occasions. The positions of particular species in (a)–(c) are constant, so that the missing components in (b) and (c) are easily identified. The base of the food chain in these tree holes is detritus; the boxes at the base trace the breakdown of detritus particles. The perceived structure of the food web varies with the temporal and spatial scale of sampling.

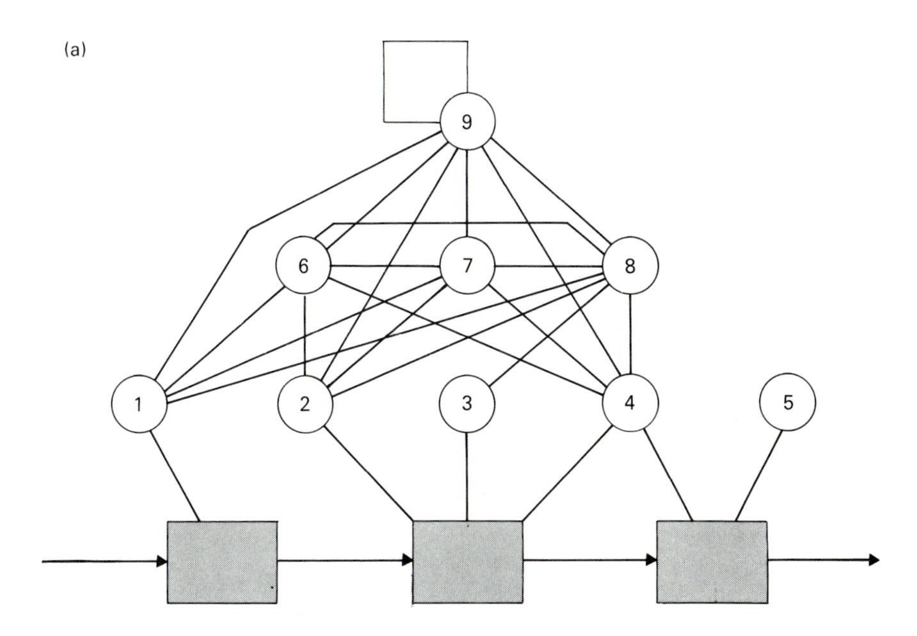
(a)
9
6
7
8
1
2
3
4
5

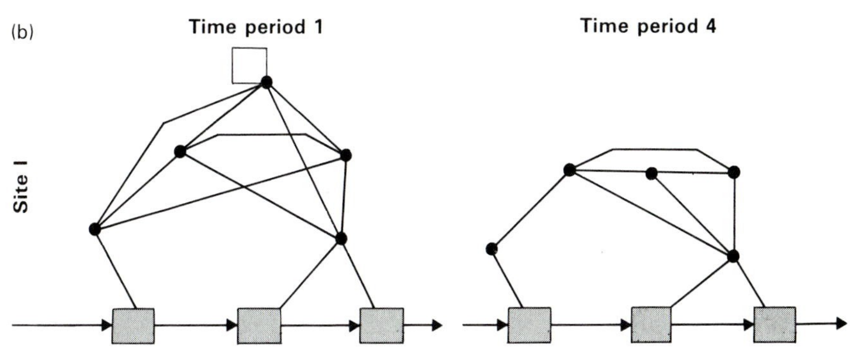
(b)
Time period 1
Time period 4
Site I

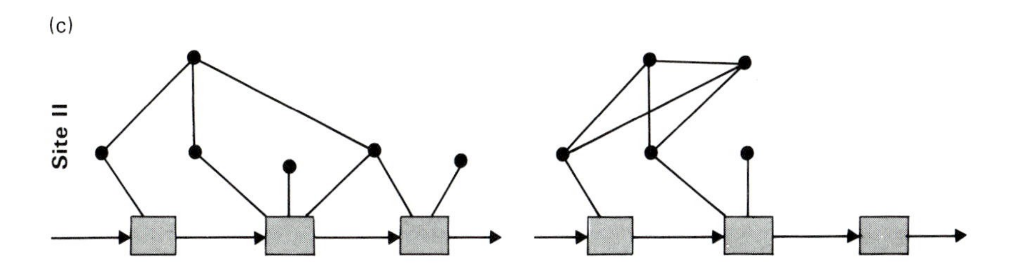
(c)
Site II

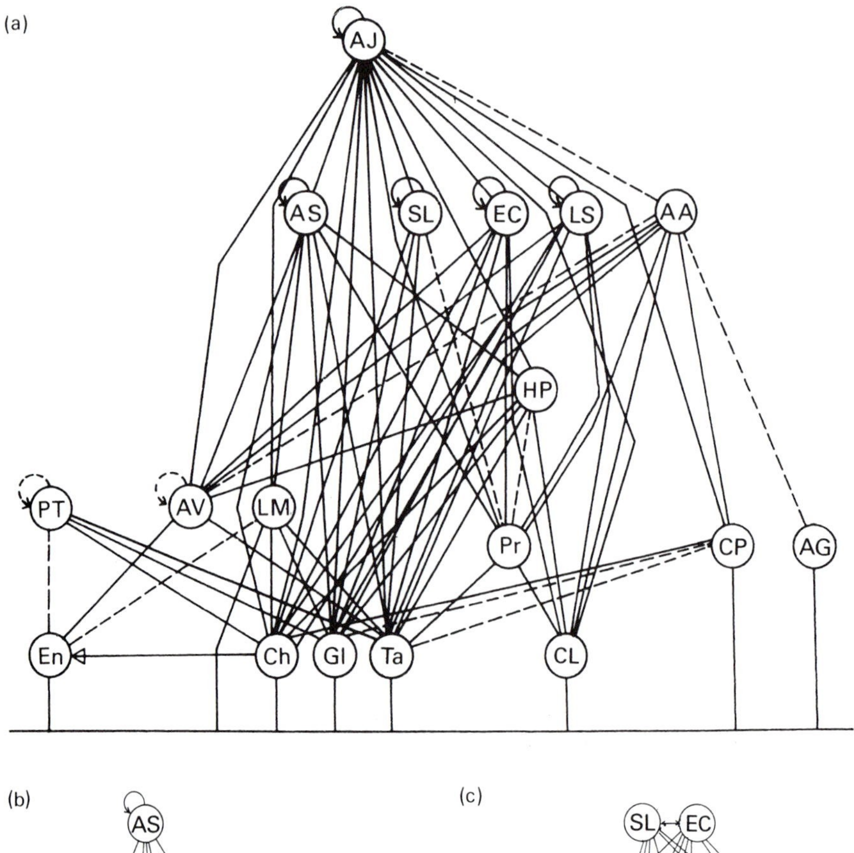

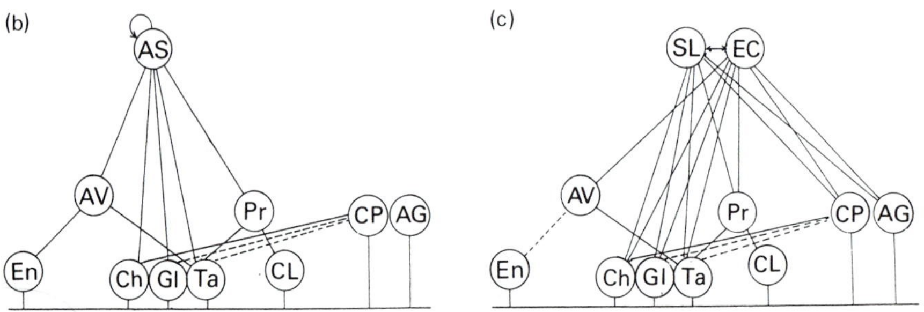

FIG. 3.2. Three versions of the same food web for animals inhabiting an acid pond at Skipwith Common, North Yorkshire (P.H. Warren, personal communication). (a) shows the full web at the weedy margin of the pond, based on all sampling dates. Also shown are webs for the open water in spring 1986 (b) and spring 1987 (c). No species occur in the open water that do not also occur at the margins. The positions of particular species, together with their coded identities, are the same in (a)–(c), so that the missing components in (b) and (c) are easily identified. Dotted lines represent probable links. Cannibalism is indicated by a small loop from each species back to itself. All links are from victims to predators at higher positions in the web, except for the link between SL and EC in (c), where each species preys on the other. As in Fig. 3.1, the structure of the food web varies spatially and temporally.

structure in the crudest data imaginable. My intuition therefore says that several of the patterns in published food webs are generated by the cascade model, with species body sizes acting as the key, organizing variable. In other words, Elton got it more or less right.

Where does this leave predictions based on dynamic models? Undoubtedly there are strong interactions in food webs, and for this subset, Lotka–Valterra models may make the right predictions for more or less the right reasons. Both static and dynamic processes then march together to yield observed patterns. [Parenthetically, note that it would be interesting to construct dynamic models of food webs explicitly incorporating species of a range of sizes, with the parameters scaled accordingly, e.g. Peters (1983)]. In comparing dynamic models and the cascade model, notice also that the density of links per species (pattern ii on p. 46) must be fixed at observed levels before the cascade model yields the right answers. It is far from clear why each species in a food web should, on average, feed on the same number of other species, irrespective of the number of potential victim species (May 1986; Warren & Lawton 1987). Elementary biological constraints must often restrict species' diets, but is it as simple as that? Too much connectance destabilizes model food webs (p. 50), and hence may impose limits on connectance in the real world. If so, a key parameter in the 'static' cascade model may be a product of dynamic interactions between species. A pioneering theoretical attack on this problem is provided by Cohen & Newman (1988). Of course, these are highly speculative arguments. If nothing else, they serve to remind us how little we still know about the structure of species assemblages.

As the quotations in the Introduction make plain, Elton believed that several processes contributed to the structure of food webs. Sixty years on, Table 3.1 suggests that he was correct. The effects of body size are clearly important, but they are unlikely to be the only processes involved. The spotlight must now focus on the relative contributions of static assembly, body size, dynamics, energy flow and natural history, as they meld to create patterns in the food cycle.

WHERE NEXT ?

I cannot see that ecologists have much more to learn from staring even harder at the compendium of published webs. As Paine (1988) points out: '. . . these qualitative descriptions were never intended to be data, to serve as fundamental grist for the theoretician's mill'. Fortunately, I sense a growing interest in gathering more and better data, in doing experiments on food webs, and in thinking about new approaches to understanding their structure.

In an ideal world, what form should new data take? In no particular order, the spatial and temporal extent of each study must be defined, and the criteria for specifying that A feeds on B need to be explicit; the probability of discovering a feeding link depends on sampling effort, for example (Cohen 1978). We should avoid lumping taxa; species identities need to remain intact. At very least, such data would allow us to be confident that the patterns so far documented are real, and not artefacts. Not everybody agrees that all, or even a majority of the patterns in Table 3.1 are reliable; others, myself included think that they are sufficiently simple and robust to be worth studying, despite the data. But I would feel happier if I had more confidence in the basic information. Several recent studies stand out as models for others to follow (e.g. Beaver 1983, 1985; Hawkins & Goeden 1984; Hildrew, Townsend & Hasham 1985; Kitching & Pimm 1985; Sprules & Bowerman 1988).

No less important or interesting would be different sorts of information. Elton (1927) drew up a hypothetical food web involving predators and parasites, and shrewdly observed that food chains might be very long when traced through mixtures of both sorts of enemies, because body size constraints are relaxed. More generally, energetic or dynamic constraints should still operate in real webs that include predators, parasites and diseases, but the distribution of feeding links is unlikely to be upper triangular, violating the cascade model. Yet despite Elton's early insight, students of food webs have studiously ignored parasites and diseases (May 1983; Price *et al.* 1986), a few have studied parasitoids, and nobody has combined all classes of 'enemies' into a study of a single web. Because such 'mixed' food webs seem to offer an opportunity to tease apart the relative contributions of dynamic and static processes, they deserve immediate attention.

Ecologists have also been slow to study and model at one and the same time the full cycle of trophic links, embracing not only plants, herbivores, predators and so on, but also corpses and decomposers. The 'national anthem' of my adopted county of Yorkshire ('On Ilkley Moor baht'at') reminds us that loops in food chains are commonplace if we include decomposers in our thinking (the song specifically identifies the pathway from humans, worms, ducks and back to humans); at least superficially, such loops do not appear to be destabilizing. Indeed, a number of recently discovered and fascinating trophic interactions have yet to feature in empirical and theoretical analyses of food webs (e.g. Hanski 1987). At least 25%, sometimes more, of the net primary production of planktonic algae may appear as dissolved organic matter (DOM) in marine ecosystems (Kurihara & Kikkawa 1986). DOM is taken up and used by bacteria, which in turn are consumed by heterotrophic flagellates. Both bacteria and flagellates serve as food for consumers that also double up as 'normal' primary consumers of algae (Azam *et al.* 1983; Fenchel 1986, 1987). The 'microbial loop' adds considerable complexity, not to say several more steps, to planktonic food chains. Some freshwater food webs, particularly those in humic lakes, may also support hitherto unexpected links via DOM, heterotrophic flagellates and zooplankton (Salonen & Hammer 1986). Equally intriguing for the light they might shed on the processes structuring food webs would be detailed studies of other unconventional trophic links: planktonic ciliates that store chloroplasts from algae, and which therefore live double lives as grazers and primary producers (Hanski 1987), and insectivorous plants in terrestrial and freshwater communities that turn the tables on animals and act as predators. Mutualists and symbionts might also be usefully and routinely incorporated into our studies whenever and wherever there is exchange of energy between them.

It is unclear to me what predictions dynamic models might make about permitted structures in food webs containing these and other unconventional or at least poorly studied trophic links. If they turn out to be clearly different from food web structures expected under alternative models, they may throw a chink of light on the conflicting explanations summarized in Table 3.1.

Such questions aside, several major and more radical initiatives suggest themselves. Paine (1980, 1983) has been virtually alone in arguing that not only the presence, but also the strength of feeding links needs to be investigated, and has pioneered the use of manipulative field experiments to distinguish between strong and feeble interactions in intertidal food webs (e.g. Paine 1966, 1971, 1980). An apparent lack of compartments or 'modules' in food webs could conceivably be an artefact of using binary data (Paine 1980; Pimm & Lawton 1980). If strong interactions are grouped in some way, it has important theoretical implications for community stability (May 1979; O'Neill *et al.* 1986; and references therein). A second, related theme already touched on, is that much more thought needs to be given to spatial and temporal scales of study. Food webs in nature are nested in time and space (Harris 1985; O'Neill *et al.* 1986). Current investigations simply break into this hierarchy at arbitrary points. For sound practical reasons, an investigator may choose to focus on insect herbivores, parasitoids and predators on one or more species of food plants in a small area. Or he may choose to paint a bigger picture on a larger canvas incorporating deer, rabbits, foxes and eagles, and ignore or roughly sketch the insignificant insects. But what we actually have are a series of webs made up of small organisms, with short generation times, interacting over small spatial scales, embedded within food chains of progressively larger, longer-lived and more widely ranging beasts. This hierarchical structure may profoundly influence the way in which ecosystems work (Harris 1985; O'Neill *et al.* 1986), as well as making it more difficult for biologists to understand how they work. As Harris (1985) succinctly points out: 'a reductionist approach using simple equations (suitable for small number systems) cannot represent the hierarchical and emergent behaviour of a loosely nested middle number system.' ('Small number' systems are tractable sets of three or four species; 'large number' systems have emergent statistical properties typified by the gas laws; for better or worse, students of real food webs and communities are stuck with 'middle number' systems.)

Third, and again related, what are the principal mechanisms that filter species out from a regional pool of potential food web members, and determine who feeds where and when? (Beaver 1983, 1985; Kitching 1987; see also related discussions in Ricklefs 1987). Presumably, chance dispersal, environmental conditions and species interactions all play a part, but how do they combine to create local webs? Beaver's (1983) work on the food webs of *Nepenthes* pitcher plants in South East Asia illustrates the nature of the processes involved.

There is considerable variation between individual pitchers in insect species composition, even after the effects of pitcher age have been eliminated. At the level of the individual pitcher, chance processes and the recent past history of the patch probably play an important role in the determination of which species are available to colonise a pitcher when it first opens. There may be priority effects, probably usually mediated through intra- or interspecific competition for food. . . . The presence of predator larvae will also affect the faunal composition . . . The community within an (individual) pitcher can never reach equilibrium. At the level of the patch of *Nepenthes* pitchers, greater stability is possible.

Or consider the more specific question, touched on in this quotation from Beaver. How important are trophic interactions involving polyphagous enemies in excluding species from particular habitats and food webs — so called competition for 'enemy-free space' (Jeffries & Lawton 1984; Lawton 1986b; Mithen & Lawton 1986), or 'apparent competition' (Holt 1977, 1987)? Paradoxically, notice that because apparent competition implies that polyphagous enemies sustained by alternative prey totally eliminate particularly vulnerable victims, some of the most interesting and potent feeding links in food webs may be ones that are no longer there to investigate!

Overlapping with all that has been said up to this point is the need not only to observe and to describe food webs, but to manipulate them. Pimm & Kitching (1987) show what can be done, particularly in small aquatic communities. I would like to see many more experiments that vary habitat productivity, predictability and structure (to name but three possibilities), and then monitor what happens to food web structure. Finally, theoretical and empirical studies all make the obvious point that food webs are about more than just energy transfer. Crucial species interactions may involve exchange and cycling of limiting nutrients, for example amongst mutualists, or between plants, micro-organisms and detritus. These vital links between species are not represented in the gallery of published webs. Moreover, trophic interactions (broadly defined) are themselves played out against a background that also involves competitive contests between species. General models of community assembly usually embrace all classes of species interactions (e.g. May 1973; Lawton 1987a). Some attempt has been made to allow for them in empirical studies of food webs (e.g. Rejmanek & Stary 1979; Briand 1983); but a widespread failure properly to quantify and simultaneously study predator–prey, competitive and mutualistic links in communities significantly weakens the bonds between theory and empirical studies.

CAN IT BE DONE ?

It is embarrassingly easy to list ways in which existing empirical data on food webs could be improved, and depressingly difficult to see how to achieve these high ideals in practice. Food web studies highlight a crucial problem that confronts all areas of ecology (Lawton 1987b). There is no shortage of important theoretical ideas about how populations are regulated, or how communities of living things are put together, or about many other ecological phenomena. To test these ideas, to refine them, refute and build better hypotheses, ecologists need more than anything else teams of people, sets of willing and skilled hands to do field manipulation experiments, or to sort, identify, weigh, grind up and so on. It is trite, but true that most of the ideas outlined above are Herculean tasks that cannot be solved without great labour; there are no sophisticated machines costing millions of pounds to do the job for us. Yet precisely because a great deal of important ecology cannot be done with very expensive electronic wizardry, it is too often seen as 'soft science', not to be taken too seriously. The irony is that we now know far more about black holes and distant galaxies than we do about the communities of living organisms that make up the life-support systems of our own planet. Food webs provide a touchstone in our efforts to discern and then understand patterns in nature. Sixty years down the road pioneered by Charles Elton, ecology stands at a crossroads. Unless we can win a share of resources more equal to the tasks that confront us, progress over the next 60 years will be painfully slow. By then, many of the world's most exciting ecosystems may have been lost for ever, and with them our last chance to understand how they work.

ACKNOWLEDGMENTS

I have benefited greatly from discussions on food webs with Phil Warren, who also generously made some of his painstakingly gathered data available for my own use. Kevin Gaston, Brad Hawkins, Richard Law, Bob Paine, Stuart Pimm and Phil Warren all commented on drafts of the manuscript, and made numerous helpful suggestions.

REFERENCES

Arnold, S. J. (1972). Species densities of predators and their prey. *American Naturalist*, **106,** 220–236.

Azam, F., Fenchel, T., Field, J.G., Gray, J.S., Meyer-Reil, L.A. & Thingstad, F. (1983). The ecological role of water-column microbes in the sea. *Marine Ecology Progress Series*, **10,** 257–263.

Beddington, J.R. & Hammond, P.S. (1977). On the dynamics of host-parasite-hyperparasite interactions. *Journal of Animal Ecology,* **46,** 811–821.

Beaver, R.A. (1983). The communities living in *Nepenthes* pitcher plants: fauna and food webs. *Phytotelmata: Terrestrial Plants as Hosts for Aquatic Insect Communities* (Ed. by J.H. Frank & L.P. Lounibos), pp. 125–159. Plexus, Medford, New Jersey.

Beaver, R.A. (1985). Geographical variation in food web structure in *Nepenthes* pitcher plants. *Ecological Entomology*, **10,** 241–248.

Briand, F. (1983). Environmental control of food web structure. *Ecology*, **64,** 253–263.

Briand, F. & Cohen, J.E. (1984). Community food webs have a scale-invariant structure. *Nature*, **307,** 264–267.

Briand, F. & Cohen, J.E. (1987). Environmental correlates of food chain length. *Science*, **238,** 956–960.

Cameron, G.N. (1972). Analysis of insect trophic diversity in two salt marsh communities. *Ecology*, **53,** 58–73.

Carpenter, S.R. & Kitchell, J.F. (1987). The temporal scale of variance in limnetic primary production. *American Naturalist*, **129,** 417–433.

Carpenter, S.R. Kitchell, J.F., Hodgson, J.R., Cochran, P.A., Elser, J.J., Elser, M.M., Lodge, D.M., Kretchmer, D., He, X. & Ende, von C.M. (1987). Regulation of lake primary productivity by food web structure. *Ecology*, **68,** 1863–1876.

Caughley, G. & Lawton, J.H. (1981). Plant-herbivore systems. *Theoretical Ecology. Principles and Applications* (Ed. by R.M. May), pp.132–166. Blackwell Scientific Publications, Oxford.

Chesson, P.L. & Murdoch, W.W. (1986). Aggregation of risk: relationships among host-parasitoid models. *American Naturalist,* **127,** 696–715.

Cohen, J.E. (1977). Food webs and the dimensionality of trophic niche space. *Proceedings of the National Academy of Sciences USA*, **74,** 4533–4536.

Cohen, J.E. (1978). *Food Webs and Niche Space.* Princeton University Press, Princeton.

Cohen, J.E. (1983). Recent progress and problems in food web theory. *Current Trends in Food Web Theory. Report on a Food Web Workshop* (Ed. by D.L. DeAngelis, W.M. Post & G. Sugihara), pp. 17–24. Oak Ridge National Laboratory, Oak Ridge, Tennessee.

Cohen, J.E. & Briand, F. (1984). Trophic links of community food webs. *Proceedings of the National Academy of Sciences USA*, **81,** 4105–4109.

Cohen, J.E., Briand, F. & Newman, C.M. (1986). A stochastic theory of community food webs III. Predicted and observed lengths of food chains. *Proceedings of the Royal-Society, London B*, **228,** 317–353.

Cohen, J.E. & Newman, C.M. (1985). A stochastic theory of community food webs I. Models and aggregated data. *Proceedings of the Royal Society, London B*, **224,** 421–448.

Cohen, J.E. & Newman, C.M. (1988). Dynamic basis of food web organization. *Ecology*, **69,** 1655–1664.

Cohen, J.E., Newman, C.M. & Briand, F. (1985). A stochastic theory of community food webs II. Individual webs. *Proceedings of the Royal Society, London B*, **224,** 449–461.

Cousins, S.H. (1980). A trophic continuum derived from plant structure, animal size and a detritus cascade. *Journal of Theoretical Biology*, **82,** 607–618.

Cousins, S.H. (1987). The decline of the trophic level concept. *Trends in Ecology and Evolution*, **2,** 312–316.

Darnell, R.M. (1961). Trophic spectrum of an estuarine community, based on studies of Lake Pontchartrain, Louisiana. *Ecology*, **42,** 553–568.

DeAngelis, D.L. (1975). Stability and connectance in food web models. *Ecology*, **56,** 238–243.

DeAngelis, D.L. (1980). Energy flow, nutrient cycling, and ecosystem resilience. *Ecology*, **61,** 764–771.

DeAngelis, D.L., Gardner, R.H., Mankin, J.B., Post, W.M. & Carney, J.H. (1978). Energy flow and the number of trophic levels in ecological communities. *Nature*, **273,** 406–407.

DeAngelis, D.L. & Waterhouse, J.C. (1987). Equilibrium and nonequilibrium concepts in ecological models. *Ecological Monographs*, **57,** 1–21.

Dempster, J.P. & Pollard, E. (1981). Fluctuations in resource availability and insect populations. *Oecologia*, **50,** 412–416.

Elton, C.S. (1927). *Animal Ecology.* Sidgwick & Jackson, London.

Elton, C.S. (1958). *The Ecology of Invasions by Animals and Plants.* Chapman & Hall, London.

Evans, F.C. & Murdoch, W.W. (1968). Taxonomic composition, trophic structure and seasonal occurrence in a grassland insect community. *Journal of Animal Ecology*, **37,** 259–273.

Fenchel, T. (1986). The ecology of heterotrophic flagellates. *Advances in Microbial Ecology*, **9,** 57–97.

Fenchel, T. (1987). *Ecology – Potentials and Limitations.* Ecology Institute, Nordbunte, Federal Republic of Germany.

Fowler, C.W. & MacMahon, J.A. (1982). Selective extinction and speciation: their influence on the structure and functioning of communities and ecosystems. *American Naturalist*, **119,** 480–498.

Free, C.A., Beddington, J.R. & Lawton, J.H. (1977). On the inadequacy of simple models of mutual interference for predation and parasitism. *Journal of Animal Ecology*, **46,** 543–554.

Gallopin, G.C. (1972). Structural properties of food webs. *Systems Analysis and Simulation in Ecology*, Vol. **2** (Ed. by B.C. Patten), pp. 241–282. Academic Press, New York.

Gittleman, J.L. (1985). Carnivore body size: ecological and taxonomic correlates. *Oecologia*, **67**, 540–544.

Hanski, I. (1987). Plankton that don't obey the rules. *Trends in Ecology and Evolution*, **2,** 350–351.

Hardy, A.C. (1924). The herring in relation to its animate environment. Part 1. The food and feeding habits of the herring with special reference to the east coast of England. *Ministry of Agriculture and Fisheries, Fisheries Investigations, Series II*, **7,** 1–45.

Harris, G.P. (1985). The answer lies in the nesting behaviour. *Freshwater Biology*, **15**, 375–380.

Hassell, M.P. (1978). *The Dynamics of Arthropod Predator-Prey Systems.* Princeton University Press, Princeton.

Hassell, M.P. & May, R.M. (1985). From individual behaviour to population dynamics. *Behavioural Ecology. Ecological Consequences of Adapative Behaviour* (Ed. by R.M. Sibley & R.H. Smith), pp. 3–32. Blackwell Scientific Publications, Oxford.

Hastings, H.M. & Conrad, M. (1979). Length and evolutionary stability of food chains. *Nature*, **282**, 838–839.

Hawkins, B.A. & Goeden, R.D. (1984). Organization of a parasitoid community associated with a complex of galls on *Atriplex* spp. in southern California. *Ecological Entomology*, **9**, 271–292.

Hildrew, A.G., Townsend, C.R. & Hasham, A. (1985). The predatory Chironomidae of an iron-rich stream: feeding ecology and food web structure. *Ecological Entomology*, **10**, 403–413.

Hokkanen, H.M.T. (1986). Success in classical biological control. *CRC Critical Reviews in Plant Sciences*, **3**, 35–72.

Holt, R.D. (1977). Predation, apparent competition, and the structure of prey communities. *Theoretical Population Biology*, **12**, 197–229.

Holt, R.D. (1987). Prey communities in patchy environments. *Oikos,* **50,** 276–290.

Huffaker, C.B. (ed.) (1971). *Biological Control.* Plenum, New York.

Humphreys, W.F. (1979). Production and respiration in animal populations. *Journal of Animal Ecology*, **48**, 427–453.

Hutchinson, G.E. (1959). Homage to Santa Rosalia or why are there so many kinds of animals? *American Naturalist*, **93**, 145–159.

Jeffries, M.J. & Lawton, J.H. (1984). Enemy free space and the structure of ecological communities. *Biological Journal of the Linnean Society*, **23**, 269–286.

Jeffries, M.J. & Lawton, J.H. (1985). Predator-prey ratios in communities of freshwater invertebrates: the role of enemy free space. *Freshwater Biology,* **15**, 105–112.

Johnson, D.M., Pierce, C.L., Martin, T.H., Watson, C.N., Bohanan, R.E. & Crowley, P.H. (1987). Prey depletion by odonate larvae: combining evidence from multiple field experiments. *Ecology*, **68**, 1459–1465.

Kareiva, P. (1983). Influence of vegetation texture on herbivore populations: resource concentration and herbivore movement. *Variable Plants and Herbivores in Natural and Managed Systems* (ed. by R.F. Denno & M.S. McClure), pp. 259–289. Academic Press, New York.

Kelleher, J.S. & Hulme, M.A. (Eds) (1984). *Biological Control Programmes Against Insects and Weeds in Canada 1969–1980.* Commonwealth Agricultural Bureaux, Farnham Royal, Slough.

Kerr, S.R. (1974). Theory of size distribution in ecological communities. *Journal of the Fisheries Research Board Canada*, **31**, 1859–1862.

Kirkwood, R.S.M. & Lawton, J.H. (1981). Efficiency of biomass transfer and the stability of model food-webs. *Journal of Theoretical Biology*, **93**, 225–237.

Kitching, R.L. (1983). Community structure in water-filled treeholes in Europe and

Australia — comparisons and speculations. *Phytotelmata: Terrestrial Plants as Hosts for Aquatic Insect Communities.* (Ed. by J.H. Frank & L.P. Lounibos), pp. 205–222. Plexus, Medford, New Jersey.

Kitching, R.L. (1987). Spatial and temporal variation in food webs in water-filled treeholes. *Oikos*, **48**, 280–288.

Kitching, R.L. & Pimm, S.L. (1985). The length of food chains: phytotelmata in Australia and elsewhere. *Proceedings of the Ecological Society of Australia*, **14**, 123–140.

Krebs, C.J. (1985). *Ecology. The Experimental Analysis of Distribution and Abundance.* Harper & Row, New York.

Kurihara, Y. & Kikkawa, J. (1986). Trophic relations of decomposers. *Community Ecology, Pattern and Process* (Ed. by J. Kikkawa & D.J. Anderson), pp. 127–160. Blackwell Scientific Publications, Oxford.

Lawton, J.H. (1981). Moose, wolves, *Daphnia* and *Hydra*: on the ecological efficiency of endotherms and ectotherms. *American Naturalist*, **117**, 782–783.

Lawton, J.H. (1986a). Surface availability and insect community structure: the effects of architecture and fractal dimension of plants. *Insects and the Plant Surface* (Ed. by B.E. Juniper & T.R.E. Southwood), pp. 317–331. Edward Arnold, London.

Lawton, J.H. (1986b). The effect of parasitoids on phytophagous insect communities. *Insect Parasitoids* (Ed. by J. Waage & D. Greathead), pp.265–287. Academic Press, London.

Lawton, J.H. (1987a). Are there assembly rules for successional communities? *Colonization, Succession and Stability* (Ed. by A.J. Gray, M.J. Crawley & P.J. Edwards), pp. 225–244. Blackwell Scientific Publications, Oxford.

Lawton, J.H. (1987b). Fluctuations in a patchy world. *Nature*, **326**, 328–329.

Lawton, J.H. & Pimm, S.L. (1978). Population dynamics and the length of food chains. *Nature*, **272**, 189–190.

Leigh, E.G. Jr. (1981). The average lifetime of a population in a varying environment. *Journal of Theoretical Biology*, **90**, 213–239.

Lessells, C.M. (1985). Parasitoid foraging: should parasitism be density dependent? *Journal of Animal Ecology*, **54**, 27–41.

MacArthur, R.H. (1955). Fluctuations of animal populations, and a measure of community stability. *Ecology*, **36**, 533–536.

MacDonald, N. (1979). Simple aspects of foodweb complexity. *Journal of Theoretical Biology*, **80**, 577–588.

Macfadyen, A. (1979). The role of the fauna in decomposition processes in grasslands. *Scientific Proceedings, Royal Dublin Society, A*, **6**, 197–206.

May, R.M. (1972). Will a large complex system be stable? *Nature*, **238**, 413–414.

May, R.M. (1973). *Stability and Complexity in Model Ecosystems.* Princeton University Press, Princeton.

May, R.M. (1979). The structure and dynamics of ecological communities. *Population Dynamics.* (Ed. by R.M. Anderson, B.D. Turner & L.R. Taylor), pp. 385–407. Blackwell Scientific Publications, Oxford.

May, R.M. (1981). Patterns in multi-species communities. *Theoretical Ecology, Principles and Applications* (Ed.by R.M. May), pp.197–227. Blackwell Scientific Publications, Oxford.

May, R.M. (1983). Food web structure: some thoughts and some problems. *Current Trends in Food Web Theory. Report on a Food Web Workshop* (Ed. by D.L. DeAngelis, W.M. Post & G. Sugihara), pp. 127–129. Oak Ridge National Laboratory, Oak Ridge, Tennessee.

May, R.M. (1986). The search for patterns in the balance of nature: advances and retreats. *Ecology*, **67**, 1115–1126.

May, R.M. & Hassell, M.P. (1981). The dynamics of multiparasitoid-host interactions.

American Naturalist, **117**, 234–261.

McCormick, S. & Polis, G. (1982). Arthropods that prey on vertebrates. *Biological Reviews*, **57**, 29–58.

McNeill, S. & Lawton, J.H. (1970). Annual production and respiration in animal populations. *Nature*, **225**, 472–474.

Menge, B.A. & Sutherland, J.P. (1987). Community regulation: variation in disturbance, competition, and predation in relation to environmental stress and recruitment. *American Naturalist*, **130**, 730–757.

Mithen, S.J. & Lawton, J.H. (1986). Food-web models that generate constant predator-prey ratios. *Oecologia*, **69**, 542–550.

Moran, V.C. & Southwood, T.R.E. (1982). The guild composition of arthropod communities in trees. *Journal of Animal Ecology*, **51**, 289–306.

Morse, D.R., Lawton, J.H., Dodson, M.M. & Williamson, M.H. (1985). Fractal dimension of vegetation and the distribution of arthropod body lengths. *Nature*, **314**, 731–733.

Murdoch, W.W. & Reeve, J.D. (1987). Aggregation of parasitoids and the detection of density dependence in field populations. *Oikos*, **50**, 137–141.

Newman, J.E. & Cohen, J.E. (1986). A stochastic theory of community food webs IV. Theory of food chain lengths in large webs. *Proceedings of the Royal Society, London B*, **228**, 355–377.

Newton, I. (1986). *The Sparrowhawk.* Poyser, Staffordshire.

Odum, E.P. & Biever, L.J. (1984). Resource quality, mutualism, and energy partitioning in food chains. *American Naturalist*, **124**, 360–376.

Oksanen, L., Fretwell, S.D., Arruda, J. & Niemela, P. (1981). Exploitation ecosystems in gradients of primary productivity. *American Naturalist*, **118**, 240–262.

O'Neill, R.V., DeAngelis, D.L., Waide, J.B. & Allen, T.F.H. (1986). *A Hierarchical Concept of Ecosystem.* Princeton University Press, Princeton.

Paine, R.T. (1963). Trophic relationships of 8 sympatric predatory gastropods. *Ecology*, **44**, 63–73.

Paine, R.T. (1966). Food web complexity and species diversity. *American Naturalist*, **100**, 65–75.

Paine, R.T. (1969). A note on trophic complexity and community stability. *American Naturalist*, **103**, 91–93.

Paine, R.T. (1971). A short-term experimental investigation of resource partitioning in a New Zealand rocky intertidal habitat. *Ecology*, **52**, 1096–1106.

Paine, R.T. (1980). Food webs: linkage, interaction strength and community infrastructure. *Journal of Animal Ecology*, **49**, 667–685.

Paine, R.T. (1983). Intertidal food webs: does connectance describe their essence. *Current Trends in Food Web Theory. Report on a Food Web Workshop* (Ed. by D.L. DeAngelis, W.M. Post & G. Sugihara), pp. 11–15. Oak Ridge National Laboratory, Oak Ridge, Tennessee.

Paine, R.T. (1988). On food webs: road maps of interactions or grist for theoretical development? *Ecology*, **69**, 1648–1654.

Patten, B.C. (1985). Energy, length of food chains, and direct versus indirect effects in ecosystems. *Canadian Bulletin of Fisheries and Aquatic Sciences*, **213**, 119–138.

Peters, R.H. (1983). *The Ecological Implications of Body Size.* Cambridge University Press, Cambridge.

Pimm, S.L. (1978). Niche overlaps. *Science*, **202**, 1075–1076.

Pimm, S.L. (1979). The structure of food webs. *Theoretical Population Biology*, **16**, 144–158.

Pimm, S.L. (1980a). Properties of food webs. *Ecology*, **61**, 219–225.

Pimm, S.L. (1980b). Bounds on food web connectance. *Nature*, **284**, 591.

Pimm, S.L. (1982). *Food Webs*. Chapman & Hall, London.

Pimm, S.L. (1984). The complexity and stability of ecosystems. *Nature*, **307**, 321–326.

Pimm, S.L. & Kitching, R.L. (1987). The determinants of food chain lengths. *Oikos*, **50**, 302–307.

Pimm, S.L. & Lawton, J.H. (1977). Number of trophic levels in ecological communities. *Nature*, **268**, 329–331.

Pimm, S.L. & Lawton, J.H. (1978). On feeding on more than one trophic level. *Nature*, **275**, 542–544.

Pimm, S.L. & Lawton, J.H. (1980). Are food webs divided into compartments? *Journal of Animal Ecology*, **49**, 879–898.

Pimm, S.L. & Lawton, J.H. (1983). The causes of foodweb structure: dynamics, energy flow and natural history. *Current Trends in Food Web Theory. Report on a Food Web Workshop* (Ed. by D.L. DeAngelis, W.M. Post & G. Sugihara), pp. 45–49. Oak Ridge National Laboratory, Oak Ridge, Tennessee.

Pimm, S.L. & Rice, J.C. (1987). The dynamics of multispecies, multilife-stage models of aquatic food webs. *Theoretical Population Biology*, **32**, 303–325.

Platt, T. (1985). Structure of marine ecosystems: its allometric basis. *Canadian Bulletin of Fisheries and Aquatic Sciences*, **213**, 55–64.

Platt, T. & Denman, K. (1977). Organisation in the pelagic ecosystem. *Helgolander wiss. Meeresunters*, **30**, 575–581.

Price, P.W., Westoby, M., Rice, B., Atsatt, P.R., Fritz, R.S., Thompson, J.N. & Mobley, K. (1986). Parasite mediation in ecological interactions. *Annual Review of Ecology and Systematics*, **17**, 487–505.

Rejmanek, M. & Stary, P. (1979). Connectance in real biotic communities and critical values for stability of model ecosystems. *Nature*, **280**, 311–313.

Ricklefs, R.E. (1987). Community diversity: relative roles of local and regional processes. *Science*, **235**, 167–171.

Salonen, K. & Hammer, T. (1986). On the importance of dissolved organic matter in the nutrition of zooplankton in some lake waters. *Oecologia*, **68**, 246–253.

Saunders, P.T. (1978). Population dynamics and the length of food chains. *Nature*, **272**, 189.

Schoener, T.W. & Spiller, D.A. (1987). Effects of lizards on spider populations: manipulative reconstruction of a natural experiment. *Science*, **236**, 949–952.

Shelford, V.E. (1913). *Animal Communities in Temperate America as Illustrated in the Chicago Region. A Study in Animal Ecology*. Bulletin of the Geographical Society of Chicago, **5**. Reprinted Arno Press, New York (1977).

Sih, A. (1987). Prey refuges and predator-prey stability. *Theoretical Population Biology*, **31**, 1–12.

Sih, A., Crowley, P., McPeek, M., Petranka, J. & Strohmeier, K. (1985). Predation, competition and prey communities: a review of field experiments. *Annual Review of Ecology and Systematics*, **16**, 269–311.

Slobodkin, L.B. (1961). *Growth and Regulation of Animal Populations*. Holt, Rinehart & Winston, New York.

Sprules, W.G. & Bowerman, J.E. (1988). Omnivory and food chain length in zooplankton food webs. *Ecology*, **69**, 418–426.

Strong, D.R., Lawton, J.H. & Southwood, T.R.E. (1984). *Insects on Plants. Community Patterns and Mechanisms*. Blackwell Scientific Publications, Oxford.

Stenseth, N.C. (1985). The structure of food webs predicted from optimal food selection models: an alternative to Pimm's stability hypothesis. *Oikos*, **44**, 361–364.

Sugihara, G. (1983). Holes in niche space: a derived assembly rule and its relation to intervality. *Current Trends in Food Web Theory. Report on a Food Web Workshop* (Ed.

by D.L. DeAngelis, W.M. Post & G. Sugihara), pp. 25–35, Oak Ridge National Laboratory, Oak Ridge, Tennessee.

Sugihara, G. (1984). Graph theory, homology and food webs. *Proceedings of Symposia in Applied Mathematics*, **30**, 83–101.

Teal, J.M. (1962). Energy flow in the salt marsh ecosystem of Georgia. *Ecology*, **43**, 614–624.

Thorp, J.H. (1986). Two distinct roles for predators in freshwater assemblages. *Oikos*, **47**, 75–82.

Verezina, A.F. (1985). Empirical relationships between predator and prey size among terrestrial vertebrate predators. *Oecologia*, **67**, 555–565.

Walter, D.E. (1987). Trophic behaviour of 'mycophagous' microarthropods. *Ecology*, **68**, 226–228.

Warren, P.H. & Lawton, J.H. (1987). Invertebrate predator-prey body size relationships: an explanation for upper triangular food webs and patterns in food web structure? *Oecologia*, **74**, 231–235.

Yodzis, P. (1980). The connectance of real ecosystems. *Nature*, **284**, 544–555.

Yodzis, P. (1981). The stability of real ecosystems. *Nature*, **289**, 674–676.

Yodzis, P. (1984a). Energy flow and the vertical structure of real ecosystems. *Oecologia*, **65**, 86–88.

Yodzis, P. (1984b). The structure of assembled communities II. *Theoretical Population Biology*, **107**, 115–126.

Yodzis, P. (1984c). How rare is omnivory? *Ecology*, **65**, 321–323.

4 THE ECOLOGICAL NICHE

T. W. SCHOENER
Department of Zoology, University of California, Davis, California 95616, USA

INTRODUCTION

Nowadays students beginning ecology are taught that the niche is a multidimensional utilization distribution, giving a population's use of resources ordered along axes such as prey size or feeding height. Surely this definition seems much at odds with common parlance. Indeed, an ordinary dictionary (*Oxford Concise*) defines 'niche' in a non-ecological sense as a 'space or recess in wall, to contain statue, vase, etc.' Etymology and prior usage help little to understand the present ecological meaning. The *Oxford English Dictionary* (1971) gives 'a.F. *niche*, ad. It. *nicchia*, of doubtful origin (by Diaz connected with *nicchio*, mussel shell)', and reports the long history of figurative usage, beginning about 250 years ago. One such sense is as 'a place or position adapted to the character or capabilities, or suited to the merits of a person or thing'; thus from Jonathan Swift (1726) 'If I can but fill my Nitch, I attempt no higher Pitch', or from W. M. Rossetti (1869) 'The work fills a niche of its own and is without competition.' A second figurative sense is a 'place of retreat or retirement'; thus from Richard Bradley (1725) 'The way to destroy the Niches of Spiders in our Garden. . . .'; or from Thomas Woolner (1863) 'I told of gourmand thrushes, which to feast on morsels oozy rich, Cracked poor snails' curling niche.' These examples, however apparently ecological, are not obviously ancestral to present ecological usage.

How has the ecological definition evolved so far from even metaphor? To understand this is in large part to trace the concept's history. I will argue that when 'niche' was first used as an ecological concept, it was in the figurative sense of a 'recess' or place in the community. With time, 'niche' became identified with the occupant of that recess. Perhaps this was because niche occupants were being viewed as more and more variable, especially due to competition, which was becoming increasingly prominent as an explanatory concept. In attaching names and papers to the above transformation, I will argue unconventionally that Grinnell's and Elton's concepts were similar except in minor ways, that Hutchinson's concept was revolutionary, and that the present 'utilization' concept was forged from Hutchinson's under utilitarian pressures.

After presenting this history, I will summarize the modern theory of the niche, especially as it relates to competition, which is its major application. I will then evaluate the usefulness of the modern niche concept and ask to what extent its replacement by a set of more specialized concepts is desirable. In this latter inquiry I will be particularly interested in the relationship of competition to niche overlap, which I will treat as a decision tree rather than as a simple scientific law in the sense of Nagel (1961).

FOUR CONCEPTS OF THE NICHE

Grinnell's concept of the niche

In a series of papers begun in 1914 and continuing through the 1920s, Joseph Grinnell developed his concept of the niche. His contributions in these papers are usually acknowledged even in the newest textbooks. Yet they are often distorted or oversimplified, e.g. Grinnell's 'niche' equals 'habitat'. I was certainly surprised, reading Grinnell's papers carefully, at the extent to which a coherent niche concept was developed, to say nothing of rather modern concepts of speciation, geographic variation, adaptive radiation and quantitative habitat utilization. As I now discuss, four of Grinnell's works in particular (Grinnell 1914, 1917, 1924, 1928) show that his niche concept ranged well beyond the simple label 'habitat', at least insofar as the latter is presently construed.

Although Grinnell's first published mention of 'niche' (Griesemer, unpublished ms) is in Grinnell & Swarth (1913), the concept is first used substantially in 'An account of the mammals and birds of the lower Colorado Valley', a several-hundred-page publication of the University of California's zoological series. This work is devoted almost entirely to species-by-species accounts, and employment of the term 'niche' appears almost incidental. In comparing two finch species (p. 91), Grinnell notes that they can be found together, but 'to prove beyond doubt what is the true ecologic niche of each, a knowledge of the distribution of each species elsewhere in their respective ranges is necessary'. One species can be assigned to the mesquite vegetation association and the other to the arrowweed association. He then writes '. . . if associational analysis is carried far enough, no two species of birds or mammals will be found to occupy precisely the same ecologic niche, though they may approximately do so when their respective associations are represented fragmentarily and in intermixture.' (p. 91.) A few pages later (p. 98), he asks 'is the Colorado fauna *full*? Are all the ecologic niches, which are available in this area and which have occupants in other regions, occupied here? Probably not . . .' he continues, because of limitations to dispersal by geographic barriers. Although not giving in detail what his niche concept is, Grinnell seems to imply in this paper that every species occupies a niche, although not necessarily the reverse — some niches may not contain a species. Each niche is somehow closely related to, but not identical to, one or more vegetation associations. Determination of a species distribution over a set of associations brings one closer to ascertaining its niche, indeed, his monograph is filled with a variant of utilization distributions (Fig. 4.1). Finally, he gives a version of the Competitive Exclusion Principle, usually credited to Gause (1934; but see Udvardy 1959), as well as posing the question of ecological saturation, long before this became fashionable in the 1960s.

Three years later the word 'niche' appeared in the title of Grinnell's (1917) most cited paper on the topic, 'The niche-relationships of the California thrasher'. This species has an unusually restricted distribution, and the purpose of the paper is to seek an explanation. Grinnell writes that it is '. . . probably to be found in the close adjustment of the bird in various physiological and psychological respects to a narrow range of environmental conditions. The nature of these critical conditions is to be learned through an examination of the bird's habitat . . .' (p. 428.) He recommends that this examination should be made at as many points throughout the species' range as possible. In so doing, an hierarchial characterization is made. First, the species occurs almost entirely within the Upper Sonoran life-zone. Second, 'faunal' restrictions (p. 430), 'indicating dependence upon atmospheric humidity', are of minor importance, although arid country is not occupied. Third, associational restrictions are strong, the species being limited to California Chaparral. Grinnell then finds an explanation for these macrospatial trends *in terms of* what he calls (p. 432) 'habitat relations'. The species is '. . . a habitual forager beneath dense and continuous cover. Furthermore, probably two-thirds of its foraging is done on the ground. . .The Thrasher is relatively omnivorous in its diet. Beal . . . found that 59 per cent of the food was . . . vegetable and 41 animal . . . [It feeds] by working over the litter beneath chaparral'. (p. 432.) He then comments that its bill and its food-gathering methods are quite specialized but not so much so that chaparral should be the exclusive foraging ground, as the (p.432) 'same mode of food-getting ought to be just as useful on the forest floor, or even the meadow . . . We must look farther'. The next paragraph does so, but now with respect to anti-predator adaptations and not those for feeding. He notes how well the thrasher's colour and pattern blend it to its

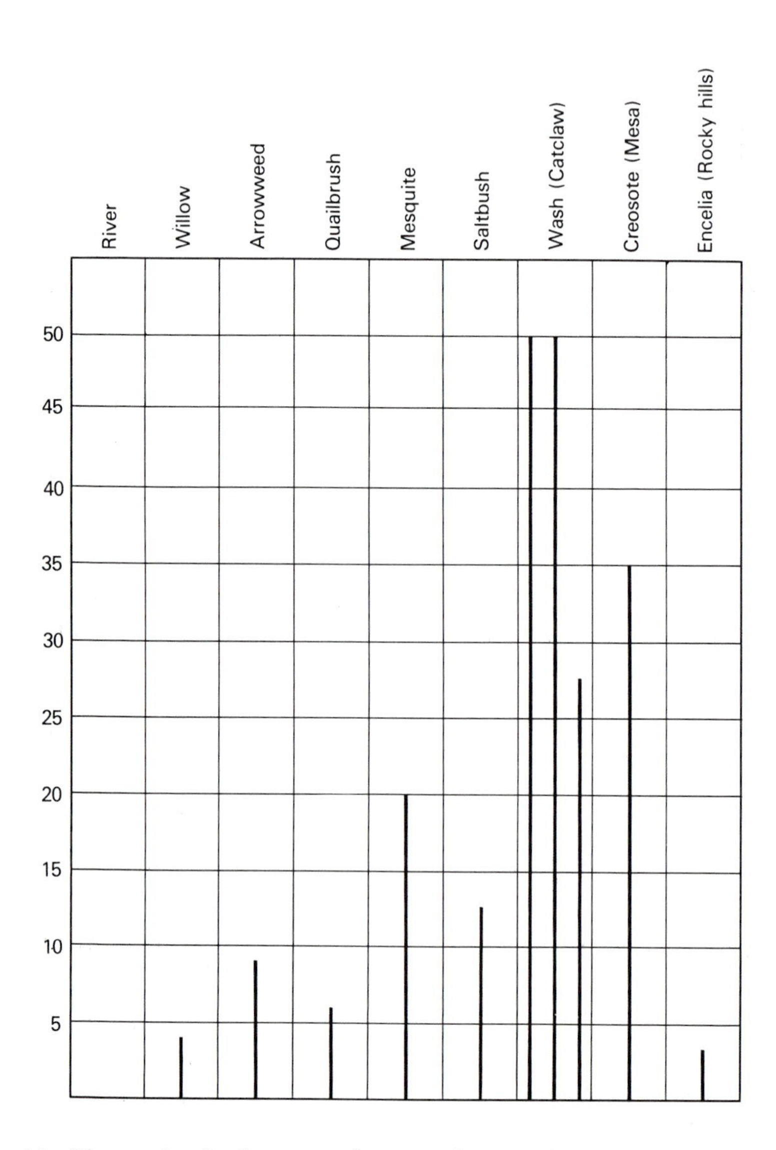

FIG. 4.1 Diagram showing frequency of capture of *Perognathus p. penicillatus* in the several associations. Two hundred and seventeen individuals were taken, of which associational occurrence was definitely recorded. This species is seen to be very widely distributed, yet exhibiting marked preference for certain associations. [After Grinnell (1914).]

background and how it shuffles quietly away through the underbrush when chased. Nest sites are similarly protected. Finally, in the penultimate paragraph of the paper, 'niche' is mentioned for the first time since the title:

> These various circumstances, which emphasize dependence upon cover, and adaptation in physical structure and temperament thereto, go to demonstrate the nature of the ultimate associational niche occupied by the California Thrasher. This is one of the minor niches which with their occupants all together make up the chaparral association. It is, of course, axiomatic that no two species regularly established in a single fauna have precisely the same niche relationships. [p. 433]

A number of points from this second paper are worth noting. First, Grinnell's niche encompasses both spatial and dietary dimensions, not just spatial ones. Second, niche characteristics are understandable both with respect to feeding and avoidance of predation. Third, the niche provides an explanation of the close correspondence of a species to a particular vegetation association, although it is in some way a deeper concept than just the fact of that correspondence, a point better understood after discussion of the next paper. Fourth, niches as 'recesses' in a community are clearly distinguished from their occupants. Fifth, a competitive exclusion principle is again asserted. Finally, it is possible from the position of the word 'habitat' in the above chain of reasoning that Grinnell did not wish this word to have an entirely spatial connotation; diet and foraging technique are classified as if part of the 'habitat relations' of the Thrasher.

Six years later, in a published address to a joint meeting of the Western Society of Naturalists and the Ecological Society of America, Grinnell (1924) further elaborated upon the relation between habitat and niche:

> Habitats have been variously classified . . . we can usefully recognize, as measures of distributional behavior, the realm, the region, the life-zone, the fauna, the subfauna, the association, and the ecologic or environmental niche. The latter, ultimate unit, is occupied by just one species or subspecies; if a new ecologic niche arises, or if a niche is vacated, nature hastens to supply an occupant . . . [p. 227]

From this passage alone, habitat could be interpreted as having a largely spatial connotation, but a much broader one than in the 1917 paper. Furthermore, it is easy to see how, from just the above words, Grinnell's concept of the niche could be mistaken as at least a subset of a habitat concept. Yet in this hierarchy 'fauna' and 'subfauna' appear, terms apparently without an exclusively spatial connotation. Using the 1917 paper, in which niche is more elaborately discussed, a plausible interpretation of the above passage is that 'habitat', in addition to 'niche', has a broader than spatial domain. Udvardy (1959) goes further and interprets the passage as implying, along with other writings, that Grinnell means niche to be 'not of a spatial nature' (p. 726).

Again, the quotations illustrate that for Grinnell, niches can exist in the absence of occupants and be 'vacated'. Also in this address, Grinnell discusses adaptive radiation, during the course of which ecological equivalents can evolve in separate places. Thus the kangaroo rat of North American deserts (p. 227) 'corresponds exactly' to the jerboa of the Sahara. Existence of such equivalents seems to imply that for Grinnell rather invariant rules determine the niches available for occupancy under a specific set of environmental conditions.

Finally, in a rather late paper, Grinnell (1928) comments on the niche of the meadowlark:

> Another example is the Meadow-lark . . . the bird is equipped to get its food safely and in adequate amount only from ground surface which is open . . . For successful food-getting. . .and for escape from its enemies, the bird is specialized for existence upon a meadow or grassy plain . . . Furthermore, within this general associational habitat, close observation shows that for its nest the Meadow-lark employs only certain types of meadow-land cover and that it chooses for its food chiefly items of a certain relatively large caliber. Other birds living in the same meadow association levy their toll upon other sources of subsistence afforded there. These different species are able compatibly to occupy the same habitat because no two of them depend upon precisely the same means of subsistence. Thus the ultimate unit in the general association occupied by each single species, Meadow-lark or any other bird, is its ecologic niche. [p. 193]

This passage confirms nicely points brought out in earlier works; the inclusion of dietary as well as habitat factors in the niche, the non-identity but close relationship of the niche to the association, a competitive exclusion principle, and consideration of both food-getting and predator-avoidance. Here, however, the term 'habitat' appears to have a more traditional meaning than in the 1917 paper, and the niche appears to include at least some macrohabitat features (see also next section).

Grinnell's early development of the niche concept was not without contemporaneous influence. W. P. Taylor, whom Hutchinson (1978, pp. 156–157) notes 'had worked with Grinnell' and whose acknowledgments in his 1916 paper thank 'especially Dr Joseph Grinnell' (p. 416), used both the word and the concept in a study of beavers. Taylor's work, perhaps even more than Grinnell's, focuses on ecological equivalents, 'Each group. . . occupies the same ecologic niche in different places. . .'(p. 474) with the proviso that '. . .no two ecologic niches can be precisely the same' (p. 479). In a section entitled 'How Have Different Ecologic Niches Been Filled?' (p. 479), Taylor, like Grinnell, clearly distinguishes the 'recess' from its occupant, 'In almost every locality there are ecologic niches which are unoccupied'. (p. 482.) Rather than repeated local adaptive radiation occurring from a single stock into the same set of niches (i.e. convergence), however, he imagined that the same stock would usually fill the same niche repeatedly in different geographic areas. Barriers would therefore be especially important in preventing some niches from being occupied.

As Gaffney (1973) was the first to point out, these early usages of the word 'niche' by Grinnell and his associate Taylor were not the first ecological usages in the literature. That precedent (quotations in the introduction aside) belongs to R. H. Johnson's *Determinate Evolution in the Color-pattern of the Lady-beetles*, published in 1910, 3 years (not 16 years as Gaffney claims) before Grinnell's first usage. I was able to find only one place in Johnson's work where 'niche' appears:

> One expects the different species in a region to occupy different niches in the environment.This is at least a corollary of the current belief that every species is as common as it can be, its numbers being limited only by its food-supply, a belief which is the result of the strong Malthusian leanings of Darwin. The major species of the coccinellids do not seem to be so distributed. [p. 87]

Not only was this usage approximately like Grinnell's, but it was related to the idea of competitive exclusion, albeit not considered applicable to the ladybird subjects of his treatise. Here, incidentally, is an early example of an arthropod biologist not finding clear niche separation in his system, a pattern to be repeated many times (see Schoener 1986a for a review). Hutchinson (1978) is 'inclined to view the use of the word niche in this statement as an ordinary example of the figurative employment of this architectural term'. (p. 156.) Certainly no more formal concept was developed, but it is especially easy to see with this example how well-suited the label 'niche' was to its ecological connotation in the early days. For those interested in immediate causes, Hutchinson (1978) writes:

> . . . I have looked through all Grinnell's earlier major works and as many of his small contributions as were available at Yale without finding any use prior to 1914. Frank A. Pitelka most kindly has examined the collections of books that Grinnell had in 1910–14 and found that Johnson's work was not among them. There are, moreover, no letters from him in Grinnell's very complete file. [p. 156]

Elton's concept of the niche

In 1927 Elton published his pioneering book *Animal Ecology*, in which an entire section is entitled 'Niches'. The concept is introduced as follows 'It is . . . convenient to have some term to describe the status of an animal in its community, 'to indicate what it is *doing* and not merely what it looks like, and the term used is "niche".' (pp. 63–64.)

Elton centres his concept about trophic properties:

The niche of an animal can be defined to a large extent by its size and food habits. . . [For example, there] is in every typical community a series of herbivores ranging from small ones (e.g. aphids) to large ones (e.g. deer). Within the herbivores of any one size there may be further differentiation according to food habits. [p. 64]

Unlike Grinnell's and subsequent concepts, Elton's niche need not be restricted to a single species:

For instance, there is *the* niche which is filled by birds of prey which eat small mammals . . . Or we might take *as a niche* all the carnivores which prey upon small mammals, and distinguish them from those which prey upon insects . . . the niches about which we have been speaking are only smaller subdivisions of the old conceptions of carnivore, herbivore, insectivore, etc., and. . .we are only attempting to give more accurate and detailed definitions of the food habits of animals. [p. 64] [Italics mine.]

Elton explicitly excludes macrohabitat factors in his characterization of niche, 'In an oak wood [the niche comprising birds of prey that eat small mammals] is filled by tawny owls, while in the open grassland it is occupied by kestrels'. (p. 64.) Inasmuch as 'oak wood' and 'open grassland' have about the same classificatory status as Grinnell's (1914) associations, e.g. 'willow' and 'creosote', there is certainly some difference in the *application* of Elton's and Grinnell's niches, although not necessarily in the *definition*; for Grinnell, primary occurrence in different associations necessarily implied different niches (see also below).

The need to understand ecological equivalents as a *raison d'être* for the concept is explicit here as with Grinnell's niche (pp. 64–65) 'There is often an extraordinarily close parallel between niches in widely separated communities'. Examples are hyenas and arctic foxes, or sand-martins and bee-eaters. Elton also cites the 'mouse niche's coming very close to Grinnell's kangaroo rat/jerboa pair. Somewhat unlike Grinnell or Taylor, he points out that equivalents may cross major taxonomic lines; for example, the niche occupied by species that pick ticks off other animals can be filled by birds (Africa, England) or land-crabs (Galápagos).

Finally, Elton does not limit his discussion to food '. . . it is convenient sometimes to include other factors than food when describing the niche of any animal.' (p. 65.) As an example, he compares the nests of sand-martins and bee-eaters, both found in sandy cliffs lining river valleys.

Elton's concept is sometimes claimed to be quite different from Grinnell's, and this notion has been enshrined in certain textbooks. For example, Ricklefs (1979) states that '... the niche has been given a variety of meanings including. . .Grinnell's (1917) use . . . to describe the habitats and habits of birds and. . .Elton's (1927) ... to describe the species' place in the biological environment — its relation to food and enemies' (p. 242). Krebs (1978) goes further 'The term *niche* was almost simultaneously defined to mean two different things . . . Grinnell in 1917 . . . viewed it as a subdivision of the habitat. . . Elton in 1927 independently defined the niche as the "role" of the species in the community'. (p. 226.) In contrast to such statements, I believe the similarities are more important than the differences, especially with a hindsight that includes more modern concepts. What are these similarities?

First, both see niches primarily as largely immutable 'places' or 'recesses' in the community. Because Elton allows niches to house larger units than individual species, this immutability is perhaps more plausible for his concept, but in both cases a rather rigid determinism is proposed that does not allow much variation in relevant traits of the occupants. In neither case is the niche primarily identified with its occupant.

Second, ecological equivalents arise in both discussions, as evidence that repeated (or nearly so) niches do indeed occur and as one of the more intriguing facts associated with the niche that requires explanation.

Third, both concepts characterize food resources as major components but neither is restricted to them; both include anti-predator traits as well.

Fourth, Grinnell's and Elton's concepts both include dietary and microhabitat factors. Several passages show that Grinnell intends the niche to include diet, a central property of Elton's niche. In addition, Grinnell uses microhabitat (foliage layer from which food is taken); while Elton does not do so explicitly, he does it implicitly, so that one has, for example, niches of species that feed on ticks located on animals, or of birds that feed primarily by hawking over rivers, or of sap-suckers, or of burrowing detritivores.

Differences between Grinnell's and Elton's concepts are less fundamental. One clear difference is that Elton's niche may include more than a single species. A second difference is harder to pin down. Elton clearly excluded macrohabitat factors from his concept; see the passage above on having the same niche in oak wood and grassland, impossible if macrohabitat is included. Grinnell's position on macrohabitat is harder to discern. On the one hand, he does give the distribution of observations for various species of animals over a set of vegetation associations (e.g. Fig. 4.1), but he does not say this is the niche; rather the niche somehow is the endpoint of a hierarchy of spatial and other levels, the penultimate of which is the association. The meadowlark passage, on the other hand, does seem to invoke certain macrohabitat factors as part of niche specification. Macrohabitat factors, however, provide an incomplete and perhaps non-fundamental niche specification; several species can occur in the same macrohabitat (association), and these can be discriminated only by dietary and microhabitat factors. Although never explicitly done by Grinnell, pursuit of the underlying relationships that result in species having a particular set of macro- and micro-habitat properties would probably reveal largely quantitative rather than qualitative differences between the two [see also discussion of Whittaker, Levin & Root (1973) below].

If the similarities are so great, how likely is it that the concepts were framed 'independently' (e.g. Krebs 1978)? Of obvious interest is whether Grinnell's classic papers are cited in Elton's book, which postdates the former by 3–13 years. In fact, none of the classic papers discussed above is cited by Elton, but he does cite another of Grinnell's works, *Animal Life in the Yosemite*, a large book co-authored with Tracey Storer (for which Storer Hall at Davis is named, where I am writing these words). This book was published in 1924, the same year as was Grinnell's address to the Western Society of Naturalists and the Ecological Society of America. The book is cited, not in reference to the niche, but to the life-zone concept and also in a methodological section. 'Niche' *is*, however, discussed in Grinnell & Storer (1924) '. . . no two species well established in a region occupy precisely the same ecologic space; each has its own peculiar place for foraging, and for securing safety for itself and for its eggs and young. These ultimate units of occurrence are called "ecologic niches".' (p. 12.)

Hutchinson's concept of the niche

Hutchinson (1978, p.158) says that his niche concept was first introduced in a footnote to a 1944 paper in *Ecology* on phytoplankton periodicity: 'The term *niche* (in Gause's sense, rather than Elton's) is here defined as the sum of all the environmental factors acting on the organism; the niche thus defined is a region of an n-dimensional hyper-space, comparable to the phase-space of statistical mechanics.'(p. 20.)

Thirteen years elapsed before his classic paper appeared as an overview of a conference on animal populations at Cold Spring Harbor. One of these 'concluding remarks', entitled 'The Formalization of the Niche and the Volterra–Gause Principle', begins thus (Hutchinson 1957):

Consider two independent environmental variables x_1 and x_2, which can be measured along ordinary rectangular coordinates. Let the limiting values permitting a species S_1 to survive and reproduce be respectively x'_1, x''_1 for x_1 and x'_2, x''_2 for x_2. An area is thus defined, each point of which corresponds to a possible environmental state permitting the species to exist indefinitely. If the variables are independent in their action on the species, we may regard this area as the rectangle ($x_1 = x'_1$, $x_1 = x''_1$, $x_2 = x'_2$, $x_2 = x'_2$, but failing such independence the area will exist whatever the shape of its sides.

We may now introduce another variable, x_3 and obtain a volume, and then further variables $x_4 \ldots x_n$ until all of the ecological factors relative to S_1 have been considered. In this way an n-dimensional hypervolume is defined, every point in which corresponds to a state of the environment which would permit the species S_1 to exist indefinitely. For any species S_1, this hypervolume N_1 will be called the *fundamental niche*[2] of S_1. Similarly for a second species S_2 the fundamental niche will be a similarly defined hypervolume N_2.

It will be apparent that if this procedure could be carried out, all X_n variables, both physical and biological, being considered, the fundamental niche of any species will completely define its ecological properties. The fundamental niche defined in this way is merely an abstract formalisation of what is usually meant by an ecological niche.' [p. 416]

The niche is defined for 'abstract space', and every point in some portion of real space (the 'biotope space'), for example a specific cubic metre of Lake Washington, will map onto a point of abstract (niche) space; the mapping may be several-to-one (i.e. more than a single point in real space can map onto a single point in abstract space). The opposite mapping need not be possible for every point in abstract (niche) space. Fig 4.2, reproduced exactly from Hutchinson (1957), illustrates a possible mapping.

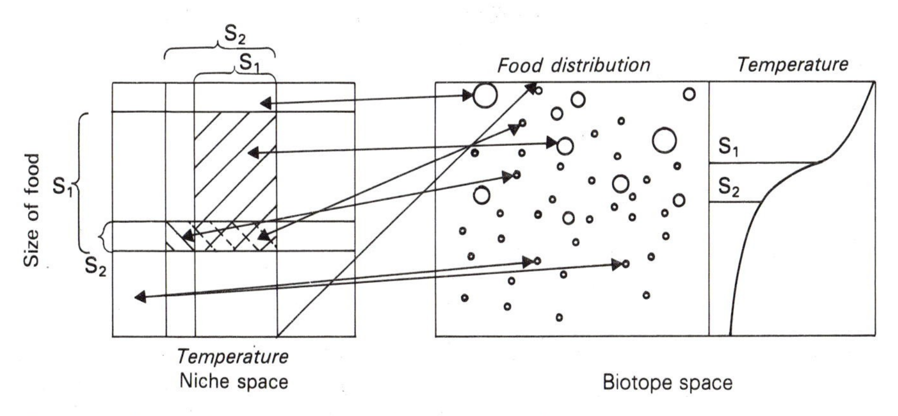

FIG. 4.2. Two fundamental niches defined by a pair of variables in a two-dimensional niche space. Only one species is supposed to be able to persist in the intersection subset region. The lines joining equivalent points in the niche space and biotope space indicate the relationship of the two spaces. The distribution of the two species involved in shown on the right hand panel with a temperature depth curve of the kind usual in a lake in summer. [After Hutchinson (1957).] (In this figure, S_1 and S_2 refer to Species 1 and Species 2.)

Hutchinson lists four 'restrictions' (p. 417) on the hypervolume representation, three of which are as follows. First, some parts of the niche may be better than others in terms of the probability that the species will persist. MaGuire (1973) extended this idea by calculating a population rate of increase for various points in the niche spaces of certain micro-organisms, thereby producing a contour diagram. Second, not all environmental variables may be linearly orderable. Examples are prey species or type of vegetation association. Third, '. . .the model refers to a single instant of time, . . .' but 'A nocturnal and diurnal species will appear in quite separate niches, even if they feed on the same food, . . ., etc.' In more recent treatments, temporal dimensions were added to cover this possibility. The competitive exclusion principle is then given informally as (p.417) '. . .two species, when they co-occur, must in some sense be occupying different niches'. To put this in his more precise, set-theoretic terminology, Hutchinson defined the realized niche of a species as that portion (if any) of the fundamental niche not overlapping fundamental niches of other species, plus that overlapping portion within which the given species can survive. Competitive exclusion is thus expressed as '. . . realised niches do not intersect.' (p. 418.)

Hutchinson's formulation was revolutionary, in that it *defined* a niche strictly with respect to its occupant, a species population, and not at all with respect to the place or 'recess' in the community, the metaphorically more suitable definition used in part by Grinnell and Elton. This shift to the occupant may have been forced for consistency's sake; Hutchinson, much more than previous workers, imagined niche properties of occupants as readily mutable, e.g. his distinction between fundamental and realized niches. If competition between species commonly caused their niches to change, then observation of an occupant's characteristics cannot be used in any simple way to determine characteristics of the 'recess', as would be perfectly possible in Grinnell's concept. Indeed, given the notion of such highly protean occupants, recesses could only be viewed as very diffuse attractors, loosely constrained but not uniquely determined by the functional rules relating species to environments.

A second major advance of Hutchinson's formulation was its precision. The niche could only be defined with respect to a set of continuous axes. Hutchinson (1957, 1978) gave as examples of such axes food size, temperature and water depth. Discrete categories, such as Grinnell's associations (Fig. 4.1), cannot be directly used to characterize the niche. Nonetheless, Hutchinson's formulation has both explicit and implicit spatial (or habitat, according to the modern meaning) axes. Water depth is explicitly such an axis, and temperature is implicitly one, as it often co-varies with habitat gradients.

Finally, both of Hutchinson's (1957, 1978) major discussions strongly emphasize that interspecific competition affects niches as they occur in nature. This emphasis on competition i.e. mainly on obtaining food rather than on avoiding predators, is somewhat different from Grinnell's treatment and very different from Elton's.

Hutchinson's concept of the niche is such a break from both Grinnell and Elton that it is of interest to probe its origins. Hutchinson himself gives most credit to a paper by Haskell (1940), although why was not obvious to me upon reading the paper. Multidimensional approaches had been part of the plant-ecology literature for some time, as both Whittaker, Levin & Root (1973) and Hutchinson (1978) note, beginning with the work of Ramensky (1924), and extending through the well-known ordination work of Ellenberg (e.g. 1950), Whittaker (e.g. 1951) and Bray & Curtis (1957). An extremely similar formalization to that of Hutchinson occurs in a 1935 book by Kostitzin; the last section, 'Stabilité de la vie,' beings as follows 'Imaginons un espace symbolique à plusieurs dimensions représentant les facteurs vitaux: p = pression, T = température, l = éclairage, etc. Dans cet espace chaque être vivant à un moment donné occupe un point, une espèce peut être représentée par un ensemble de points.' (p.43.) Of this work, Hutchinson (1978) writes 'James R. Ziegler has drawn my attention to a passage in V.A. Kostitzin . . . in which a formulation for scenopoetic factors identical to mine occurs. As I knew of the work in the early 1940s, it is almost certain that it had a direct influence, though I did not remember the passage.' (p. 158.)

The niche as a utilization distribution

That concept of the niche nearly always used in the body of concepts known as 'niche theory' (e.g. Levins 1966,1968; MacArthur & Levins 1967; MacArthur 1968; Pianka 1969; Roughgarden 1972; May 1974; Schoener 1974a) is not Hutchinson's, but rather the 'utilization distribution'. It is defined for a particular species population and gives the fractional use of resources arranged along one or more dimensions called 'niche axes'. It is thus nothing more than a frequency histogram of resource use by some population. An example is given in Fig. 4.3, where both one- and two-axis distributions are shown. More niche axes can be added, but with corresponding loss of graphical representation.

Typically, a wide variety of niche axes is allowed in this scheme. These axes may be roughly classified by food, space and time (Pianka 1969; Schoener 1974a). Food utilizations are determined by estimating the relative number of food items of say, different sizes, that individuals of a population eat over some time period; this might be determined

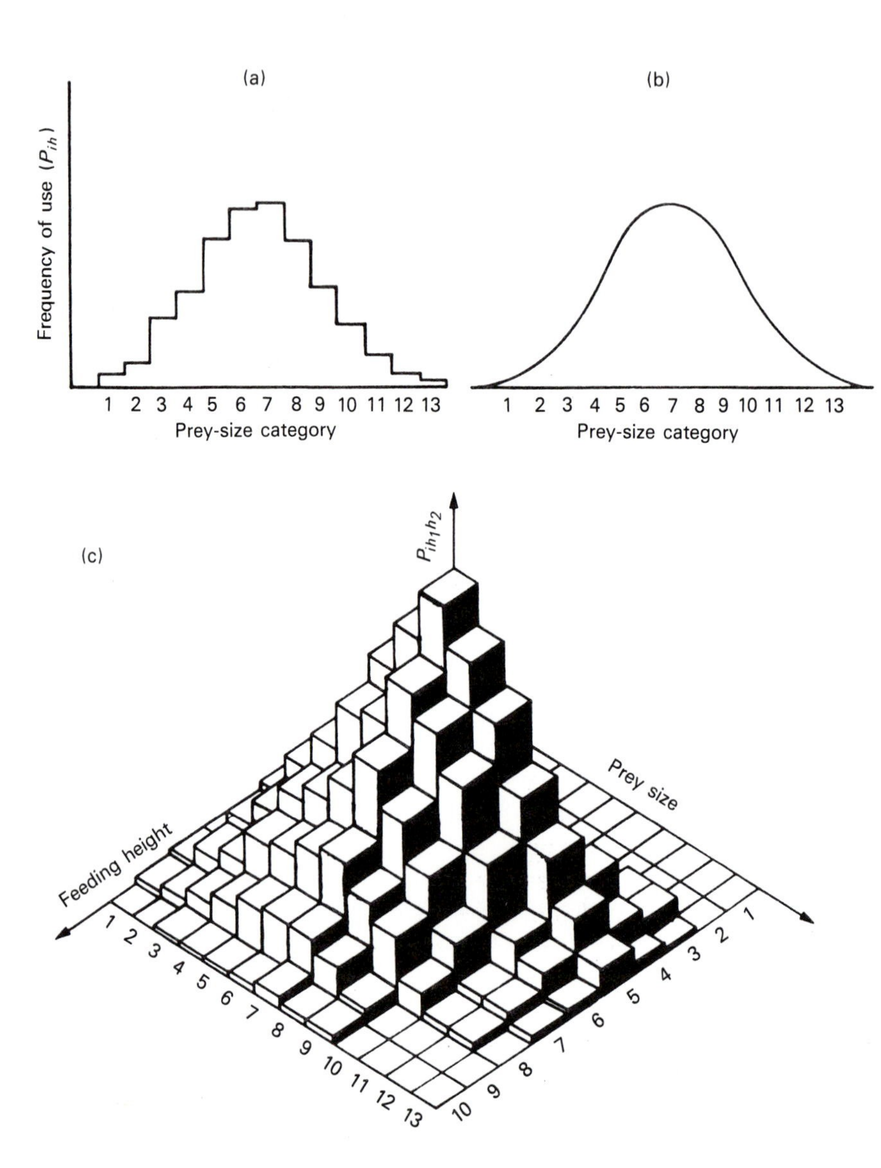

FIG. 4.3. (a) The utilization for Species i as a frequency histogram. This utilization is one-dimensional, where the dimension is prey size. Numbers refer to prey-size categories, indexed by *h*. (b) The same utilization smoothed. (c) A utilization for two resource dimensions, prey size and feeding height. [After Schoener (1986b).]

from gut contents. Space utilizations are determined by observing where individuals of a population are found over some activity period, perhaps using feeding observations alone. Spatial axes may be those of macrohabitat (coarse-grained, see below) or microhabitat (fine-grained). Temporal utilizations are determined on a diel basis by counting the relative number of 'active' individuals at various times of the day; seasonal occurrences are determined similarly, but over periods of months.

The transformation of Hutchinson's multidimensional concept into this one was doubtless spurred by the eminently operational concept of the utilization niche. One major difficulty with Hutchinson's niche is the empirical determination of dimensional states permitting the species to survive. As mentioned, MaGuire (1973) had made some progress here with micro-organisms, but in general, the criterion of population survival is too difficult to ascertain for most organisms, especially in the field. A second set of difficulties, in part logical and in part logistical, has to do with the representation of the niche as a range of values, and the mapping of points from niche (abstract) space into real space. The difficulty is that populations do not survive (or for that matter occur) at single values of axes such as food size, or even depth or ambient temperature. Although any single food item in Hutchinson's biotope (see Fig. 4.2) maps into a location on the food–size axis, no population, let alone individual of that population, can be characterized by exposure to food of just that particular size. The same can be said to a lesser extent for spatially varying axes such as temperature and depth. Hutchinson seems to have realized the problem in part, because in his 1978 (p. 159, fig. 99) treatment (as opposed to the original 1957 one), he uses 'mean food size', rather than food size *per se*. Even this modification is not without difficulty, because a large number of prey-size distributions have the same mean, some of which will be more favourable for consumers than others. In short, Hutchinson's niche is a set of ranges which, even if enriched with MaGuire's contours, allows little representation of distributional properties of resource use.

Because of these difficulties, Hutchinson's formulation gives immediate trouble with concepts both of niche breadth and of niche overlap. Hutchinson (1978) says that '. . . the concept of niche breadth is initially due to Levins (1968), who used the term to indicate the length of that part of any niche axis that included all the points defining viable values of the variable that is measured along the axis.' (p. 169.) This definition is cast in Hutchinson's terminology, but what Levins (1968) actually said was 'Any measure of spread could be used to quantify niche breadth.' (p. 41.)

The measures Levins himself proposes are two:

$$\log B = -\Sigma p_i \log(p_i)$$

or

$$B = 1/\Sigma p_i^2,$$

where p_i is the proportion of the species which is found in Environment i, which selects Environment i, or in the case of a viability measure $p_i = v_i/\Sigma v_i$, where v_i is the viability in Environment i. Both of Levins's measures are sensitive to the variance and other distributional properties of resource use or viability. Similarly for niche overlap, Hutchinson's (1957) realized niches cannot overlap where species coexist, provided competition is occurring. Yet (see next section), modern niche theory relates niche overlap to the competition coefficient, which is in turn related to coexistence. A little overlap allows coexistence, but somewhat more overlap does not. Hutchinson's formulation of niche overlap acts as if competing species are placed together in arenas having single values of such niche dimensions as food size or temperature. Should such arenas exist, modern niche theory would support Hutchinson's contention that only one species could survive, as resource use here will perforce be identical. But real arenas where populations interact are characterized by distributions of values over axes of resource availability, not by single values.

The utilization concept of the niche originated soon after Hutchinson's development. The first theoretical paper employing resource-utilization distributions was that on limiting similarity by MacArthur & Levins (1967); the actual word 'utilization' appears to have been coined here, although with a somewhat more specific meaning than the current one. Levins (1966) however, had defined niche breadth a year earlier. Furthermore, numerous authors had been presenting data on resource use as utilization distributions for some time (see papers in Schoener 1974a; Toft 1985; Ross 1986). Perhaps this already compiled store of data helped establish the utilization concept of the niche.

A critique of 'niche' usage

A major critique of the route that the usage of the word 'niche' was taking was written by Whittaker, Levin & Root in 1973. Their paper strongly urges the separation of 'niche' and 'habitat', a plea that in retrospect seems to have been largely ignored. Their justification, largely historical, runs as follows. They identified three senses of 'niche': (a) the 'functional' sense — the position or role of the species in the community; (b) the habitat niche or 'place niche concept'; and (c) the niche as (a) + (b). These senses can be identified with earlier concepts as follows. First, they argue that Grinnell's niche concept is closer to (c), although they also say that it is sometimes misinterpreted or simplified to (b). I agree with them on this. Second, they argue that Elton's concept is (a); here, I do not agree, insofar as microhabitat is concerned; one needs to read Elton's examples, not merely the words he uses to describe his concept (see above). They believe it was Elton's concept more than Grinnell's that was adopted in most discussions of the 1930s to the late 1950s. Then they state 'The current difficulty derives primarily from different interpretations of a single, seminal paper by Hutchinson (1958[sic]).' (p. 322.) After describing his concept, they write 'A number of other writers (including two authors of this paper) have considered that Hutchinson's fundamental niche formulated concept (c)... However... His formulation was of... concept (a), with the niche defined purely *intensively* — that is, by variables as they apply within the biotope (Hutchinson, 1967, p. 232). Factors of habitat, in contrast, have spatial extension.' (p. 323.) Put another way, they argue that Hutchinson's niche is defined *within* a particular community. Turning to Hutchinson's (1967) statements, certain portions seem at first glance to support their argument 'The *habitat* of a species, within the geographic range, may be regarded as appropriately defined by specifying those parts of the ecosystem that must be present in a biotope in order for the species to occur. The habitat is regarded as having spatial extension. The *niche* of a species is defined purely intensively.' (p. 232.). But then he continues 'It is assumed that all the variation of the factors required to define a habitat can be ordered linearly on the axis of an n-dimensional coordinate system.' My interpretation of these passages is that Hutchinson considers habitat, like biotope, to be some particular real place, albeit at a smaller scale. For example, the chapparal of the western Sierra Nevada slope is such a place. 'Chapparal' in contrast, would not be any particular place, but rather would be a *kind* of habitat that might occur *inter alia* in California or in Spain or in Chile. Hutchinson's 'habitat' is a proper term, whereas the usage of habitat in most modern niche studies is as a kind of place, the 'place niche.' Whittaker, Levin & Root (1973) seem to interpret Hutchinson's discussion as excluding the 'place niche' from the hypervolume, whereas in fact Hutchinson says that spatial factors can be ordered on the latter's axes. About this issue incidentally, Hutchinson (1978) himself comments 'The original multidimensional concept of the niche, initially elaborated for the plankton, certainly was conceived to extend over more than one community insofar as in a stratified lake several communities are likely to be present one above the other.' (p. 197.)

Aside from the historical precedent argument, Whittaker, Levin & Root give scientific reasons why they prefer Concept (a). Most of these seem to argue against macrospatial axes only, and boil down to the difficulty that if too large a habitat gradient is used, species may not be in spatial contact and certainly would not belong to the same community, so that use of niche concepts as means toward understanding competitive exclusion would be impossible. I agree that this distinction, which can also be labelled 'coarse-grained' (the macrospatial case) vs 'fine- grained' (the microspatial case), can be quite important (see below). Hutchinson (1978) has commented '. . . the original multidimensional niche was not properly distinguished from what would now be called an ecotope, . . .' (p. 197) which latter is defined as a hypervolume that includes broad-scale habitat axes. However, it is difficult *a priori* to know which case one has, and there would appear to me to be a continuum of spatial scales from the finest microhabitat to the coarsest macrohabitat, where fine- or coarse-grained would depend in part on the kind of organism being studied (see also Kulesza 1975). Exactly where the line should be drawn could be moot. To illustrate, a year or two after the 1973 paper was written Whittaker visited Harvard, and I had a substantial discussion with him about this issue with particular reference to the *Anolis* lizards I was then studying. I noted that perch height and diameter were important dimensions separating species, and I asked him if he would consider them valid niche dimensions. He said yes (see also Whittaker, Levin & Root 1975). I then asked him about climatic differences, those of sun versus shade, for example. He felt these should be in the non-niche, 'habitat' category. I then pointed out that these 'climatic' differences could operate over the space of a few metres, and that they interacted (in a multiway-contingency-table or ANOVA sense) with perch height and perch diameter; thin perches are preferred in the sun, for example (Schoener 1970a; Schoener & Schoener 1971a, b). He agreed that the distinction could be difficult.

MODERN THEORY OF THE NICHE

The modern theory of the niche is largely directed towards systems in which resource competition predominates. It attempts to explain *inter alia* the number of species in a community and their abundances, the degree of invasibility (or inversely, saturation) of a community, and changes in niche properties of particular species from one community to another. It has been the subject of several reviews, some quite recent (Schoener 1974a, 1977, 1986b, 1988a; Roughgarden 1979; Abrams 1983; Giller 1984). In this essay I have emphasized the concept's historical development. Here I give the barest précis of the theory and discuss a few applications and criticisms. Readers are referred to the above references for further discussion.

Models

1 Coexistence of species with static niches

Ecological models of species co-occurence do not allow evolution; they ask, given an established equilibrium community, what species if any can (i) invade, and (ii) persist? Largely because evolution is excluded, utilizations remain constant throughout the time span of interest (behavioural shifts are ignored), and such immutable species either fit together or not. In the latter case, at least one becomes extinct. MacArthur & Levins (1967), the founding paper of the modern theory, presents a model of this general kind. It asks whether the utilization of an invading species can be sandwiched between those of two established species. The model assumes a single important niche axis and normally distributed utilizations, equally spaced and equally shaped. A crucial assumption is that a formula for niche overlap can be used to compute the competition coefficient in the Lotka–Volterra model, itself assumed to describe competition dynamics appropriately. The formula in discrete form (MacArthur & Levins give it in continuous form) is

$$\alpha_{12} = \sum_h p_{ih} p_{2h} / \sum_h p_{1h}{}^2,$$

where α_{12} is the competition coefficienct for Species 2 on 1, and p_{ih} is the utilization frequency of Resource Kind h by Species i. Fixing the K values in the Lotka–Volterra model, we can compute those α values just small enough to allow coexistence, thereby giving the so-called 'limiting similarity'. These α values can be converted into units measuring properties of the utilizations by back-calculating from the above equation. In particular, if d is the distance between utilizations (measured as the distance between the modes of adjacent distributions, and w is utilization width (here the standard deviation of a distribution), limiting similarity can be measured in units of d/w. May (1973, 1974) and Roughgarden (1974) expanded the MacArthur–Levins analysis to include persistence as well as invasion. May (1973, 1974) also considered more than three species; the more species the larger the limiting d/w, because of greater summed or 'diffuse' competition from all the species (see also MacArthur 1972).

This sort of model is particularly sensitive to five departures. First, as niche shape varies from normal to leptokurtic (thick-tailed, peaked) distributions, limiting d/w diminishes (Roughgarden 1974). Second, if α is calculated from utilizations in other sensible ways than the above formula, limiting d/w diminishes (Abrams 1975). Third, under increasing degrees of frequency-dependent predation, limiting d/w diminishes (Roughgarden & Feldman 1975). Fourth, a two-level model of competing consumers and resources (MacArthur 1968) leads to an α that includes total consumption rates and resource availabilities (Schoener 1974b); this formula can give very different, even opposite results to that of MacArthur and Levins. Spiller (1986), for example, found that only the more complicated formula agreed with experimental results on spiders. Fifth, competition models giving concave zero-isoclines (rather than linear ones, as does the Lotka–Volterra model) generally increase the likelihood of coexistence (Schoener 1974c, 1976, 1978).

The above treatment is deterministic. In contrast, environmental stochasticity, modelled as 'white noise' affecting particular parameters, was analysed by May & MacArthur (1972) and Turelli (1981). The treatments are somewhat at variance. The latter, apparently correct one, shows (i) that slightly fluctuating environments give results very similar to deterministic ones (models for the latter are therefore approximately valid for the former; see also Ågren & Fagerström 1980); and (ii) that only for large fluctuations will environmental stochastic variation markedly affect limiting d/w, and then the effect can be either destabilizing or stabilizing, depending on the particular model used.

2 Coexistence of species with dynamic niches

Although MacArthur & Levins (1967) briefly treated the evolution of niche properties, the first major papers on the topic were by Roughgarden (1972, 1976), on changes in niche breadth and niche position, respectively. Most such models give d/w values that at evolutionary equilibrium are substantially greater than limiting d/w values calculated from ecological models (review in Schoener 1988a). In Roughgarden's (1976) treatment, (i) the smaller the number of species the larger will be the 'realized' (Roughgarden 1979) d/w, and (ii) the wider the available resource spectrum, the further apart will utilizations be. This model is automatically a model of character displacement (Brown & Wilson 1956) as well, and it was soon joined by a number of others. Slatkin's (1980) treatment allowed both niche position and between-phenotype variation to evolve. Case (1982) allowed niche position and within-phenotype variation to evolve. The latter was recently expanded by Taper & Case (1985) to allow niche position and both within- and between-phenotype variation to evolve; the general model also included, explicitly rather than implicitly, resource dynamics, as in the Lotka–Volterra-based treatments. Taper & Case's model provides a strong basis for expecting character divergence, as well as predicting that (i) d/w should decrease as species number increases, (ii) within-phenotype variation should be larger than between-phenotype variation, and (iii) utilization separation should be smaller than trait separation. Milligan (1985), in a quite different analysis, pointed out that divergence is more likely where species utilizations match the available resource spectrum less precisely. Finally, another recent treatment (Abrams 1986) includes resource dynamics as well as other density-dependent processes; here convergence can occur relatively easily.

Character-displacement models, although usually exemplified by phenotypic changes in body size which may reflect utilization changes in food size, are also applicable to shifts in microhabitat, although details may vary. Shifts at a greater habitat scale were treated by Schoener (1974b, 1975) as if habitat types were separate arenas of competition. Competition along very large-scale habitat gradients has only been considered recently; an important initial paper by Shigesada, Kawasaki & Teramoto (1979) was followed by Shigesada & Roughgarden (1982) and Pacala & Roughgarden (1982a). Most recently, Brew (1984, 1987) modelled competition in coarse-grained habitats, defined as those in which encounter rates are under the control of individuals through their active choice (see also Rosenzweig 1979). A major result was that niche separation along coarse- grained dimensions implies strong competition, the opposite of the conclusion for the fine-grained dimensions considered above. Finally, habitat shifts resulting from very short-term changes in behaviour, as well as dietary shifts, were modelled using optimal foraging approaches. This collection of effects is known as the compression hypothesis (MacArthur & Wilson 1967; Schoener 1974d). Predictions are that species habitats should diverge in early periods of sympatry (and perhaps expand as well), whereas food types should not diverge and should typically expand (except where species are largely parapatric; Schoener, Huey & Pianka 1979).

Applications and tests of modern niche theory

The theory just described was usually applied to data already gathered, although some new data collection was also stimulated.

1 Niche separation along a single dimension

Shortly after publishing his classical presentation of the niche, Hutchinson (1959) published his equally famous 'Santa Rosalia' paper, in which he tabulated a few sets of congeneric sympatric species whose trophic appendages differed by a ratio of *c* 1.3. Other cases showing Hutchinson's ratio were subsequently reported, but exceptions also began to accumulate (reviews in Abrams 1983; Schoener 1986b). As discussed, niche theory does not predict a single value for limiting similarity, much less a single value for realized similarity. Nonetheless (with the proviso that niches are constrained by the available resource spectrum), some sort of regular spacing of niches is expected, although d may not be constant. For example, among prey size utilizations d may increase as w increases (Schoener 1974a).

Simberloff & Boecklen (1981) proposed several tests for regular spacing and minimal differences, and only about 25% of the data they examined could be so characterized. Their treatment generated various criticisms (review in Schoener 1986b, 1988a), the most severe problem being that the test for minimal differences is highly dependent on the actual data. One consequence of this is that very small but similar ratios support a minimal difference while large but less similar ratios do not (Schoener 1984; Tonkyn & Cole 1986). Another approach, testing actual size differences against those produced by constructing communities randomly from the species in question, was proposed by Strong, Szyska & Simberloff (1979); this stimulated a whole set of papers on similar Monte Carlo techniques (reviews in Schoener 1986b, 1988a, 1988b). The method of Strong, Szyska & Simberloff, although also drawing substantial criticism (e.g. Colwell & Winkler 1984), is probably the best for analysing size differences in the aggregate, and it is more powerful than, as well as sometimes giving very different results from, the method of Simberloff & Boecklen (1981) (Schoener 1984). However, it looks only for size differences which statistically are unusually large; it does not ask if niches are spaced in a non-random way along an axis. Inspection of the expected versus actual distribution of ratios permits at least a qualitative statement to be made about how constant ratios are. Proceeding in this way, Schoener (1984) found little such constancy in bird-eating hawks. Eldridge & Johnson (1988) however, found rather striking support for

Hutchinson's generalization among North American shorebirds; at least eight unique pairs had ratios between 1.2 and 1.3!

2 Niche separation along several dimensions

Although little theory exists (see however, Yoshiyama & Roughgarden 1977; Pacala & Roughgarden 1982b), patterns of niche separation in several dimensions have also been examined. For example, complementarity in niche separation (Schoener 1974a), whereby substantial overlap along one dimension necessitates small overlap along another, was described for grassland birds (Cody 1968), *Anolis* lizards (Schoener 1968), robber flies (Shelly 1985), and a number of other groups. Most recently, DuBowy (1988) found that complementarity in diet and microhabitat among species of ducks was strong in winter, when food was scarce, but was practically absent in summer, when food was abundant.

3 Niche changes in particular species

The previous discussions deal with community patterns in the aggregate; in addition, niche theory predicts patterns with respect to particular species. Most examples of character displacement were presented without particular reference to niche formalism, probably because most have no data on resource utilizations. A major exception is the study by Fenchel (1975) (see also Schoener, Roughgarden & Fenchel 1986; Levinton 1987) on the deposit-feeding molluscs *Hydrobia*. This group shows repeated, independent character displacement, to about the same degree of divergence, in body sizes of species inhabiting Danish fjords. Moreover, body size correlates well with mean prey size, and prey-size utilizations differ by about $d/w = 1$. This value was expected from May & MacArthur's (1972) early stochastic model, which is now questioned. It is also however, well within the range expected from deterministic models, e.g. May (1974). A study on *Hydrobia* in Britain, however, did not find statistically significant displacement (Cherrill & James 1987). On the other hand, the sometimes-expressed view that character displacement is seldom found is probably not correct, as most studies that consider all legitimate comparisons among species of a widely distributed group have fairly commonly found it (e.g. Schoener 1970b; Dunham, Smith & Taylor 1979; Fjeldså 1983; Schluter 1988). Habitat shifts are also rather frequently found, and niche formalism is sometimes used to characterize them, e.g. shifts in perch height and diameter among *Anolis* lizards (Schoener 1975). The presence or absence of these latter shifts is consistent with competition, e.g. similarly sized species affect one another more than dissimilarly sized species, and larger species affect smaller ones more than the reverse. The distribution of lizard species over microhabitat types on satellite islands of the Greater Antilles is also consistent with competition, according to the most reasonable Monte Carlo analysis (Schoener 1988b). The behavioural compression hypothesis has been tested experimentally by Werner & Hall (1979) for sunfish habitats. Shifts occurred as would be expected from food-energetic considerations. A variety of other experimentally produced habitat shifts has also been reported from the field, twenty-two such experiments being noted through 1982 (Schoener 1983).

In addition to fairly general aspects, specific predictions of niche theory have sometimes been tested. Pianka (1974) found that desert lizard species are more loosely packed (d/w larger), the greater the total number of species in the community; this supports diffuse competition. Working in Thailand, Inger & Colwell (1977) found tighter packing of the herpetofaunas in evergreen forest, climatically the most predictable; this supports certain stochastic models. Case (1979) predicted the sizes of coexisting *Cnemidophorus* lizards in the Gulf of California using co-evolutionary formalism. More recent special models by Roughgarden, Heckel & Fuentes (1983) explained the co-evolution of sizes among *Anolis* lizards of the Lesser Antilles. Taper & Case's (1985) model of character displacement is consistent with many available data, e.g. that between-phenotype variation should be greater than within-phenotype variation. Brew's (1984) predictions about macro versus microhabitat overlap in relation to competition agree with most experimental field studies (Schoener 1983). However, many specific predictions of niche theory remain to be tested. As reviewed in the next section, this testing has slowed recently, perhaps because of mounting criticism of the theory.

Criticism of modern niche theory

At its most ambitious, the theory of the niche helps us understand fundamental questions of ecology: what determines the number of species in a community and their abundances, or inversely, what determines both the extent of species distributions and the variability from community to community in how particular species utilize resources? Yet we are now less optimistic than in the euphoric 1960s and 1970s; as Gilpin (1986) laments '. . . I must confess that I am saddened by [an example of modern ecology's] honest realism, its unabashed pluralism. Something of a romantic, I long for the heady days of an earlier decade. . .'. Oddly, the downslide began well before the theory was tested with any kind of thoroughness; indeed, as Roughgarden (1983) pointed out, much criticism was levelled without a detailed understanding of the theory. I now briefly review some of the more sensible of the criticisms.

First, the theory was regarded by some botanists as being inappropriate, or at least of very limited domain, for plants. In particular, all autotrophs require light, water and similar minerals; hence, substantial resource partitioning would seem impossible. Motivated by this difficulty, Grubb (1977) advocated a rather expanded niche definition, including aspects of habitat, phenology, life-form and regeneration. The first two of these mesh well with 'utilization' concepts, but the latter two do not. The third would probably be viewed by most zoologists as 'indicator traits', i.e. morphological properties reflecting utilizations. The fourth aspect, or 'regenerative niche', while not unimportant for mobile animals (e.g. Varley 1949), is especially important for space-occupying organisms when most space is filled and reproductive capabilities greatly exceed what is necessary to populate the empty space. The regenerative niche concept was made numerical by Fagerström & Ågren (1979). They showed that differentiation of species with respect to mean diaspore production, temporal variability in that production, or specific reproductive phenology, can allow species to coexist which would otherwise competitively exclude one another. While thus appropriate for certain organisms, especially terrestrial plants, the regenerative theory is substantially more inclusive than traditional niche theory, incorporating as it does life-history variation. Incidentally, it shows some major similarities, to recent theory for coexistence of coral-reef fish (Chesson & Warner 1981; Abrams 1984). Countering this trend, a major programme to explain plant diversity in terms of resource partitioning was embarked upon by Tilman (e.g. 1986), but here Lotka–Volterra-based models are explicitly rejected (Tilman 1987).

Second, 'friendly' mathematical critics argued that the theory was not as general as claimed, even for highly resource-partitioned systems. Each new theoretical treatment seemed to produce unexpected new results, and it was not always clear that the abundant modelling activity was converging (review in Schoener 1982). It might pessimistically be argued that the resulting proliferation showed the theory was too protean and too complex to be useful. For example, a decade separated the appearance of the first models, which gave roughly a single value of limiting similarity (e.g. May & MacArthur 1972), from the full realization that different models give quite different values (Abrams 1983). A second example of multifarious theoretical predictions is given by the character-displacement models discussed above. Because this complexity can be expressed in biological terms, a diverse theory for a diverse world is not necessarily unacceptable (Schoener 1986c). The question is, however, just how diverse the eventual theory will be in relation to the number of communities available on Earth for falsification. The same can be said for other aspects of community theory, and we have little indication as yet of the answer (see also Schoener 1986c).

A third cluster of criticisms is concerned with factors outside the theory that can invalidate its assumptions. In particular, niche characteristics can be affected *inter alia* by predation, physical factors and mutualism. Hence it is argued that competition must be shown experimentally before knowing that the phenomena of interest are in the domain of niche theory; after that, however, of what further use is the theory? I return to this issue below.

The latter two types of criticism come together in what has perhaps been the most serious of specific problems for niche theory; the relation of competition to niche overlap. To see just how convoluted an issue it can be, consider the following two empirical results on niche overlap and competition, both supported by experiment. Dunham (1983) studied two species of iguanid lizards in Texas and experimentally found competition during dry but not wet years. During the former, dietary and microhabitat niche overlap was lower than during the latter (non-competing) years. The conclusion here is that competition is associated with lower niche overlap. Pacala & Roughgarden (1985) studied two species of iguanid lizards in the Lesser Antilles. They showed experimentally that species of similar size competed but those of dissimilar size did not; the former overlap much more in diet than the latter. The conclusion here is that competition is associated with higher niche overlap. In addition to these empirical results, plausible conceptual arguments can be made on both sides of the issue (see also Colwell & Futuyma 1971; Vandermeer 1972). On the one hand, species might be zoned in different macrohabitats (have less niche overlap) where competition is high. On the other, niche theory typically assumes that competition coefficients vary directly with niche overlap, as in the equation given on pp. 97.

To lead us out of this tangle, I have constructed a dichotomous decision tree (Fig. 4.4) for relating niche overlap to competition. Decisions are needed at three levels. First, does or did competition occur? If competition is not an issue, the degree of niche overlap is not specified by niche theory; it will be high if all species are concentrating on the same superabundant food resources (Smith *et al.* 1978; Schoener 1982). Second, is competition present or past? Past competition has caused niche separation and/or fixed it evolutionarily, so that strong resource partitioning implies minimal present competition: one finds little disease once the patient is cured. Third, given present competition,

1 *A* Competition occurs or has occurred **2**
B Competition not an issue; relation indeterminate from niche theory (if a superabundant resource is profitably selected, as in optimal foraging theory, then species will overlap greatly)

2 *A* Competition present-day **3**
B Competition past, niche overlap low

3 *A* Niche dimension(s) fine-grained (food type, microhabitat)
competition high, niche overlap high; or
competition low, niche overlap low
B Niche dimension coarse-grained (macrohabitat)
competition high, niche overlap low; or
competition low, niche overlap high

FIG. 4.4. Dichotomous key relating niche overlap to competition, according to niche theory.

are the relevant niche dimensions fine- or coarse- grained? If the former, we have ordinary niche theory (e.g. MacArthur 1968), which assumes consumers and resources are in a well-stirred soup. Here, the greater the overlap (at least as modified in Schoener 1974c), the greater the competition. If the latter, we have Brew's (1984) model, in which lower overlap is associated with greater competition. Going back to the two lizard examples, Dunham's (1983) case splits at Level 1; in some years no competition occurs because resources are superabundant and the species do the same thing (other examples in Schoener 1982). When competition does occur, overlap is lower, presumably because niche separation evolved during such times (however, we might not expect competition to be *extremely* great then; see above). Pacala & Roughgarden's (1985) case does not split until Level 3. Here, resources are always limiting, and niche dimensions are fine-grained, so greater overlap implies greater competition. Thus it seems there is nothing wrong with the logic of the theory, but the decision tree requires rather detailed information.

SYNTHESIS AND PROSPECT

The meaning of 'niche' has changed substantially over the past three-quarters of a century. It began by reflecting the figurative usage of a place or 'recess' in the community. The concepts of both Grinnell and Elton were of this sort, seeking to understand ecological equivalents, i.e. the parallel types of opportunities which occur for species to use resources and avoid predators from one locality to another. Both Grinnell and Elton considered diet and microhabitat as part of the niche. Grinnell's concept extended to at least the finest macrohabitat level, whereas

Elton's did not, but even for Grinnell, macrohabitat was not emphasized inasmuch as it did not fully discriminate species. Elton but not Grinnell allowed the niche to include more than one species, e.g. a trophic type. But generally the concepts were similar. Hutchinson's formulation was quite different, as he shifted the emphasis from the recess to its occupant. The notion that niche properties were quite variable, especially when affected by interspecific competition, would seem to have necessitated this reversal. Hutchinson phrased his concept quantitatively as a set of ranges along multiple dimensions creating a hypervolume within which a species population could survive. Operational difficulties with this concept caused Hutchinson's niche to be succeeded by the utilization distribution, which simply describes resource use by a population, much as studies of resource partitioning had been doing for some time. The utilization, the most easily measured of the four 'niches', is appropriately the subject of the modern theory of the niche. This theory has been almost entirely about competition. Table 4.1 summarizes the main differences between these four major niche concepts.

For competitively controlled communities, the modern theory of the niche predicts their number of species, their invasibility or saturation, and the degree of specialization and similarity of their species' niches. With the waxing of pluralism in various areas of ecology, niche theory was seen as of more limited use, at least in terms of the kinds of organisms to which the original theory could be applied. Parallel theories

TABLE 4.1. Comparison between the four principal niche concepts

	Grinnell	Elton	Hutchinson	Utilization
Recess- or occupant-based	Recess	Recess	Occupant	Occupant
Extent of spatial scale	Mostly microhabitat*	Microhabitat	Micro- and macrohabitat†	Micro- and Macrohabitat
Mainly food resources or both food and predators included	Both	Both	Mainly food	Mainly food
Breadth of taxonomic group	One species only	Sometimes broader	One species only	One species only
Survival or occurrence	Not specified	Not specified	Survival	Occurrence

* Some broader scale habitat factors used to specify the association.

† Originally included both (Hutchinson 1957), but some retraction later (Hutchinson 1978; see text).

for how predation might affect niche properties are now emerging (Holt 1987; Martin 1988), earlier such ideas being exemplified by Clarke (1962), Moment (1962) and Rand (1967). In this respect we are coming full circle back to Grinnell and Elton. Even within the competitive domain, niche theory rapidly became more diverse as it was realized that predictions varied strongly with the detailed assumptions of a model. Inasmuch as such assumptions reflect biological variation, the theory may be enriched by this diversification, provided the number of distinguishable theoretical cases is not on the scale of the number of independent communities. Furthermore, where model details matter, a strong biological justification for the particular model used is especially important. For this purpose the mechanistic approach (Price 1986; Schoener 1986d; Tilman 1987) has already proved valuable and will probably continue to do so. This approach is largely a theoretical programme, in which community- and population-level theory is derived mainly from theory at individual levels, i.e. behavioural ecology, physiological ecology and ecomorphology.

A possible long-term consequence of the particularization of niche theory that might occur, for instance, with a mechanistic approach, is that many present niche concepts may no longer play useful roles in explanatory chains. For example, we showed above that niche overlap could be related to competition, although considerable special knowledge and perhaps experimentation was required to ascertain how. It could be argued that the very act of attaining such specificity obviates the need for niche concepts. I would counter argue that this overlooks at least two major points. First, niche properties, because they are so easily measured, will at least provide an important way of describing data, as is shown by the many resource-partitioning studies performed before the modern theory originated. Second and more important, observation and experiment should interplay many times over the course of a research programme. Eventually, correspondences should be established between observed niche (and other) patterns on the one hand and the processes revealed by experimental manipulation on the other. It is possible, however, that with more mechanistic approaches a more detailed set of niche properties, which may not involve much present niche terminology, will replace the set presently in use. At the most optimistic, these detailed correspondences will then allow equivalencies to be set up between pattern and process, so that the two become interchangable in theory and practice. Such correspondences are presently few, and that is why we argue so much.

ACKNOWLEDGMENTS

I thank J. M. Cherrett, J. Griesemer, W. C. Wimsatt and two anonymous reviewers for comments on a previous draft; F. A. Pitelka for discussion and bibliographical information; and the United States National Science Foundation for support.

REFERENCES

Abrams, P. A. (1975). Limiting similarity and the form of the competition coefficient. *Theoretical Population Biology,* **8,** 356–375.

Abrams, P. A. (1983). The theory of limiting similarity. *Annual Review of Ecology and Systematics,* **14,** 359–376.

Abrams, P. A. (1984). Recruitment, lotteries and coexistence in coral fish. *American Naturalist,* **123,** 44–55.

Abrams, P. A. (1986). Character displacement and niche shift analyzed using consumer-resource models of competition. *Theoretical Population Biology,* **29,** 107–160.

Ågren, G. I. & Fagerström, T. (1980). On environmental variability and limits to similarity. *Journal of Theoretical Biology,* **82,** 401–404.

Bray, J. R. & Curtis, J. T. (1957). An ordination of the upland forest communities of southern Wisconsin. *Ecological Monographs,* **27,** 325–349.

Brew, J. S. (1984). An alternative to Lotka–Volterra competition in coarse-grained environments. *Theoretical Population Biology,* **25,** 265–288.

Brew, J. S. (1987). Competition and niche dynamics from steady-state solutions of dispersal equations. *Theoretical Population Biology,* **32,** 240–261.

Brown, W. L. & Wilson, E. O. (1956). Character displacement. *Systematic Zoology,* **5,** 49–64.

Case, T. J. (1979). Character displacement and coevolution in some *Cnemidophorus* lizards. *Fortschritte der Zoologie,* **25,** 235–282.

Case, T. J. (1982). Coevolution in resource-limited competition communities. *Theoretical Population Biology,* **21,** 69–91.

Cherrill, A. J. & James, R. (1987). Character displacement in *Hydrobia. Oecologia,* **71,** 618–623.

Chesson, P. L. & Warner, R. R. (1981). Environmental variability promotes coexistence in lottery competitive systems. *American Naturalist,* **117,** 923–943.

Clarke, B. (1962). Balanced polymorphism and the diversity of sympatric species. *Systematics Association Publication No. 4, Taxonomy and Geography,* pp. 47–70.

Cody, M. L. (1968). On the methods of resource division in grassland bird communities. *American Naturalist,* **102,** 107–147.

Colwell, R. K. & Futuyma, D. J. (1971). On the measurement of niche breadth and overlap. *Ecology,* **52,** 567–576.

Colwell, R. K. & Winkler, D. W. (1984). A null model for null models in biogeography. *Ecological Communities: Conceptual Issues and the Evidence* (Ed. by D.R. Strong, Jr., D. Simberloff, L.G. Abele & A.B. Thistle), pp.344–359. Princeton University Press, Princeton, New Jersey.

DuBowy, P. J. (1988). Waterfowl communities and seasonal environments; temporal variability in interspecific competition. *Ecology,* **69,** 1439–1453.

Dunham, A. E. (1983). Realized niche overlap, resource abundance and the intensity of interspecific competition. *Lizard Ecology: Studies of a Model Organism* (Ed. by R.B. Huey, E.R. Pianka & T.W. Schoener), pp. 261–280. Harvard University Press, Cambridge, Massachusetts.

Dunham, A. E., Smith, G. R. & Taylor, J. N. (1979). Evidence for ecological character displacement in western American catostomid fishes. *Evolution,* **33,** 877–896.

Eldridge, J. L. & Johnson, D. H. (1988). Size differences in migrant sandpiper flocks: ghosts in ephemeral guilds. *Oecologia,* **77,** 433–444.

Ellenberg, H. (1950). *Landwirtschaftliche Pflanzensoziologie. I. Unkrautgemeinschaften als Zeiger für Klima und Boden.* Ulmer, Stuttgart.

Elton, C. (1927). *Animal Ecology.* Sidgwick & Jackson, London.

Fagerström, T. & Ågren, G. I. 1979). Theory for coexistence of species differing by regeneration properties. *Oikos,* **33,** 1–10.

Fenchel, T. (1975). Character displacement and coexistence in mud snails. *Oecologia,* **20,** 19–32.

Fjeldså, J. (1983). Ecological character displacement and character release in grebes Podicipedidae. *Ibis,* **125,** 463–481.

Gaffney, B. M. (1973). Roots of the niche concept. *American Naturalist,* **109,** 490.

Gause, G. F. (1934). *The Struggle for Existence.* Reprinted 1969. Hafner Publishing Company, New York.

Giller, P. J. (1984). *Community Structure and the Niche.* Chapman & Hall, London.

Gilpin, M. E. (1986). Review of predation. *American Scientist,* **74,** 202–203.

Grinnell, J. (1914). An account of the mammals and birds of the Lower Colorado Valley with especial reference to the distributional problems presented. *University of California Publication in Zoology,* **12,** 51–294.

Grinnell, J. (1917). The niche-relationships of the California Thrasher. *Auk,* **34,** 427–433.

Grinnell, J. (1924). Geography and evolution. *Ecology,* **5,** 225–229.

Grinnell, J. (1928). Presence and absence of animals. *University of California Chronicle,* **30,** 429–450. (Reprinted in Grinnell, J. (1943). *Joseph Grinnell's Philosophy of Nature,* pp. 187–208. University of California Press, Berkeley and Los Angeles).

Grinnell, J. & Storer, T. I. (1924). *Animal Life in the Yosemite.* University of California Press, Berkeley, California.

Grinnell, J. & Swarth, H. (1913). An account of the birds and mammals of the San Jacinto area of Southern California. *University of California Publications in Zoology,* **10,** 197–406.

Grubb, P. J. (1977). The maintenance of species-richness in plant communities: the importance of the regeneration niche. *Biological Reviews,* **52,** 107–145.

Haskell, E. F. (1940). Mathematical systematization of 'environment', 'organism' and 'habitat'. *Ecology,* **21,** 1–16.

Holt, R. D. (1987). Prey communities in patchy environments. *Oikos,* **50,** 276–290.

Hutchinson, G. E. (1944). Limnological studies in Connecticut. VII. A critical examination of the supposed relationship between phytoplankton periodicity and chemical changes in lake waters. *Ecology,* **25,** 3–26.

Hutchinson, G. E. (1957). Concluding remarks. *Cold Spring Harbor Symposia on Quantitative Biology,* **22,** 415–427.

Hutchinson, G. E. (1959). Homage to Santa Rosalia, or why are there so many kinds of animals? *American Naturalist,* **93,** 145–159.

Hutchinson, G. E. (1967). *A Treatise on Limnology. Volume* **II.** *Introduction to Lake Biology and the Limnoplankton.* John Wiley & Sons, New York.

Hutchinson, G. E. (1978). *An Introduction to Population Ecology.* Yale University Press, New Haven, Connecticut.

Inger, R. F. & Colwell, R. K. (1977). Organization of contiguous communities of amphibians and reptiles in Thailand. *Ecological Monographs,* **47,** 229–253.

Johnson, R. H. (1910). *Determinate Evolution in the Color-Pattern of the Lady-Beetles.* Carnegie Institution of Washington, Washington, DC.

Kostitzin, V. A. (1935). *Evolution de l'Atmosphère.* Exposés de Biomètrie et de Statistique

Biologique. **VIII.** Hermann, Paris, France.

Krebs, C. J. (1978). *Ecology: The Experimental Analysis of Distribution and Abundance,* 2nd edition. Harper & Row, New York.

Kulesza, G. (1975). Comment on 'Niche, habitat and ecotope'. *American Naturalist,* **109,** 476–479.

Levins, R. (1966). The strategy of model building in population biology. *American Scientist,* **54,** 421–431.

Levins, R. (1968). *Evolution in Changing Environments.* Princeton University Press, Princeton, New Jersey.

Levinton, J. W. (1987). The body size-prey size hypothesis and *Hydrobia. Ecology,* **68,** 229–231.

MacArthur, R. H. (1968). The theory of the niche. *Population Biology and Evolution* (Ed. by R.C. Lewontin), pp. 159–176. Syracuse University Press, Syracuse, New York.

MacArthur, R. H. (1972). *Geographical Ecology: Patterns in the Distribution of Species.* Harper & Row, New York.

MacArthur, R. H. & Levins, R. (1967). The limiting similarity, convergence and divergence of coexisting species. *American Naturalist,* **101,** 377–385.

MacArthur, R.H. & Wilson, E. O. (1967). *The Theory of Island Biogeography.* Princeton University Press, Princeton, New Jersey.

MaGuire, B., Jr. (1973). Niche response structure and the analytical potentials of its relationship to the habitat. *American Naturalist,* **107,** 213–246.

Martin, T. E. (1988). On the advantage of being different: nest predation and the coexistence of bird species. *Proceedings of the National Academy of Sciences,* **85,** 2196–2199.

May, R. M. (1973). *Stability and Complexity in Model Ecosystems.* Princeton University Press, Princeton, New Jersey.

May, R. M. (1974). On the theory of niche overlap. *Theoretical Population Biology,* **5,** 297–332.

May, R. M. & MacArthur, R. H. (1972). Niche overlap as a function of environmental variability. *Proceedings of the National Academy of Sciences,* **69,** 1109–1113.

Milligan, B. G. (1985). Evolutionary divergence and character displacement in two phenotypically-variable, competing species. *Evolution* **39,** 1207–1222.

Moment, G. B. (1962). Reflexive selection: a possible answer to an old puzzle. *Science,* **136,** 262–263.

Nagel, E. (1961). *The Structure of Science: Problems in the Logic of Scientific Explanation.* Harcourt, Brace and World, New York.

Pacala, S. W. & Roughgarden, J. (1982a). Spatial heterogeneity and interspecific competition. *Theoretical Population Biology,* **21,** 92–113.

Pacala, S. W. & Roughgarden, J. (1982b). The evolution of resource partitioning in a multidimensional resource space. *Theoretical Population Biology,* **22,** 127–145.

Pacala, S. W. & Roughgarden, J. (1985). Population experiments with *Anolis* lizards of St. Maarten and St. Eustatius. *Ecology,* **66,** 129–141.

Pianka, E. R. (1969). Sympatry of desert lizards (*Ctenotus*) in western Australia. *Ecology,* **50,** 1012–1030.

Pianka, E. R. (1974). Niche overlap and diffuse competition. *Proceedings of the National Academy of Sciences,* **71,** 2141–2145.

Price, M. V. (1986). Introduction to the symposium: Mechanistic approaches to the study of natural communities. *American Zoologist,* **26,** 3–4.

Ramensky, L. G. (1924). Die Grundgesetsmässigkeiten im Aufbau der Vegetationsdecke. *Vêstnik Opytnogo Dêla, Voronezh,* 37–73. (Abstr. in Bot. Centralblatt, n.s. 7, 453–455, 1926).

Rand, A. S. (1967). Predator–prey interactions and the evolution of aspect diversity. *Atas do Simpósio sôbre a Biota Amazônica* **5,** (Zool.) 73–83.

Ricklefs, R. E. (1979). *Ecology,* 2nd edition. Chiron, New York.

Rosenzweig, M. L. (1979). Optimal habitat selection in two-species competition systems. *Population Ecology* (Ed. by U. Halbach & J. Jacobs), pp. 283–293. Gustav Fischer Verlag, Stuttgart, West Germany.

Ross, S. T. (1986). Resource partitioning in fish assemblages: review of field studies. *Copeia,* **1986,** 352–387.

Roughgarden, J. (1972). Evolution of niche width. *American Naturalist,* **106,** 683–719.

Roughgarden, J. (1974). Species packing and the competition function with illustrations from coral reef fish. *Theoretical Population Biology,* **5,** 163–186.

Roughgarden, J. (1976). Resource partitioning among competing species — a coevolutionary approach. *Theoretical Population Biology,* **9,** 388–424.

Roughgarden, J. (1979). *Theory of Population Genetics and Evolutionary Ecology: An Introduction.* Macmillan, New York.

Roughgarden, J. (1983). Competition and theory in community ecology. *American Naturalist,* **122,** 583–601.

Roughgarden, J. & Feldman, M. (1975). Species packing and predation pressure. *Ecology,* **56,** 489–492.

Roughgarden, J., Heckel, D. & Fuentes, E. (1983). Co-evolutionary theory and the biogeography and community structure of *Anolis. Lizard Ecology: Studies of a Model Organism* (Ed. by R.B. Huey, E.R. Pianka & T.W. Schoener), pp.371–410. Harvard University Press, Cambridge, Massachusetts.

Schoener, T. W. (1968). The *Anolis* lizards of Bimini: resource partitioning in a complex fauna. *Ecology,* **45,** 704–726.

Schoener, T. W. (1970a). Nonsynchronous spatial overlap of lizards in patchy habitats. *Ecology,* **51,** 408–418.

Schoener, T. W. (1970b). Size patterns in West Indian *Anolis* lizards. II. Correlations with the sizes of particular sympatric species—displacement and convergence. *American Naturalist,* **104,** 155–174.

Schoener, T. W. (1974a). Resource partitioning in ecological communities. *Science,* **185** 27–39.

Schoener, T. W. (1974b). Some methods for calculating competition coefficients from resource-utilization spectra. *American Naturalist,* **108,** 332–340.

Schoener, T. W. (1974c). Competition and the form of habitat shift. *Theoretical Population Biology,* **6,** 265–307.

Schoener, T. W. (1974d). The compression hypothesis and temporal resource partitioning. *Proceedings of the National Academy of Sciences,* **71,** 4169–4172.

Schoener, T. W. (1975). Presence and absence of habitat shift in some widespread lizard species. *Ecological Monographs,* **45,** 233–258.

Schoener, T. W. (1976). Alternatives to Lotka–Volterra competition: models of intermediate complexity. *Theoretical Population Biology,* **10,** 309–333.

Schoener, T. W. (1977). Competition and the niche. *Biology of the Reptilia* (Ed. by C. Gans & D.W. Tinkle), pp.35–136. Academic Press, London.

Schoener, T. W. (1978). Effect of density-restricted food encounter on some single-level competition models. *Theoretical Population Biology,* **13,** 365–381.

Schoener, T. W. (1982). The controversy over interspecific competition. *American Scientist,* **70,** 586–595.

Schoener, T. W. (1983). Field experiments on interspecific competition. *American Naturalist,* **122,** 240–285.

Schoener, T. W. (1984). Size differences among sympatric, bird-eating hawks: a worldwide survey. *Ecological Communities: Conceptual Issues and the Evidence* (Ed. by D.R. Strong, D. Simberloff, L.G. Abele & A.B. Thistle), pp.254–281. Princeton University Press, Princeton, New Jersey.

Schoener, T. W. (1986a). Patterns in terrestrial vertebrate versus arthropod communities: do systematic differences in regularity exist? *Community Ecology* (Ed. by J. Diamond & T.J. Case), pp. 556–586. Harper & Row, New York.

Schoener, T. W. (1986b). Resource partitioning. *Community Ecology — Pattern and Process* (Ed. by J. Kikkawa & D. Anderson), pp. 91–126. Blackwell Scientific Publications, Oxford.

Schoener, T. W. (1986c). Kinds of ecological communities — ecology becomes pluralistic. *Community Ecology* (Ed. by J. Diamond & T.J. Case), pp. 467–479. Harper & Row, New York.

Schoener, T. W. (1986d). Mechanistic approaches to community ecology: a new reductionism? *American Zoologist,* **26,** 81–106.

Schoener, T. W. (1988a). Ecological interactions and biogeographic patterns. *Analytical Biogeography: an Integrated Approach to the Study of Animal and Plant Distribution* (Ed. by A.A. Myers & P.S. Giller), Chapman & Hall, London.

Schoener, T. W. (1988b). Testing for non-randomness in sizes and habitats of West Indian lizards: choice of species pool affects conclusions from null models. *Evolutionary Ecology,* **2,** 1–26.

Schoener, T. W. & Schoener, A. (1971a). Structural habitats of West Indian *Anolis* lizards. I. Jamaican lowlands. *Breviora Museum of Comparative Zoology,* **368,** 1–53.

Schoener, T. W. & Schoener, A. (1971b). Structural habitats of West Indian *Anolis* lizards. II. Puerto Rican uplands. *Breviora Museum of Comparative Zoology,* **375,** 1–39.

Schoener, T. W., Huey, R. B. & Pianka, E. R. (1979). A biogeographic extension of the compression hypothesis: competitors in narrow sympatry. *American Naturalist,* **113,** 295–298.

Schoener, T. W., Roughgarden, J. & Fenchel, T. (1986). The body-size–prey-size hypothesis: a defense. *Ecology,* **67,** 260–261.

Schluter, D. (1988). Character displacement and the adaptive divergence of finches on islands and continents. *American Naturalist,* **131,** 799–824.

Shelly, T. E. (1985). Ecological comparisons of robber fly species (*Diptera: Asilidae*) coexisting in a neotropical forest. *Oecologia,* **67,** 57–70.

Shigesada, N. & Roughgarden, J. (1982). The role of rapid dispersal in the population dynamics of competition. *Theoretical Population Biology,* **21,** 353–372.

Shigesada, N., Kawasaki, K. & Teramoto, E. (1979). Spatial segregation of interacting species. *Journal of Theoretical Biology,* **79,** 83–99.

Simberloff, D. & Boecklen, W. (1981). Santa Rosalia reconsidered: size ratios and competition. *Evolution,* **35,** 1206–1228.

Slatkin, M. W. (1980). Ecological character displacement. *Ecology,* **61,** 163–177.

Smith, J. N. M., Grant, P. R., Grant, B. R., Abbott, I. J. & Abbott, L. K. (1978). Seasonal variation in feeding habits of Darwin's ground finches. *Ecology,* **59,** 1137–1150.

Spiller, D. A. (1986). Consumptive-competition coefficients: an experimental analysis with spiders. *American Naturalist,* **127,** 604–614.

Strong, D. R., Szyska, L. A. & Simberloff, D. S. (1979). Tests of community-wide character displacement against null hypotheses. *Evolution,* **33,** 897–913.

Taper, M. L. & Case, T. J. (1985). Quantitative genetic models for the coevolution of character displacement. *Ecology,* **66,** 355–371.

Taylor, P. (1916). The status of the beavers of Western North America, with a consideration of the factors in their speciation. *University of California Publications in Zoology,* **12,** 413–495.

Tilman, D. (1986). Evolution and differentiation in terrestrial plant communities: The importance of the soil resource: light gradient. *Community Ecology* (Ed. by J. Diamond & T.J. Case), pp. 359–380. Harper & Row, New York.

Tilman, D. (1987). The importance of mechanisms of interspecific competition. *American Naturalist,* **129,** 769–774.

Toft, C. A. (1985). Resource partitioning in amphibians and reptiles. *Copeia,* **1985,** 1–20.

Tonkyn, D. W. & Cole, B. J. (1986). The statistical analysis of size ratios. *American Naturalist,* **128,** 66–81.

Turelli, M. (1981). Niche overlap and invasion of competitors in random environments. I. Models without demographic stochasticity. *Theoretical Population Biology,* **20,** 1–56.

Udvardy, M. F. D. (1959). Notes on the ecological concepts of habitat, biotope and niche. *Ecology,* **40,** 725–728.

Vandermeer, J. H. (1972). Niche theory. *Annual Review of Ecology and Systematics,* **3,** 107–132.

Varley, G. C. (1949). Population changes in German forest pests. *Journal of Animal Ecology,* **18,** 117–122.

Werner, E. E. & Hall, D. J. (1979). Foraging efficiency and habitat switching in competing sunfishes. *Ecology,* **60,** 256–264.

Whittaker, R. H. (1951). A criticism of the plant association and climatic climax concepts. *Northwest Science,* **25,** 17–31.

Whittaker, R. H., Levin, S. A. & Root, R. B. (1973). Niche, habitat and ecotope. *American Naturalist,* **107,** 321–338.

Whittaker, R. H., Levin, S. A. & Root, R. B. (1975). On the reasons for distinguishing 'Niche, habitat and ecotope'. *American Naturalist,* **109,** 479–482.

Yoshiyama, R. M. & Roughgarden, J. (1977). Species packing in two dimensions. *American Naturalist,* **111,** 107–121.

5 DIVERSITY AND STABILITY

D. WALKER
Research School of Pacific Studies, Australian National University, Canberra, ACT 2601, Australia

CONCEPTS, DEFINITIONS AND MEASURES

The world is alive with organisms, 1.5×10^6 kinds, if a kind is defined as a biological species, and several exponents more if distinctive populations within such species are counted. Except where specific qualification or implication of context indicates otherwise, in what follows it is *species* diversity which is under discussion. Species are not evenly distributed over the globe and I am concerned with the relationship of this patchiness, in space and time, with the variety of ideas subsumed under 'stability'.

The *diversity* of an assemblage is usually broken down into the number of species in it (species richness), and the distribution of the number of individuals amongst the species (equitability); for the most part I shall be concerned with the former, to which I shall restrict the term diversity. For many purposes it is useful to define the geographical scales of diversities and their differences, perhaps the best system being that of Whittaker (1977) (Table 5.1). Numerical statements of diversity are complicated by the desirability, for some purposes, of confounding species richness, equitability and the area from which the sample is drawn (Preston 1962; Williams 1964; May 1965; Peet 1974; Pielou 1975;

TABLE 5.1. Types of species diversity, after Whittaker (1977)

Inventory diversities	Differentiation diversities
1 Point, subsample or internal, diversity; small sample of a 'homogenous' assemblage.	2 Pattern, or internal β, diversity; difference between samples of a 'homogenous' assemblage.
3 α, or within-habitat, diversity; total from all samples of a 'homogenous' assemblage.	4 β, diversity; differences along a gradient or between the assemblages in a landscape.
5 γ, or landscape, diversity; total from all assemblages in a landscape.	6 δ, or geographic, diversity; differences along climatic or 'geographical' gradients or between landscapes in a geographical region.
7 ϵ, or regional, diversity; total for differing landscapes in a large geographic region.	

Taylor 1978; Grassle *et al.* 1979; Kikkawa 1986) but many of the problems referred to below are sufficiently gross to be relatively unaffected by such sophistications. Yet insufficient attention has been paid to the dimensioning of the samples between which diversities are compared; similarly-sized fish gills make neat units for parasites (Rohde 1978), phytophagous insects may sometimes be sufficiently choosey for their numbers to be reasonably related to the number of attractive locations on host plants (Eastop 1978; Gilbert & Smiley 1978; Lawton 1978) but in the greater world of a complex forest, a reference to ground area, the most commonly used unit, can only have meaning for a minority of the biota. Another difficulty in enumeration leads to an important philosophical deficiency. This is that the very size of even modest assemblages, together with the specialist taxonomic and enumerating requirements for dealing with each major group of organisms, discourages studies of total diversity in favour of those dealing with the diversity of one or a very few major groups. In relation to some problems, of course, such numbers can be assumed to be additive (or multiplicative) but in other instances they leave the question of total diversity virtually untouched.

Complexity, despite its useful everyday connotations, is best reserved to express trophic differentiation, the connectance (King & Pimm 1983; Pimm 1984) of food webs and so on, with which diversity and stability have been related.

Stability is a greatly over-worked word, expressing concepts more difficult to deal with than diversity because they involve time. Of all the proposed expressions about the changeability of the biotic components of ecosystems (Lewontin 1969; Botkin & Sobel 1975; Pimm 1982; King & Pimm 1983), those of Holling (1973) have proved to be generally the most useful; *stability* 'is the ability of a system to return to an equilibrium state after a temporary disturbance' whilst *resilience* is the ability of a system to absorb changes in all its inputs and still persist. Because diversity measurements are often made for a single taxonomic group, rather than a whole system, it is useful to extend these stability definitions to such groups and to populations of individual species. Stability defined in this way, or in terms (e.g. measures of fluctuations about means) more appropriate to particular needs, can be estimated by experiment and predicted, for closely defined boundary conditions, by mathematical analysis and modelling. Even so, experiments accurately reflecting natural conditions and meeting the requirements of initial equilibrium are extremely difficult to design. In the majority of field studies of biotas through time, what can actually be measured are changes in the taxonomic identity, diversity, equitability, production, biomass, population age structures and so on. The first step in the analysis of observed changes in a biota must be an attempt to allocate the total change, or lack of it, to the variety of forces which may have contributed to it; we are all familiar with this kind of thing in the process of hypothesis generation and experimental design.

Observations of changes through time are usually geographically located and, particularly where long periods are concerned, it may be difficult to discriminate between change within an assemblage and the displacement of the assemblage and the occupation of its former space by another one. This is particulary the case in palaeoecological reconstructions in which instability is sometimes used to describe biotic change at a site irrespective of whether this is *in situ* or the result of displacement. Of course, if a displacement is oscillatory, a feature common on some scales in the Quaternary period, the length of oscillation is a measure of stability but stability of the landscape, not of the biota.

The stability of an assemblage and its resilience may be affected by purely biotic, particularly density-dependent, relationships within it. Indeed, such effects and related notions of niche differentiation gave great impetus to the highly influential attempts to explain diversity differences, through field observation and mathematical analysis, by laws applicable throughout the ecological world (Hutchinson 1959; MacArthur 1960, 1965; MacArthur & Wilson 1967). Increasingly, however, the effects of disturbance resulting from environmental change have begun to seem more common, and theoretically more interesting (White 1979; Thiery 1982). This tension between explanations of purely biological and of environmental kinds is one that runs through the whole topic. The definition of disturbance (or, where experimentally applied, perturbation) has almost become an intellectual field in its own right. Many, if not all, ecosystems can evidently persist within varying physical environments (most obviously those of seasonal climates) and have life cycle traits convincingly selected as adaptations to this. There are arguments that for many ecosystems some disturbance is essential to their survival (e.g. Connell 1978) so that fire, beginning as a destructive disturbance, might select species and interactions which eventually need it (Walker 1982a). Unpredictability, either of frequency or magnitude, seems to be the essence of disturbance, although in evolutionary time some species and systems may even be selected for resilience to this. An important, but almost mentally disabling matter, is that the organisms in an assemblage are differentially sensitive to any particular kind, magnitude or frequency of disturbance by virtue of differences in their structures, physiologies, demographies and life cycle characteristics. An applied force may be ‘perceived’ as a disturbance by some but not by others of an assemblage, which brings us dangerously close to teleology. It is essential to define the unit which is thought to be disturbed and the property of the unit by which the effects of disturbance are measured, not always easy tasks.

Investigations of diversity/stability relationships fall into two main groups. The first concerns the degree to which diversity is the product of environmental stability. The second seeks explanations of differential stability in differing degrees of diversity.

I will first give some account of opinions about periodicities in the development of organic diversity through geological time. Next I will examine environmental stability in relation to the latitudinal biogeographic diversity gradient and then the reciprocal relationship between diversity and stability on the ecological scale. Finally I will seek convergences in the development of thoughts on these matters which might point ways ahead. The literature of these topics is rich in literary allusion and in metaphor, particularly if we grant the metaphorical nature of differential equations.

ACCUMULATION OF DIVERSITY

Palaeontological units, a unit being a group of morphologically more-or-less identical, organically-derived, bits, are not species; no bit is a whole organism and nothing can be known about the former reproductive relationships between the units. For convenience, however, palaeontologists recognize grades of difference between fossil units analogous to those found amongst living species, genera and families. The probability with which fossil bits can be attributed to living taxa increases as their geological ages decrease and as the amount of morphological detail increases, so that in most applications of Quaternary research such identity is taken as a working hypothesis. By now, about 250 000 palaeontological units are known (Raup 1987) from the whole 3.5 Ga of life's history (Walter 1987), enough to give meaning to changes in this number with time despite the geographically and temporally patchy distribution of the data and the fact that it probably represents less than 0.001% of all taxa that have lived.

It seems to be generally accepted that the palaeontological record can be divided into alternating periods of geologically rapid increase in diversity (of lower taxonomic units) and of little such diversity change; punctuation and stasis (Eldredge & Gould 1972; Gould and Eldredge 1977; Raup 1987). Some periods seem to have witnessed more-or-less simultaneous diversification in many kinds of organisms but some major groups have at other times behaved more independently, witness their use in the division of the geological column. Within a higher taxonomic group, diversity often increased dramatically soon after the group's first emergence, sometimes, though not always, associated with a major environmental event or the conquest of a hitherto unoccupied realm (as plants of the land and insects of the air). Sometimes the explosive diversification of one group coincided with the crash of another and, even within a group, high rates of origination and extinction often went together.

The palaeontological world is now fizzing with discussion rivalling in its energy early Darwinmania and, like it, not totally devoid of vituperation. The two main issues have clear parallels in studies of living organisms. First, does the world have a finite carrying capacity? Opinions differ from those who think that, ideally at least, it does (Simpson 1952; Darlington 1957; discussion by Schopf 1977) to those for whom either its attainment has always been such a remote possibility as to render it a virtually pointless question (Simberloff 1981; Walker & Vallentine 1984, but see Herbold & Moyle 1986), or who have observed still vacant niches (Lawton 1982).

Second, do punctuation and stasis imply different evolutionary styles, particularly in the relationship between taxon generation and extinction mechanisms? There are those who see the genic and selection systems as we know them in population genetics today as quite adequate mechanisms for explaining all the palaeontological data (Milligan 1986; Carson 1987). Others credit the genetic machine with the capacity to produce sufficient variety but doubt whether population genetics has yet identified the crucial steps in genomic and developmental processes leading to species' fixation (Schopf 1977, 1984; Franklin 1987). Others again, incline to the view that changes in the environmental architecture of the world (e.g. through plate tectonics, sea-level changes and temperature drifts) are selectors of extinction and origination sufficiently varied in intensity and kind to account for all apparent differences of species numbers from time to time (Valentine 1977). At the extreme, Gould (1985) identifies three 'tiers' of evolutionary processes, the first appropriate to 'ecological moments' being quite different from the other two, namely those involved in punctuation and in rapid recruitment after recurrent mass extinctions which undo or over-ride anything accumulated by the first tier.

It is abundantly clear that mass extinctions have repeatedly afflicted the world's animals on scales that are difficult to attribute to competitive or predator–prey mechanisms. What kinds of environmental disturbances have been at work is still hotly argued and range from temperature changes (e.g. Stanley 1984) to 26 Ma cycles of astronomical phenomena including meteoric impacts (Raup & Sepkoski 1984; Gould 1985). Land plants, however, have a different kind of history without world-wide mass extinctions, new groups replacing old in a manner perhaps explicable without recourse to major environmental disturbances (Knoll 1984; Truswell 1987).

For present purposes, the importance of the palaeontological record is twofold. First it exposes a record of diversity changes which shows that, for animals at least, 'stability' is not the prime generator of diversity in evolutionary time. Secondly it helps to put the present into a relevant evolutionary perspective; are any of the major groups living today in a 'punctuating' mode? Following this last question further, we may take as an example the state of the mammals. We are rightly accustomed to thinking of mammals as generally a Cenozoic phenomenon. However, if the numbers of genera are standardized to a time base of a million years, they show a ten-fold rise between about 5 and 1.5 Ma ago, and a steadying or small fall thereafter. However, this trend was not one of passive accumulation, because extinctions ran parallel with originations. In short, the last 5 million years, arguably including the present, have been a tumultuous period in mammalian evolution (Gingerich 1975). For example, losses in total mammalian diversity have resulted from extinction of residents and their replacement by invaders in the interchange between North and South America (Marshall 1981) and from the environmentally (Guilday 1984; Webb 1984) or culturally (Martin 1984; Whittington & Dyke 1984) induced preferential loss of big species. The resulting biogeographical reassortment of the mammal faunas may be part of the processes of the high turnover already noted (Gingerich 1984). If so, are the genetic mechanisms governing changes in mammal populations at present comparable with those controlling other groups? If not, what are the implications for living ecological relationships? The kind of question which arises is exemplified by the shift in diet of the ground sloth (*Nothrotheriops shastense*) as shown from coproliths, from many to two plant species about 13 ka ago. The sloth became extinct about 2 ka later and, whether the events were related or not, the facts illustrate changes in ecological balances (Hansen 1978). How many of the organisms we study today are in such states of longish-term disequilibrium?

LATITUDINAL DIVERSITY GRADIENT

On the biogeographical scale, the latitudinal gradient in species richness has held central place in the interests of ecologists for more than a century, at once an unresolved problem for some and a point of reference for others. When all the caveats have been entered and components of the diversity accounted for by area, altitude, incident energy, soil and predator effects (e.g. Osman & Whitlatch 1978; Schopf, Fisher & Smith 1978; Wright 1983) to name but a few, the fact remains that, on any sampling basis so far used, humid lowland tropical forest has more species than does, say, mesic temperate forest or subarctic scrub. For all that has been written on this subject, however, there is still a need for a critical reappraisal which treats the 'exceptions' (e.g. low diversity forests at low latitudes and high diversity heaths at middle latitudes) as cells of equal initial significance to those more usually selected. Diversity will still centre on the tropical rain forest but the latitudinal diversity gradient will perhaps emerge as a slightly less special case. It is also high time to distinguish primary diversity, residing in the organisms which create the basic architecture of an assemblage, from contingent diversity which results from the niche variety created by that architecture. At present, however, it is a problem that won't go away; a feature of whole biomes and most large taxonomic groups, terrestrial and marine, animal and plant. Indeed, it is refreshing to read of a group which does *not* have most species in the tropics, even if it turns out to be the environmentally myopic aphids (Dixon *et al.* 1987).

The dominating explanations of the latitudinal diversity gradient, which were hardly challenged until the late sixties, are that : (i) the humid tropics have a less variable and therefore more predictable physical environment than do higher latitudes, and (ii) the lowland humid tropics have not been subjected to so much environmental disturbance (e.g. glaciation) in the past as have higher latitudes.

The physical environment, particularly the heat component, is less of a challenge to fundamental physiological tolerances in the tropics and this, together with the greater predictability of such variability as there is, selects for fine niche differentiation there and for wider environmental tolerances at high latitudes. The lack of historical disturbance has minimized extinctions in the tropics which have therefore become living museums of biotic diversity contrasting with the temporary collections of temperate regions. (For much fuller accounts see Fischer 1960; MacArthur 1972; Pianka 1978.) Although there is some validity in both these explanations, their application has been far too simplistic and anecdotal. Flying from the floods and landslips of the humid tropics to an English spring, one wonders why organisms have not been selected for finer subdivision of the additional niches which seasonality provides. However, in the context of this paper, the appropriate subject for further discussion is whether tropical diversity can be accounted for by long freedom from physical disturbance.

By the late sixties and the seventies, enough information had accumulated about Quaternary changes in glaciation, climate and the relative distributions of land and sea in the tropics, as well as explicit evidence of major vegetation displacements on their mountains (e.g. Livingstone 1967; Walker 1970; van der Hammen 1974; Flenley 1979), to challenge the proposition of environmental changelessness. These oscillations are now seen to be superimposed on a longer trend of change which from 50 Ma ago had accentuated the differences between low- and high-latitude temperatures (Shackleton 1984). There had also been just enough argument about mechanisms of speciation to have polarized attitudes preparatory to a more satisfactory synthesis; allopatry reigned in the biogeographical state. Against this background, Haffer (1969) proposed that the pattern of distribution of some Amazonian bird species could best be explained by repeated dissection and coalescence of the rain forest in which they lived during the Quaternary. The argument was enthusiastically applied to patterns of other groups, particularly of animals (e.g. many of the papers in Prance 1982). The next step in the argument was to attribute the diversity of humid tropical forest to repeated environmental change in the last 2 Ma. The older explanation attributing tropical diversity to environmental changelessness was stood on its head. What can be said about this now?

There can be no doubt that the last two, perhaps even the last seven, million years have witnessed recurrent, geologically rapid and large, environmental changes which have caused enormous displacements and other modifications of biotas the world over, the present humid tropics included. The latitudinal diversity/stability question is now divisible into two. Was there a gradient before these disturbances? Whether there was one or not, to what extent was diversity *differentially* affected at low and high latitudes by environmental fluctuations resulting directly or indirectly from the same basic cause, that is changes in the distribution of heat over the Earth's surface? I shall leave the first question unanswered. In tackling the second question two facts are important. First, the major climatic fluctuations were world-wide, though regionally different in their expression, whilst minor oscillations may have resulted from more local modifications. Second, the climatic, and therefore gross biotic, patterns of today's world are analogues, and rather poor ones at that, for only about 10% of Quaternary time. In thinking about speciation and extinction over the eminently appropriate period of a million years or so, we must consider forces appropriate to the patterns of the longer 'glacial' periods (of the order of 100 ka), fleetingly interrupted by much shorter periods (of the order of 10 ka) roughly of the kind we live in now. Of course, in a sense, it is reasonable to speak of 'glacial refugia' for plants and animals but it might be better to coin a description of the converse such as 'interglacial forward positions'.

Explicit fossil evidence of the location of lowland humid tropical biotas during the greater part of the Quaternary is not very abundant (Walker & Chen 1987) but it is clear that climatic oscillations affected the main tropical regions differently. In Amazonia, despite all today's biotic patterning, there is practically no solid, dated, evidence for its climatic dissection in the Quaternary. However, it might reasonably be supposed that the total forest biota was for most of that time confined below an altitudinal limit about 700 m lower and by temperatures 4.5 to 6° C cooler than today's (Colinvaux 1987). In Africa, tropical humid rain forest occupied three small areas, a strip along the east coast, a block west of the east African mountains and discontinuous areas on the west coast (Hamilton 1976; Maley 1987), the whole comprising about 25% of today's area and 10% of its greatest extent about 8 ka ago. The constraining geographies were climatic cold and drought and the absence of a large unflooded continental shelf. New Guinea, a rather special case of rapid orogeny and environmental diversification (Walker 1982b; Walker & Hope 1982), nevertheless held its forest biota between cold mountains and deep sea to the north, and arid plains to the south within something like 75% of its present area and 60% of its present altitudinal range. South East Asia was much bigger by reason of the availability of now flooded shelves and islands (Chappell & Thom 1977). Major parts almost certainly supported tropical forest, despite arguments favouring some less humid areas (Medway 1972; Verstappen 1975); bigger than Amazonia, it also had more topographic variety. The major disturbances of these patterns, have been the usually rapid climatic warmings and coolings at the beginning and end of each 'interglacial' (to use temperate zone terminology) (Flenley 1984; Walker & Sun 1988). The warmings occurred together with geologically temporary, but large, expansions of available land in Africa, contractions in South East Asia and probably rather little change in Amazonia. Clearly, there is a great deal to be investigated here at a much more detailed level but one might expect, *inter alia*, that the longer periods of occurrence of geographical restriction (refugia) in Africa and the shorter ones in South East Asia might have resulted, in the latter, in greater diversity and proportionately greater evolutionary activity (both origination and extinction) amongst *K*-selected species there. Amazonia, probably not affected by such big changes in available geography, shared in the speciation-accelerating environmental events [described vividly for modern Amazonia by Colinvaux (1987)] but might have been less prone to extinctions. In those tropical rain forests which have occupied their present positions for only 10 ka or so, the main question relating to their diversity is how it can have been achieved in such a short time at such locations. It seems to be established that a variety of diversity-disturbance interactions occurred in the terrestrial lowland tropics; how does this compare with temperate latitudes?

In the two temperate regions for which historical data are sufficiently numerous, namely Europe and North America, reconstructions of Quaternary vegetational events show similarities and some differences (Davis 1983). Judging by the pollen record from the very beginning of the present interglacial, the normal Quaternary location for mixed temperate forest of Europe is in an arc running from Spain to south Russia with a marked concentration of taxa in the Balkans (Huntley & Birks 1983). It is constrained to this region by cold and aridity elsewhere. In North America the analogous region lies in the southern Mississippi basin, and perhaps further west, whilst Florida, currently the location of the most rich biota, is a mystery at that time (Watts 1980). Northward toward the ice some conifers grew, perhaps interspersed in peculiarly favoured localities with very small and isolated patches of deciduous species (Watts & Stuiver 1980; Davis 1983; Bennett 1985; Delcourt & Delcourt 1987). The size of the 'core' area is unknown for either continent but its edges were probably more diffuse in America than in Europe. The opening of an interglacial, if the events of the present one are anything to go by, caused substantial changes in the colonization of the newly available ground. Many of the now important trees of the northern mixed conifer-hardwood forests of North America advanced to that position by different routes and at varied paces (Davis 1976), so synthesizing present forest in piecemeal fashion. The most diverse occurred in the Appalachians where there were no deciduous trees during the last glacial (Davis 1983). In Europe, mountain, and, as the ocean rose, sea barriers greatly modified the northward and the westward tree migrations although, once through the mountains, the movement of some species was probably facilitated by northwesterly flowing rivers (Huntley & Birks 1983). The fact that some tree taxa reached Britain whilst others did not, and that they arrived at different times are evidence of the differential impact of barriers and of the apparent progressive loss of cold-tolerant biotypes throughout the Quaternary (West 1980). The end of each interglacial episode brought other kinds of disturbance before mechanisms of long-term evolutionary significance had had time to take effect. Probably more variety was lost than was gained on both continents but, for simple geographical reasons, this was more extreme in Europe than in North America. In both places, only evolutionary novelties arising and remaining in the small southern areas had much chance of survival.

The effects of disturbances imposed by the same basic environmental fluctuations on major parts of low and high latitude biotas now seem much more complicated than hitherto. Each region has been subjected to different selective forces even when consideration is restricted to the simplest geographical features. Of the five regions considered, the effects of Quaternary fluctuations were probably most similar in tropical Africa and North America.

It may be that further understanding will place the latitudinal diversity gradient more conformably in the pattern of total regional biogeographic diversity. This will hardly be enough, however, to establish the contribution of disturbance on this scale to the generation and maintenance of diversity. A whole series of questions now arise. How fast do plants move? What types of selection do the various disturbance regimes imply? How much truncation of tolerances is there? How much 'flush' effect is there (Carson 1968)? How much flip-flop arises between density-dependent and density-independent processes? How much co-evolutionary strain is there as organisms are differently affected not only by the magnitude of the disturbance but by the time over which it is applied? Already, there are encouraging developments in answering these questions , emanating not only from ecologically oriented palynologists (e.g. Davis 1981, 1984; Solomon, West & Solomon 1981; Wright 1984; Bennett 1986; Ritchie 1986; Delcourt & Delcourt 1987; Dexter, Banks & Webb 1987; Walker & Sun 1988) but also from geneticists (Parsons 1987). An immediate task is to compare the directions, magnitudes and frequencies of well-specified environmental changes with the response times of organisms, in both behavioural and selectional senses. However, whilst most of the above discussion might make sense for forest-tree assemblages, has it any significance at all for the diversities of other organisms apart from any entailment through such trees? It seems that it might be only marginal for some, such as European Coleoptera (Coope 1978, 1979) and nobody can deny that there are lots of beetles in the world.

Latitudinal diversity gradients in unattached marine organisms are in some ways simpler conceptually than any similar feature on land, if only because the watery medium is likely to dampen environmental instabilities. Their major contribution to the subject has been to demonstrate that, despite varied temperature tolerances, the latitudinal range limits of various organisms may coincide because of a geographical feature such as the intrusion of a cape into the flow of currents. They have also led to the suspicion that temperature tolerances may be biproducts of selection for unrelated processes at the biochemical level (Jablonski, Flessa & Valentine 1985). Here, the genetic variability which underlies diversity is only indirectly related to the temperature gradient, as well as to other factors, through variations in the resource base (Valentine & Ayala 1978; Valentine 1984).

The concentration of diversity of stony corals in the tropics obscures the fact that they grow between 65°N and 40°S. Species without symbiotic zooxanthellae have relatively broad temperature ranges and their diversity increases with decreasing latitude until between 40° and 25°N, beyond which their numbers generally decline as the zooxanthellate species with narrow temperature ranges enter the scene carrying total species diversities to their highest values in the tropics (Rosen 1981). Within the zooxanthellates' range in the Indo-Pacific region, maximum generic diversity rises with sea surface temperature from 16 to 26° C but most reef sites in fact contain many fewer genera than the maxima for their temperatures. Rosen (1984) has explained this, and other aspects of reef relative diversity, in terms of changing barriers to larval dispersal and to changing occurrences and sizes in the past, of areas suited to coral growth as determined by geology and climate. The significance of Quaternary sea level changes for shallow reef coral evolution is that such organisms, and the enormous numbers of others which live on and amongst them, must either die or become restricted to a very narrow and ocean-battered zone along the continental shelf margin every time a shelf is exposed. For the Indo-Pacific continental shelves, Potts (1984) estimated that the duration of periods at which most shelf and margin levels were within the uppermost 20 m of water varied between 2 and 4 ka, with an isolated case of 13 ka for the two events affecting the present 10 m depth. Potts argued from this that the longer-lived corals experienced only ten to 100 generations between major changes in their environments so that there was no time for selection to produce either much speciation or extinction. On the other hand, reef animals with shorter generation times experienced enhanced speciation as a result of the same disturbances. Once again we are returned to the kinds, magnitudes and frequencies of disturbances, as they are related to organisms' life histories, being crucial to any understanding of the interplay between diversity and stability. It brings us to the ecological level.

THE ECOLOGICAL SCALE

At the ecological scale it should be possible to minimize the divinational quality of much diversity/stability discussion; after all, it seems to deal with definable organisms, to offer better opportunities for data collection and to allow real experimentation. To a degree such expectations are realized but not with the crisp results we might like, largely because of the time factor, problems in the definition of disturbance and the extreme difficulty of even planning adequately controlled experiments involving substantial numbers of units with only approximately known properties. These difficulties can only be partially offset by mathematical modelling. Despite appearances to the contrary, I suspect that in this field modelling is easier than experimentation and that this is one of the reasons why the one seems to have run way ahead of the other in producing printed pages. Nevertheless, modelling is useful for sorting out ideas in the predictive language of mathematics, even if the mathematicians are sometimes carried away to formulate patterns of great abstract beauty but little earthy use; playing Aerial to the experimentalist's Caliban. It is also at the ecological scale that we might hope to examine the practically important question of whether diversity, whatever its origin, actually confers stability on biotic assemblages.

Primary and secondary plant succession studies show that species diversity does not increase continually with time, but reaches a maximum from which it frequently falls to a more-or-less maintained level (Viereck 1966; Auclair & Goff 1971; Hanes 1971; Shafi & Yarranton 1973; Yarranton & Morrison 1974; Brackenhielm 1977; Walker 1982a). To a degree, however, this might depend on how the counts are made. Thus, whilst shrubland plant diversity, counting only photosynthesizing species, decreased after each initial post-fire flush, there was no such decrease if deposits in the seed bank were included; differing life stages partition the time (Christensen 1985). In the Hubbard Brook study, maximum diversity was attained only a few years after clear cutting and then decreased but might possibly have increased slightly afterwards as natural tree-fall disturbance began (Bormann & Likens 1979). In the Breckland heath, two sequences of diversity change were separated by mole-hilling which maintained environments otherwise expunged (Watt 1974). Perturbation by fertilizer application in the herbaceous stages of old field recolonization suggested that stability was increased by greater diversity (Mellinger & McNaughton 1975). It could be argued from all this that maximizing diversity on a plot requires disturbance which prevents the attainment of the ultimate simpler stage (Horn 1974). Sufficiently frequent local disturbance may do this at the plot level (α

diversity) but disturbance which is patchy in either space or time might only achieve high diversity on the β- or γ-scales, successions being part of the variety of the landscape (Aubreville 1938; Watt 1947; Whittaker 1975; Shugart 1984). Alternatively, it might be the increasing productivity throughout a succession, leading to the dominance of those most effective in competing for light that, through competition, decreases diversity (Tilman 1982).

The removal of a top predator from a food web often leads to loss of diversity at 'lower' trophic levels, by competitive exclusion by species formerly culled by the predator (Jones 1948; Watt 1960; Paine 1966; Connell 1971; Strong 1984). High diversity, therefore, seems to be encouraged when competition, at least within a trophic level or part of a food web, is prevented from taking its full course. This reduced competition might well result from the intervention of physical forces of a size, frequency or unpredictability for which the affected assemblages cannot be selected, except in the sense of the regional occurrence, and therefore it permits the continuous availability, of all the species (sometimes called the 'fugitives') appropriate to post-disturbance conditions. Spatial and temporal heterogeneity of the environment therefore induces biotic diversity (Loucks *et al.* 1981; Sousa 1985).

These kinds of argument, appropriate to plants on the one hand and animals on the other, came to the fore in the late sixties and are still actively sustained. They fitted well with the growing recognition of the changeability of many environmental factors on a continuum of time scales from millions of years to seasons. They challenge the notions that diversity derives from finely regulated biological relationships which selection has maximized in relation to the predictable elements of the physical environment. This received wisdom of the past had gained credibility from demonstrations of trophic connectedness and the idea that the connections were so specific, every species having a unique and necessary role in the system, that one link severed would bring all to crumpled chaos. As usual, scale was a muddling factor. With hindsight, it is odd that, fired by Alec Watt's gap regeneration and pattern and process (Watt 1923, 1924, 1925, 1947) we did not recognize a flaw in all this when scaled down from the biome to the local assemblage. This was the scale on which it might have been expected to be most strongly evidenced, yet it was the scale at which continuous change in species composition through time was most evident. However, the greater the number of species, the greater is the number of theoretically possible links, and most imagined that evolution had produced something ecologically efficient which used all the potential biotic web. The alternative viewpoint, that complex biotic webs provided many alternative pathways for energy flow so that the disruption of some could be compensated, within limits, by increased flow through others, allowed that diversity, through complexity, facilitated stability. Carefully bounded mathematical calculations, however, suggested that, within their limits, diversity might not confer stability on a perturbed biological system (May 1973, 1976). The initial specifications were frankly unrealistic, but no more so than the less explicit or less analysable ones which had gone before, and the exercise led to a thought of great potential consequence which, for mental convenience, I have tagged 'May's Remark'. It is to the effect that, at least in the selection of stability mechanisms, evolution might have taken very special courses which cannot be matched by statistically probable mathematical descriptions (May 1973, p. 173); it is a great argument for deriving models from that which *is*, and not from pure thought (Lawlor 1978).

There followed a rich development of analytical modelling which largely ignored 'the Remark' but proceeded by introducing more species, modifying their interactions, expanding from local to global stability, relaxing some of the dictates of niche theory and even allowing for some non-linear relationships (May 1974; Roberts 1974; Levandowsky & White 1977). A fair number of these exercises were, I think, ecologically irrelevant (cf. McNaughton 1977) but together they provide the field naturalist with a new set of thought patterns to help structure observations and design experiments. Amongst these are the ideas of local and global stability (Lewontin 1969), fitness sets (Levins 1968), the implications of non-linearity (May & Oster 1976; May 1986) and so on. In due course they will stimulate better field observations and experiments.

If high diversity is sustained exclusively by disturbance, tropical rain forests and coral reefs must be disturbed. It is not difficult to point to agents of physical disturbance, particularly those which lead to local extinction of some or all components of an assemblage (e.g. cyclones, landslides, fire) and to features of the life histories of the architectural elements of such assemblages (e.g. tree fall) which clearly disturb populations of others (Whitmore 1974; Grubb 1977; Strong 1977; Hubbell 1979, 1980; Denslow 1980; Hartshorn 1980; Clark & Clark 1984; Brokaw 1985). The questions are whether these events, undoubtedly destroying local biotic stability, are nevertheless necessary to ensure regional ecosystem resilience, and how their immediate effects are incorporated into longer population trajectories (Hopkins & Graham 1987). The answer must lie in understanding the processes which take place in spaces the sizes of patches immediately affected by specific kinds of disturbance. These processes will depend upon the surrounding reservoir of species and the frequency and location of the disturbing event. How does a system adapted to one set of disturbances respond to a new one?

In many tropical forests, at least during the last few thousand years, swidden agriculture has imposed a pattern which is of considerable importance for their persistence. How does this disturbance compare with the disturbance of natural tree-fall? Is the return time of clearance at any one place long enough to allow the return of slowly recovering species yet short enough to prevent their rise to diversity-reducing dominance, and is the distance between like-aged patches small enough to keep the system 'open' to necessary local immigration? How do these situations compare quantitatively with others having different diversities and environmental limits, such as the mountain horticulturalists of New Guinea (Walker 1966; Walker & Hope 1982), the North American Indians (McAndrews 1976; Delcourt *et al.* 1986), the Australian 'fire-stick' Aborigines (Jones 1975; Nicholson 1981) and the Neolithic farmers of Britain (Smith 1970)? If swidden agriculture is crucial to the diversity of some tropical rain forests, what provides analogous disturbance in Australian rain forest? In the New Guinea highlands evidence of changes in forest composition from pollen analysis have been attributed to human activity (Powell 1976; Walker & Flenley 1979), but the changes were in the quantities and distribution patterns of plants (the 'fugitives') already in the flora which become more consistent and widespread. Their availability secured the forests' resilience, in a very broad sense of that term, but they were already pre-adapted before man intervened. A similar story appears to have applied in the tropical forest of Costa Rica (Herwitz 1981). Swidden might therefore be a threat to the rainforest's persistence which has not yet crossed the threshold of its stability. In the prehistoric forests of temperate Europe, Neolithic clearances were sometimes followed by nearly total recovery, but in other circumstances encouraged the development of heaths, perhaps with greater diversity than the forests they replaced. Resilience may have been threatened everywhere for as long as humans have disturbed fellow organisms by their fecundity, their environmental tolerance and their constantly increasing capacity for destruction, permitted by their flexible resource demands.

At present many see the intermediate disturbance hypothesis, grown from several roots (e.g. Grime 1973) but most cogently stated by Connell (1978) and put in wider context by Connell & Keough (1985), as the best statement of the contribution of environmental, and to a degree biotic, disturbance to diversity. It supposes that for a given regional species pool, diversity at a site will be maximized by disturbances which are sufficient to prevent the development of dominance, yet insufficient to cause regional destruction. Equilibrium systems cannot develop under such conditions (Wiens 1984) and Connell & Sousa (1983) have shown that few long-term studies demonstrate them. Much depends on the nature of the disturbance and how it affects the components of an assemblage; indeed apparently similar results may be attained through different courses. Tilman (1982) has pointed out that disturbance may be a process which, among others, determines the rates at which resources are available to assemblages and that, for sessile organisms, space may otherwise be limiting for some and light for others. Physical disturbance would relieve the limitations of both, but for different reasons. This leads to the conclusion that maximum diversity should be correlated with 'moderately low disturbance rates'. Similarly, Huston (1979) has developed a theory based on the rates 'at which differences in competitive abilities are expressed'.

Almost all these discussions are about sessile organisms and usually about those which provide the main spatial structures of assemblages. For this reason they are evidently appropriate to plants but perhaps less so to vagile animals around which equilibrium ideas were developed, although not without challenge (Birch 1979). Rarely are the two groups considered together except in a predator–prey context. It is entirely possible that local stability of short-lived animal populations, circumscribed by the spatial or temporal presence of architectures on which they depend, have diversity/stability relationships more in keeping with equilibrium system predictions.

Happily, there are more long-term data sets slowly accumulating which begin to expand on those of Watt's Breckland (see Greig-Smith 1982) and the Park Grass plots (Williams 1978; Silvertown 1980; Tilman 1982) (e.g. Connell, Tracey & Webb 1984). There are also a wealth of experimental data on the shorter term interactions between herbaceous species in particular (e.g. Harper 1977). But there is surely an opening for work on the disturbance dynamics of assemblages of short and very-short-lived organisms living together, just as there is for the application of refined pollen analysis of sediments to the disturbance histories of long-lived plant assemblages (Walker 1982a; Green *et al.* 1988). In addition, where living organisms are concerned, the genetic processes, particularly their speeds, directions, dependence on neighbouring populations and responses to selective forces are just as important as are counts of individuals.

It is at the ecological level that we come closest to the question of the identities of the most significant units in organism–environment relationships. At this level we can begin to identify the populations through the genetics of which responses to disturbance take place and which determine, through their reservoirs of genetic polymorphisms and the rates at which their component individuals respond to selection, whether a breeding population persists or becomes locally extinct. In 1967, Waddington (1968) could protest at studies of 'the phenotypeless genotype in a uniform environment' and Slobodkin (1968) point to the futility of fitness measures 'without explicit demonstration of how that property relates to the animal's environment'. Since then the selectionist–neutralist arguments have waxed and waned and, I think, left two ecologically important conclusions. The first is that populations have enormous reservoirs of genetic polymorphisms on which environmental fluctuation at the phenotype level can operate. The very number of such polymorphisms argues that they respond only weakly to selection. They are thought to be associated with niche competition processes (Mani 1984) which only strong environmental change ('environment' being everything but the breeding population under consideration) can disrupt. The stability of an ecological assemblage is therefore strongly affected by the relative sizes of these reservoirs in the species of an assemblage and the rates at which their potentials can be expressed in phenotypic response. These rates are probably determined by sets of modifier genes more immediately susceptible to environmental change than the polymorphisms they control. Wills (1981), writing of such genetic variety, asks the clearly ecological question '. . . what sort of population structure is likely to maximize the rate of adaptation of organisms to new and unexpected environmental conditions?' (p. 12). The second important conclusion is that, even allowing for possible biases in the data, a number of experimental populations of animals tend to stable fixed points and, in general, populations in fluctuating environments retain more genetic variability than do those of more constant ones (Wills 1981).

Evidently the ecology of alleles has some of the problems of the ecologies of demes and species. It is an encouraging sign that the two fields are beginning to interact, and with increasing interest in the processes which connect them (Ennos 1983; Rice & Jain 1985; Thompson 1985; Vrijenhoek 1985; Sultan 1987).

THE STATE OF THE SCIENCE

My justification for this rather eclectic treatment of diversity and stability was that by looking at things which some still find exciting, convergences of ideas might come to notice in a topic which one recent book describes as amongst those 'rapidly being forgotten' (Price, Gaud & Slobodchikoff 1984). Analogies between trends on time and space scales as different as those I have chosen are dangerous, but can be useful if not pushed too far. One of the great waves which has swept through ecology during the latter half of this Society's existence has surely been that associated with niche theory and resource-limited processes and all that was propagated therefrom. In that equilibrium-dominated context the connection between diversity and resilience was a convenient, if sometimes uncomfortable, assumption whilst that between diversity and stability seemed to be implied by the early experiments. The data which accumulated around niche, competition and island biogeography theory demanded a more penetrating evaluation of the role of disturbance in directing ecological processes. As yet its greatest attainment has probably been the enormous amount of natural history which it has stimulated and the opportunity it has given to experiment with the interplay between correlated field observations and mathematical modelling at both ecological and biogeographical levels. It has already forced the re-recognition of the importance to ecological theory of time as well as space scales (e.g. Dayton & Tegner 1984), something which passed out of mind together with the Clementsian climax. By luck, rather than good management, this resurgent interest found the reconstruction, through Quaternary research, of environmental and biological events on scales from 10^2 to 10^6 years well developed and heading confidently for finer resolution, having already successfully demonstrated the ubiquity of change in the environmental determinants of ecological and biogeographic patterns. We can now say 'determinants' because it is inconceivable that any ecological theory could now hold sway that did not give environmental determinism an equal, though not always coincident, role with biotic interactions. In this regard, the wheel has been rediscovered somewhat rounder, smoother running and with many more spokes, than before. Palaeontology, meanwhile, has asserted the importance of massive disturbance, often associated with high levels of extinction, in the very creation of species, and has questioned whether what happens over the greater part of evolutionary time is anything more than refining the products of bursts of species turnover. In this, some have commiserated with ecologists whose 'focal range is restricted to . . . short-term consequences of process' and who are continually plagued by the problem of 'the apportionment of cause; how much to present-day effects, how much to historical constraints' (Kitchell 1985). With the confidence born of enlightenment in their own material, some palaeontologists have called in question the significance of population genetics for evolutionary processes, and even whether there might be two kinds of species with different kinds of genetic control (Van Valen 1982). Ecologists, most of whom are closer to genetics, have not been quite so uppish. Yet most now question whether equilibrium-ecology is truly the stage on which evolution is played out (Hutchinson 1965) and whether we have really understood the rules of the game (Slobodkin & Rapoport 1974). We clearly live in a world subject to changes which at some scales are biotically catastrophic yet at others are biotically regenerative. For me, the great convergence is that, through processes which we have yet to define, disturbance of biotically determined relationships is essential to the generation and maintenance of diversity.

Another convergence is faltering of confidence that a single law might uniquely relate diversity to stability or to anything else, for all assemblages and all geographies (Whittaker, 1977). Palaeontologists, biogeographers, ecologists and geneticists are all quickly moving toward the realization that diversity is not only the product of complex and largely unreconstructable histories but is maintained at 'preferred' levels by a complexity of factors the components of which have their own ups and downs, predictable and unpredictable, interrelated or independent. In any particular system, therefore, an early task is to guess at the probable relative importance of the contributors to diversity and to sort out those which are measurable from those which are not. Modellers are now becoming more helpful in providing hints for doing this most effectively (e.g DeAngelis *et al.* 1985; Gerritsen & Patten 1985).

These first two general conclusions are clearly leading to renewed interest in the ways in which organisms are linked in assemblages. Food web linkages have played an important part in diversity/stability discussions, but emphasis is now shifting to the significance of life-history characteristics in assemblage relationships and the genetic processes associated with them (Istock 1984). How do the cogs mesh together, what are the gear ratios and what constitutes a spanner in the works, to use pre-electronic metaphors?

All of which brings us to the question of appropriate units. Palaeontologists use necrological bits, biogeographers use species and genera, ecologists use demes tagged with species names, and population geneticists use those fragments of DNA for which adequate maps are available. In general, mathematicians' units can correspond with anything living which behaves more-or-less exactly to specification. Whilst theorists were busy doing sums which need not have concerned the naturalists, they also did some, the importance of which has only recently penetrated into general ecological thinking, despite the fact that, in a fuzzy sort of way, they were part of the old 'community' controversy. These demonstrated that different stabilities result, depending on whether connectance is unevenly distributed amongst the individuals of a system. In fact, stability is optimized for a given diversity when a particular differential distribution of connectance is achieved (Allen & Starr 1982; O'Neill *et al.* 1986). The important numbers of correlation with stability, therefore, are not the numbers for species but the numbers for closely linked groups of species.

A very disappointing feature is that, despite some promising beginnings, little field progess has been made in answering the question of whether diverse systems resist disturbance better than simple ones. The complexity of the maintenance processes suggests that a simple relationship will not emerge, but perhaps there is hope in the selection of more appropriate units. Diversity/stability relationships might be a great deal clearer if they were sought in sub-systems rather than species.

Finally, another convergence is in the incompleteness of data. Palaeontologists, to their credit, squeeze all the fossils they can (to the casts left by absent organisms) from their samples, even if each then works on only a favourite taxonomic group. At the biogeographical level, the size of the spaces considered makes recourse to individual group studies (e.g. plants, mammals, beetles) extremely seductive. Ecologists in the diversity/stability field usually cross groups but rarely, if ever, attempt total enumeration of the systems they study. It is very difficult to do so, which is one reason why trophic levels were such popular simplifications. I think, however, that we have had just about enough of this stuff, and that a start on some total system descriptions, and their dynamics, is called for, alert to the fundamental differences between major groups, such as animals and plants! The specification for success will lie in the selection of systems which are large enough to be representative yet small enough to handle.

It could, perhaps only glibly, be argued that the recent history of diversity/stability studies resulted from a perturbation which thankfully did not destroy the resilience of natural history, whilst pushing it into a new degree of connectance. Where did British ecology stand in all this? Glibly again, one might say that two British ecologists, Elton and the expatriate Hutchinson, kicked it off then retired to the sidelines whilst the teams played exuberantly in search of the rules. Throughout the game, a few British, some with 'Colonial Experience', ran on and off the field with oddly shaped balls, rules about rules and sensible suggestions (Smith 1974; Grubb 1977) but, by and large, the most consistent players were Americans and the odd fugitive from the Australian beaches. British ecologists meanwhile got on with the workaday business of relating plant growth to soil chemistry, of devising ways of making quantitative descriptions and then ordering them, of discovering what really happens as plant populations interfere with one-anothers' increase, of estimating production and its correlates, of measuring what animals eat, of how they interfere with one another, what directly affects their reproductive rates and so on. I am reminded that Verona Conway described the ecological scene in the late 'forties in terms of 'European formalism, American idealism and British empiricism'. Now, as the game begins to slacken and the spectators turn away, I think that British ecology is in an excellently pre-adapted condition to steal the ball. What is worth doing with it?

I doubt if there is much more to be gained, in the immediate future, by continuing to study diversity/stability as an end in itself. It is time to accept its failures and successes, and there is certainly room for a much more detailed appraisal of these than I have attempted. Amongst the exploitable directions which it has illuminated is studying the mechanics of assemblages within the interactions that occur between biologically dominated processes and the continuously changing, and periodically unstable, physical environment. Biological and environmental periodicities are essential parts of such studies and these are now much more readily comprehended and measured than of old. A good deal of the work published in the last few years (e.g. papers in Price, Slobodchikoff & Gaud 1984; Pickett & White 1985; Kikkawa & Anderson 1986; O'Neill *et al.* 1986) seems implicitly or explicitly to recognize this. Diversity will stay with us as perhaps the most fascinating aspect of the naturalist's world but increasingly, as organisms and environment become conceptually re-united, the very notion of disturbance, as distinct from variation in the forces acting on and in the system, should disappear. Eventually it may become an experimental field, but first there is a lot of correlative observation to be made, particularly of the population fluctuations of different species. Refined palynological and related techniques for reconstructing past biotas and environments have an important place both in setting contexts and testing hypotheses relating to long time periods (10–10^3 a). Probably many specialists will be needed working on each system for a long time; in this field at least, individuals just cannot cope with the complexity.

There is also a more philosophical question which nevertheless may have practical implications. Today's Quaternary world is one of such environmental change that selection of attributes contributing to stability and resilience must be increasing. Whether total diversity will increase or decrease as a consequence is a most interesting question but it is certain that some kinds of life histories will be selected in preference to others. Can the selection process operate fast enough to maintain stability in *any* configuration, particularly as the human species becomes ever more rapacious?

REFERENCES

Allen, T. F. H. & Starr, T. B. (1982). *Hierarchy: Perspectives for Ecological Complexity.* University of Chicago Press, Chicago.

Aubreville, A. (1938). La foret colonial: les forets de l'Afrique occidentale francaise. *Annales de l'Academie Scientifique Colonial de Paris,* **9,** 1–245.

Auclair, A. N. & Goff, F. G. (1971). Diversity relations of upland forests in the western Great Lakes area. *American Naturalist,* **105,** 499–528.

Bennett, K. D. (1985). The spread of *Fagus grandifolia* across eastern North America during the last 18 000 years. *Journal of Biogeography,* **12,** 147–164.

Bennett, K. D. (1986). The rate of spread and population increase of forest trees during the postglacial. *Philosophical Transactions of the Royal Society of London B,* **314,** 523–531

Birch, L. C. (1979). The effect of species of animals which share common resources on one another's distribution and abundance. *Fortschritte der Zoologie,* **25,** 197–221.

Bormann, F. H. & Likens, G. E. (1979). *Pattern and Process in a Forested Ecosystem.* Springer-Verlag, New York.

Botkin, D. W. & Sobel, M. J. (1975). Stability in time-varying ecosystems. *American Naturalist,* **109,** 625–646.

Brackenhielm, S. (1977). Vegetation dynamics of afforested farmland in a district of south-eastern Sweden. *Acta Phytogeographica Suecica,* **63,** 1–106.

Brokaw, N. V.L. (1985). Gap-phase regeneration in a tropical forest. *Ecology,* **66,** 682–687.

Carson, H. L. (1968). The population flush and its genetic consequences. *Population Biology and Evolution* (Ed. by R.C. Lewontin), pp. 123–137. Syracuse University Press, Syracuse.

Carson, H. L. (1987). Population genetics, evolutionary rates and Neo-Darwinism. *Rates of Evolution* (Ed. by K.W.S. Campbell & M.F. Day), pp. 209–217. Allen & Unwin, London.

Chappell, J. & Thom, B. G. (1977). Sea levels and coasts. *Sunda and Sahul: Prehistoric Studies in Southeast Asia, Melanesia and Australia* (Ed. by J. Allen, J. Golson & R. Jones), pp. 275–291. Academic Press, London.

Christensen, N. L. (1985). Shrubland fire regimes and their evolutionary consequences. *The Ecology of Natural Disturbance and Patch Dynamics* (Ed. by S.T.A. Pickett & P.S. White), pp. 85–100. Academic Press, Orlando.

Clark, D. A. & Clark, D. B. (1984). Spacing dynamics of a tropical tree. *American Naturalist,* **124,** 769–788.

Colinvaux, P. (1987). Amazon diversity in light of the paleoecological record. *Quaternary Science Reviews,* **6,** 93–114.

Connell, J. H. (1971). On the role of natural enemies in preventing competitive exclusion in some marine animals and in rain forest trees. *Dynamics of Populations* (Ed. by P.J. den Boer & G. Gradwell), pp. 298–312. Wageningen Centre for Agricultural Publishing and Documentation, Wageningen.

Connell, J. H. (1978). Diversity in tropical rain forest and coral reefs. *Science,* **199,** 1302–1310.

Connell, J. H. & Keough, M. J. (1985). Disturbance and patch dynamics of subtidal marine animals on hard substrata. *The Ecology of Natural Disturbance and Patch Dynamics* (Ed. by S.T.A. Pickett & P.S. White), pp. 125–151. Academic Press, Orlando.

Connell, J. H. & Sousa, W. P. (1983). On the evidence needed to judge ecological stability or persistence. *American Naturalist,* **121,** 789–824.

Connell, J. H., Tracey, J. G. & Webb, L. J. (1984). Compensatory recruitment, growth and mortality as factors maintaining rain forest tree diversity. *Ecological Monographs,* **54,** 141–164.

Coope, G. R. (1978). Constancy of insect species versus inconstancy of Quaternary environments. *Diversity of Insect Faunas* (Ed. by L.A. Mound & N. Waloff), pp. 176–187. Blackwell Scientific Publications, Oxford.

Coope, G. R. (1979). Late Cenozoic fossil Coleoptera: evolution, biogeography and ecology. *Annual Review of Ecology and Systematics,* **10,** 247–267.

Darlington, P. J. (1957). *Zoogeography: the Geographical Distribution of Animals.* John Wiley & Sons, New York.

Davis, M. B. (1976). Pleistocene biogeography of temperate deciduous forests. *Geoscience and Man,* **13,** 13–26.

Davis, M. B. (1981). Quaternary history and the stability of forest communities. *Forest Succession: Concepts and Application* (Ed. by D.C. West, H.H. Shugart & D.B. Botkin), pp. 154–177. Springer-Verlag, New York.

Davis, M. B. (1983). Quaternary history of deciduous forests of eastern North America and Europe. *Annals of the Missouri Botanical Gardens,* **70,** 550–563.

Davis, M. B. (1984). Climatic instability, time lags and community disequilibrium. *Community Ecology* (Ed. by J. Diamond & T.J. Case), pp. 269–284. Harper & Row, New York.

Dayton, P. K. & Tegner, M. J. (1984). The importance of scale in community ecology: a kelp forest example with terrestrial analogs. *A New Ecology* (Ed. by P.W. Price, C.N. Slobodchikoff & W.S. Gaud), pp. 457–481. John Wiley & Sons, New York.

DeAngelis, D. L., Waterhouse, J. C., Post, W. M. & O'Neill, R. V. (1985). Ecological modelling and disturbance evaluation. *Ecological Modelling,* **29,** 399–419.

Delcourt, P. A. & Delcourt, H. R. (1987). *Long-term Forest Dynamics of the Temperate Zone, a Case Study of Late Quaternary Forests in Eastern North America. Ecological Studies Series 63.* Springer-Verlag, New York.

Delcourt, P. A., Delcourt, H. R., Cridelbaugh, P. A. & Chapman, J. (1986). Holocene ethnobotanical and palaeoecological record of human impact on vegetation in the Little Tennessee River Valley, Tennessee. *Quaternary Research,* **25,** 330–349.

Denslow, J. S. (1980). Patterns of species diversity during succession under different disturbance regimes. *Oecologia,* **46,** 18–21.

Dexter, F., Banks, H. T. & Webb, T. (1987). Modelling Holocene changes in the location and abundance of beech populations in eastern North America. *Review of Palaeobotany and Palynology,* **50,** 273–292.

Dixon, A. F.G., Kindlman, P., Leps, J. & Holman, M. (1987). Why there are so few species of aphids, especially in the tropics. *American Naturalist,* **129,** 580–592.

Eastop, V. F. (1978). Diversity of the Sternorrhyncha within major climatic zones. *Diversity of Insect Faunas* (Ed. by L.A. Mound & N. Waloff), pp. 71–88. Blackwell Scientific Publications, Oxford.

Eldredge, W. & Gould, S. J. (1972). Punctuated equilibria: an alternative of phyletic gradualism. *Models in Paleobiology* (Ed. by T.J.M. Shopf), pp. 82–115. Freeman, Cooper, San Francisco.

Ennos, R. A. (1983). Maintenance of genetic variation in plant populations. *Evolutionary Biology,* **16,** 129–155.

Fischer, A. G. (1960). Latitudinal variation in organic diversity. *Evolution,* **14,** 64–81.

Flenley, J. R. (1979). *The Equatorial Rainforest: a Geological History.* Butterworths, London.

Flenley, J. R. (1984). Late Quaternary changes of vegetation and climate in the Malesian mountains. *Erdwissenschaftliche Forschung,* **18,** 261–267.

Franklin, I. R. (1987). Population biology and evolutionary change. *Rates of Evolution* (Ed. by K.W.S. Campbell & M.F. Day), pp. 156–174. Allen & Unwin, London.

Gerritsen, J. & Patten, B. C. (1985). System theory formulation of ecological disturbance. *Ecological Modelling,* **29,** 383–397.

Gilbert, L. E. & Smiley, J. T. (1978). Determinants of local diversity in phytophagous insects: host specialists in tropical environments. *Diversity of Insect Faunas* (Ed.by L.A. Mound & N. Waloff), pp. 89–104. Blackwell Scientific Publications, Oxford.

Gingerich, P. D. (1975). Patterns of evolution in the mammalian fossil record. *Patterns of Evolution as Illustrated by the Fossil Record* (Ed. by A. Hallam), pp. 469–500. Elsevier, Amsterdam.

Gingerich, P. D. (1984). Pleistocene extinctions in the context of origination-extinction equilibria in Cenozoic mammals. *Quaternary Extinctions, a Prehistoric Revolution* (Ed. by P.S. Martin & R.G. Klein), pp. 211–222. University of Arizona Press, Tucson.

Gould, S. J. (1985). The paradox of the first tier: an agenda for paleobiology. *Paleobiology,* **11,** 2–12.

Gould, S. J. & Eldredge, W. (1977). Punctuated equilibria: the tempo and mode of evolution reconsidered. *Paleobiology,* **3,** 115–151.

Grassle, J. F. Patil, G. P., Smith, W. & Taillie, C. (1979). *Ecological Diversity in Theory and Practice.* International Co-operative Publishing House, Fairland, Maryland.

Green, D. G., Singh, G., Polach, H., Moss, D., Banks, J. & Geissler, E. A. (1988). A fine-resolution palaeoecology and palaeoclimatology from south-eastern Australia. *Journal of Ecology,* **76,** 790–806.

Greig-Smith, P. (1982). A.S. Watt, F.R.S.: a biographical note. *The Plant Community as a Working Mechanism,* (Ed. by E.I. Newman), pp. 9–10. Blackwell Scientific Publications, Oxford.

Grime, J. P. (1973). Competitive exclusion in herbaceous vegetation. *Nature.* **242,** 344–347.

Grubb, P. J. (1977). The maintenance of species richness in plant communities: the importance of the regeneration niche. *Biological Reviews,* **52,** 107–145.

Guilday, J. E. (1984). Pleistocene extinction and environmental change : case study of the Appalachians. *Quaternary Extinctions, a Prehistoric Revolution* (Ed. by P.S. Martin & R.G. Klein), pp. 250–258. University of Arizona Press, Tucson.

Haffer, J. (1969). Speciation in Amazonian forest birds. *Science,* **165,** 131–137.

Hamilton, A. (1976). The significance of patterns of distribution shown by forest plants and animals in tropical Africa for the reconstruction of Upper Pleistocene palaeoenvironments: a review. *Palaeoecology of Africa,* **9,** 63–97.

Hanes, T. L. (1971). Succession after fire in the chaparral of southern California. *Ecological Monographs,* **41,** 27–52.

Hansen, R. M. (1978). Shasta ground sloth food habits, Rampart Cave, Arizona. *Paleobiology,* **4,** 302–319.

Harper, J. L. (1977). *Population Biology of Plants.* Academic Press, London.

Hartshorn, G. H. S. (1980). Neotropical forest dynamics. *Biotropica,* **12 (Supplement),** 23–30.

Herbold, B. & Moyle. P. B. (1986). Introduced species and vacant niches. *American Naturalist,* **128,** 751–760.

Herwitz, S. R. (1981). *Regeneration of Selected Tropical Tree Species in Corcovado National Park,* Costa Rica. University of California Press, Berkeley.

Holling, C. S. (1973). Resilience and stability of ecological systems. *Annual Review of Ecology and Systematics,* **4,** 1–23.

Hopkins, M. S. & Graham, A. W. (1987). Gregarious flowering in a lowland tropical rainforest: a possible response to disturbance by Cyclone Winifred. *Australian Journal of Ecology,* **12,** 25–29.

Horn, H. S. (1974). The ecology of secondary succession. *Annual Review of Ecology and Systematics,* **5,** 25–37.

Hubbell, S. P. (1979). Tree dispersion, abundance, and diversity in a tropical dry forest. *Science,* **203,** 1299–1308.

Hubbell, S. P. (1980). Seed predation and the coexistence of tree species in tropical forests. *Oikos,* **35,** 214–229.

Huntley, B. & Birks, H. J. B. (1983). *An Atlas of Past and Present Pollen Maps for Europe: 0–13,000 Years Ago.* Cambridge University Press, Cambridge.

Huston, M. (1979). A general hypothesis of species diversity. *American Naturalist,* **113,** 81–101.

Hutchinson, G. E. (1959). Homage to Santa Rosalia, or Why are there so many kinds of animals? *American Naturalist,* **93,** 145–159.

Hutchinson, G. E. (1965).*The Ecological Theater and the Evolutionary Play.* Yale University Press, New Haven.

Istock, C. A. (1984). Boundaries to life history variation and evolution. *A New Ecology : Novel Approaches to Interactive Systems* (Ed. by P.W. Price, C.N. Slobodchikoff & W.S. Gaud), pp. 143–168. John Wiley & Sons, New York.

Jablonski, D., Flessa, K. W. & Valentine, J. W. (1985). Biogeography and paleobiology. *Paleobiology,* **11,** 75–90.

Jones, N. S. (1948). Observations and experiments on the biology of *Patella vulgata* at Port St. Mary, Isle of Man. *Proceedings and Transactions of the Liverpool Biological Society,* **56,** 60–77.

Jones, R. (1975). The Neolithic Palaeolithic and hunting gardeners: man and land in the Antipodes. *Quaternary Studies* (Ed. by R.P. Suggate & M.M. Cresswell), pp. 21–34. Royal Society of New Zealand, Wellington.

Kikkawa, J. (1986). Complexity, diversity and stability. *Community Ecology : Pattern and Process* (Ed. by J. Kikkawa & D.J. Anderson), pp. 41–62. Blackwell Scientific Publications, Melbourne.

Kikkawa, J. & Anderson, D. J. (1986). *Community Ecology: Pattern and Process.* Blackwell Scientific Publications, Melbourne.

King, A. W. & Pimm, S. L. (1983). Complexity, diversity and stability: a reconciliation of theoretical and empirical results. *American Naturalist,* **122,** 229–239.

Kitchell, J. A. (1985). Evolutionary paleoecology: recent contributions to evolutionary theory. *Paleobiology,* **11,** 91–104.

Knoll, A. H. (1984). Patterns of extinction in the fossil record of vascular plants. *Extinctions* (Ed. by M.H. Nitecki), pp. 21–68. University of Chicago Press, Chicago.

Lawlor, L. R. (1978). A comment on randomly constructed model ecosystems. *American Naturalist,* **112,** 445–447.

Lawton, J. H. (1978). Host-plant influences on insect diversity: the effects of space and time. *Diversity of Insect Faunas* (Ed. by L.A. Mound & N. Waloff), pp. 105–125. Blackwell Scientific Publications, Oxford.

Lawton, J. H. (1982). Vacant niches and unsaturated communities: a comparison of bracken herbivores at sites on two continents. *Journal of Animal Ecology,* **51,** 573–595.

Levandowsky, M. & White, B. S. (1977). Randomness, time scales and the evolution of biological communities. *Evolutionary Biology,* **10,** 69–161.

Levins, R. (1968). *Evolution in Changing Environments: Some Theoretical Explorations.* Princeton University Press, Princeton, New Jersey.

Lewontin, R. C. (1969). The meaning of stability. *Brookhaven Symposia on Biology,* **22,** 13–24.

Livingstone, D. A. (1967). Postglacial vegetation of the Ruwenzori mountains in equatorial Africa. *Ecological Monographs,* **37,** 25–52.

Loucks, O. L., Ek, A. R., Johnson, W. C. & Monserud, R. A. (1981). Growth, aging and succession. *Dynamic Properties of Forest Ecosystems* (Ed. by D.E. Reichle), pp. 37–85. Cambridge University Press, Cambridge.

MacArthur, R. H. (1960). On the relative abundance of species. *American Naturalist,* **94,** 25–36.

MacArthur, R. H. (1965). Patterns of species diversity. *Biological Reviews,* **40,** 410–533.

MacArthur, R. H. (1972). *Geographical Ecology.* Harper & Row, New York.

MacArthur, R. H. & Wilson, E. O. (1967). *The Theory of Island Biogeography.* Princeton University Press, Princeton.

Maley, J. (1987). Fragmentation de la fôret dense-humide africaine et extension des biotopes montagnardes au Quaternaire recent: nouvelles données polliniques et chronologiques. Implications paleoclimatiques et biogeographiques. *Palaeoecology of Africa and the Surrounding Islands,* **18,** 307–334.

Mani, G. S. (1984). Genetic diversity and ecological stability. *Evolutionary Ecology* (Ed. by B. Shorrocks), pp. 363–396. Blackwell Scientific Publications, Oxford.

Marshall, L. G. (1981). The great American interchange–an invasion induced crisis for South American mammals. *Biotic Crises in Ecological and Evolutionary Time* (Ed. by M.H. Nitecki), pp. 133–229. Academic Press, New York.

Martin, P. S. (1984). Prehistoric overkill : the global model. *Quaternary Extinctions, a Prehistoric Revolution* (Ed. by P.S. Martin & R.G. Klein), pp. 354–403. University of Arizona Press, Tucson.

May, R. M. (1965). Patterns of species abundance and diversity. *Ecology and Evolution of Communities* (Ed. by M.L. Cody & J.M. Diamond), pp. 81–120. Belknap Press, Cambridge, Massachusetts.

May, R. M. (1973). *Stability and Complexity in Model Ecosystems.* Princeton University Press, Princeton, New Jersey.

May, R. M. (1974). Ecosystem patterns in randomly fluctuating environments. *Progress in Theoretical Biology* (Ed. by R. Rosen & F. Snell), pp. 1–50. Academic Press, New York.

May, R. M. (1976). Patterns in multi-species communities. *Theoretical Ecology. Principles and Applications* (Ed. by R.M. May), pp. 142–162. Blackwell Scientific Publications, Oxford.

May, R. M. (1986). When two and two do not make four: nonlinear phenomena in ecology. The Croonian lecture 1985. *Proceedings of the Royal Society of London B,* **228,** 241–266.

May, R. M. & Oster, G. (1976). Bifurcations and dynamic complexity in simple ecological models. *American Naturalist,* **110,** 573–599.

McAndrews, J. H. (1976). Fossil history of man's impact on the Canadian flora: an example from southern Ontario. *Canadian Botanical Association Bulletin, Supplement,* **9,** 1–6.

McNaughton, S. J. (1977). Diversity and stability of ecological communities: a comment on the role of empiricism in ecology. *American Naturalist,* **111,** 515–524.

Medway, Lord. (1972). The Quaternary mammals of Malesia: a review. *The Quaternary Mammals of Malesia Transactions of the 2nd Aberdeen - Hull Symposium on Malesian Ecology,* (Ed. by P. & M. Ashton), pp. 63–82. Department of Geography, University of Hull, Hull.

Mellinger, M. V. & McNaughton, S. J. (1975). Structure and function of successional vascular plant communities in central New York. *Ecological Monographs,* **45,** 161–182.

Milligan, B. G. (1986). Punctuated equilibrium induced by ecological change. *American Naturalist,* **127,** 522–532.

Nicholson, P. H. (1981). Fire and the Australian Aborigine — an enigma. *Fire and the Australian Biota* (Ed. by A.M. Gill, R.H. Groves & I.R. Noble), pp. 54–76. Australian Academy of Science, Canberra.

O'Neill, R. V., DeAngelis, D. L., Waide, J. B. & Allen, T. F. H. (1986). *A Hierarchical Con-*

cept of Ecosystems. Princeton University Press, Princeton, New Jersey.

Osman, R. W. & Whitlatch, R. B. (1978). Patterns of species diversity: fact or artifact? *Paleobiology,* **4,** 41–54.

Paine, R. T. (1966). Food web complexity and species diversity. *American Naturalist,* **100,** 65–75.

Parsons, P. A. (1987). Evolutionary rates under environmental stress. *Evolutionary Biology,* **21,** 311–347.

Peet, R. K. (1974). The measurement of species diversity. *Annual Review of Ecology and Systematics,* **5,** 285–307.

Pianka, E. R. (1978). *Evolutionary Ecology,* 2nd edition. Harper & Row, New York.

Pickett, S. T. A. & White. P. S. (1985). *The Ecology of Natural Disturbance and Patch Dynamics.* Academic Press, Orlando.

Pielou, E. C. (1975). *Ecological Diversity.* Wiley-Interscience, New York.

Pimm, S. L. (1982). *Food Webs.* Chapman & Hall, London.

Pimm, S. L. (1984). The complexity and stability of ecosystems. *Nature,* **307,** 321–326.

Potts, D. C. (1984). Generation times and the Quaternary evolution of reef building corals. *Paleobiology,* **10,** 48–58.

Powell, J. M. (1976). Ethnobotany. *New Guinea Vegetation* (Ed. by K. Paijmans), pp. 106–183. Australian National University Press, Canberra.

Prance, G. T. (1982). *Biological Diversification in the Tropics.* Columbia University Press, New York.

Preston, F. W. (1962). The canonical distribution of commonness and rarity. *Ecology,* **43,** 185–215.

Price, P. W., Gaud, W. S. & Slobodchikoff, C. N. (1984). Introduction: Is there a new ecology? *A New Ecology : Novel Approaches to Interactive Systems* (Ed. by P.W. Price, C.N. Slobodchikoff & W.S. Gaud), pp. 1–11. John Wiley & Sons, New York.

Price, P. W., Slobodochikoff, C. N. & Gaud, W. S. (eds) (1984). *A New Ecology: Novel Approaches to Interactive Systems.* John Wiley & Sons, New York.

Raup, D. M. (1987). Major features in the fossil record and their implications for fossil rate studies. *Rates of Evolution* (Ed. by K.W.S. Campbell & M.F. Day), pp. 1–14. Allen & Unwin, London.

Raup, M. M. & Sepkoski, J. J. (1984). Periodicity of extinctions in the geological past. *Proceedings of the National Academy of Sciences,* **81,** 801–805.

Rice, K. & Jain, S. (1985). Plant population genetics and evolution in disturbed environments. *The Ecology of Natural Disturbance and Patch Dynamics* (Ed. by S.T.A. Pickett & P.S. White), pp. 287–303. Academic Press, Orlando.

Ritchie, J. C. (1986). Climatic change and vegetation response. *Vegetatio,* **67,** 65–74.

Roberts, A. (1974). The stability of a feasible random ecosystem. *Nature,* **251,** 607–608.

Rohde, K. (1978). Latitudinal gradients in species diversity and their causes II. Marine parasitological evidence for a time hypothesis. *Biologisches Zentralblatt,* **97,** 405–418.

Rosen, B. R. (1981). The tropical high diversity enigma – the corals' eye view. *Chance, Change and Challenge : the Evolving Biosphere* (Ed. by P.H. Greenwood & P.L. Forey), pp. 103–129. Cambridge University Press and British Museum (Natural History), Cambridge.

Rosen, B. R. (1984). Reef coral biogeography and climate through the late Cainozoic: just islands in the sun or a critical pattern of islands? *Fossils and Climate* (Ed. by P.J. Brenchley), pp. 201–262. John Wiley & Sons, Chichester.

Schopf, T. J. M. (1977). Patterns of evolution : a summary and discussion. *Patterns of Evolution as Illustrated by the Fossil Record* (Ed. by A. Hallam), pp. 547–561. Elsevier, Amsterdam.

Schopf, T. J. M. (1984). Rates of evolution and the notion of 'living' fossils. *Annual Review of Earth and Planetary Sciences,* **12,** 245–292.

Schopf, T. J. M., Fisher, H. B. & Smith, C. A. F. (1978). Is the marine latitudinal diversity gradient merely another example of the species area curve? *Marine Organisms: Genetics, Ecology and Evolution* (Ed. by B. Battaglia & J.A. Beardmore), pp. 365–385. Plenum, New York.

Shackleton, N. J. (1984). Oxygen isotope evidence for Cenozoic climatic change. *Fossils and Climate* (Ed. by P.J. Brenchley), pp. 27–34. John Wiley & Sons, Chichester.

Shafi, M. I. & Yarranton, G. A. (1973). Diversity, floristic richness and species evenness during a secondary (post-fire) succession. *Ecology,* **54,** 897–902.

Shugart, H. H. (1984). *A Theory of Forest Dynamics.* Springer-Verlag, New York.

Silvertown, J. (1980). The dynamics of a grassland ecosystem: botanical equilibrium in the Park Grass experiment. *Journal of Applied Ecology,* **17,** 491–504.

Simberloff, D. (1981). Community effects of introduced species. *Biotic Crises in Ecological and Evolutionary Time* (Ed. by T.H. Nitecki), pp. 53–81. Academic Press, New York.

Simpson, G. G. (1952). How many species? *Evolution,* **6,** 342.

Slobodkin, L. B. (1968). Toward a predictive theory of evolution. *Population Biology and Evolution* (Ed. by R.C. Lewontin), pp. 187–205. Syracuse University Press, Syracuse, New York.

Slobodkin, L. B. & Rapoport, A. (1974). An optimal strategy of evolution. *Quarterly Review of Biology,* **49,** 181–200.

Smith, A. G. (1970). The influence of mesolithic and neolithic man on British vegetation: a discussion. *Studies in the Vegetational History of the British Isles* (Ed. by D. Walker & R.G. West), pp. 81–96. Cambridge University Press, Cambridge.

Smith, J. M. (1974). *Models in Ecology.* Cambridge University Press, Cambridge.

Solomon, A. M., West, D. C. & Solomon, J. A. (1981). Simulating the role of climate change and species immigration in forest succession. *Forest Succession: Concepts and Application* (Ed. by D.C. West, H.H. Shugart & D.B. Botkin), pp. 154–177. Springer-Verlag, New York.

Sousa, W. P. (1985). Disturbance and patch dynamics on rocky intertidal shores. *The Ecology of Natural Disturbance and Patch Dynamics* (Ed. by S.T.A. Pickett & P.S. White), pp. 101–124. Academic Press, Orlando.

Stanley, S. M. (1984). Marine mass extinctions: a dominant role for temperatures. *Extinctions* (Ed. by M.H. Nitecki), pp. 69–117. Chicago University Press, Chicago.

Strong, D. R. (1977). Epiphyte loads, treefalls and perennial disruption: a mechanism for maintaining higher species richness in the tropics without animals. *Journal of Biogeography,* **4,** 215–218.

Strong, D. R. (1984). Exorcising the ghost of competition past: phytophagous insects. *Ecological Communities; Conceptual Issues and the Evidence* (Ed. by D.R. Strong, D. Simberloff, L.G. Abele & A.B. Thistle), pp. 28–41. Princeton University Press, Princeton, N.J.

Sultan, S. E. (1987). Evolutionary implications of phenotypic plasticity in plants. *Evolutionary Biology,* **21,** 127–178.

Taylor, L. R. (1978). Bates, Williams, Hutchinson — a variety of diversities. *Diversity of Insect Faunas* (Ed. by L.A. Mound & N. Waloff), pp. 1–18. Blackwell Scientific Publications, Oxford.

Thiery, R. G. (1982). Environmental instability and community diversity. *Biological Reviews,* **57,** 671–710.

Thompson, J. N. (1985). Within-patch dynamics of life histories, populations and interactions: selection over time in small spaces. *The Ecology of Natural Disturbance and Patch Dynamics* (Ed. by S.T.A. Pickett & P.S. White), pp. 253–264. Academic Press, Orlando.

Tilman, D. (1982). *Resource Competition and Community Structure.* Princeton University Press, Princeton, New Jersey.

Truswell, E. M. (1987). The initial radiation and rise to dominance of the angiosperms. *Rates of Evolution* (Ed. by K.W.S. Campbell & M.F. Day), pp. 101–128. Allen & Unwin, London.

Valentine, J. W. (1977). General patterns of metazoan evolution. *Patterns of Evolution as Illustrated by the Fossil Record* (Ed. by A. Hallam), pp. 27–57. Elsevier, Amsterdam.

Valentine, J. W. (1984). Climate and evolution in the shallow sea. *Fossils and Climate* (Ed. by P.J. Brenchley), pp. 265–277. John Wiley & Sons, Chichester.

Valentine, J. W. & Ayala, F. J. (1978). Adaptive strategies in the sea. *Marine Organisms: Genetics, Ecology and Evolution* (Ed. by B. Battaglia & J.A. Beardmore), pp. 323–345. Plenum, New York.

van der Hammen, T. (1974). The Pleistocene changes of vegetation and climate in tropical South America. *Journal of Biogeography,* **1,** 3–26.

Van Valen, L. M. (1982). Integration of species: stasis and biogeography. *Evolutionary Theory,* **6,** 99–112.

Verstappen, H. T. (1975). On palaeoclimates and landform development in Malesia. *Modern Quaternary Research in Southeast Asia,* **1,** 3–35.

Viereck, L. A. (1966). Plant succession and soil development on gravel outwash of the Muldrow glacier, Alaska. *Ecological Monographs,* **36,** 181–199.

Vrijenhoek, R. C. (1985). Animal population genetics and disturbance: the effects of local extinctions and recolonizations on heterozygosity and fitness. *The Ecology of Natural Disturbance and Patch Dynamics* (Ed. by S.T.A. Pickett & P.S. White), pp. 265–285. Academic Press, Orlando.

Waddington, C. H. (1968). The paradigm for the evolutionary process. *Population Biology and Evolution* (Ed. by R.C. Lewontin), pp. 37–45. Syracuse University Press, Syracuse New York.

Walker, D. (1966). Vegetation of the Lake Ipea region, New Guinea highlands 1. Forest, grassland and 'garden'. *Journal of Ecology,* **54,** 503–533.

Walker, D. (1970). The changing vegetation of the montane tropics. *Search,* **1,** 217–221.

Walker, D. (1982a). The development of resilience in burned vegetation. *The Plant Community as a Working Mechanism* (Ed. by E.I. Newman), pp. 27–43. Blackwell Scientific Publications, Oxford.

Walker, D. (1982b). Speculations on the origin and evolution of Sunda – Sahul rain forests. *Biological Diversification in the Tropics* (Ed. by G.I. Prance), pp. 554–575. Columbia University Press, New York.

Walker, D. (in press). Directions and rates of tropical rain forest processes. *Proceedings of the ANZAAS Conference, Townsville, Australia, 1987,* on Australian Tropical Rainforests: Science, Values and Meaning. CSIRO, Melbourne.

Walker, D. & Chen, Y. (1987). Paylnological light on tropical rainforest dynamics. *Quaternary Science Reviews,* **6,** 77–92.

Walker, D. & Flenley, J. R. (1979). Late Quaternary vegetational history of the Enga Province of upland Papua New Guinea. *Philosophical Transactions of the Royal Society of London B,* **286,** 265–344.

Walker, D. & Hope, G. S. (1982). Late Quaternary vegetation history. *Biogeography and Ecology of New Guinea* (Ed. by J.L. Gressitt), pp. 263–285. Junk, The Hague.

Walker, D. & Sun, X. (1988). Vegetational and climatic changes at the Pleistocene-Holocene transition across the eastern tropics. *The Palaeoenvironment of East Asia from the Mid-Tertiary,* Centre for Asian Studies, Hong Kong University, Hong Kong.

Walker, T. D. & Valentine, J. W. (1984). Equilibrium models of evolutionary species diversity and the number of empty niches. *American Naturalist,* **124,** 887–899.

Walter, M. R. (1987). The timing of major evolutionary innovations from the origin of life to the origins of the Metaphyta and Metazoa: the geological evidence. *Rates of Evolution* (Ed. by K.W.S. Campbell & M.F. Day), pp. 15–38. Allen & Unwin, London.

Watt, A. S. (1923). On the ecology of British beechwoods with special reference to their regeneration. *Journal of Ecology,* **11,** 1–48.

Watt, A. S. (1924). On the ecology of British beechwoods with special reference to their regeneration. Part II. The development and structure of beech communities on the Sussex Downs. *Journal of Ecology,* **12,** 145–204.

Watt, A. S. (1925). On the ecology of British beechwoods with special reference to their regeneration. Part II, sections II and III. The development and structure of beech communities on the Sussex Downs (continued). *Journal of Ecology,* **13,** 27–73.

Watt, A. S. (1947). Pattern and process in the plant community. *Journal of Ecology,* **35,** 1–22.

Watt, A. S. (1960). The effect of excluding rabbits from acidiphilous grassland in Breckland. *Journal of Ecology,* **48,** 601–604.

Watt, A. S. (1974). Senescence and rejuvenation in ungrazed chalk grassland (grassland B) in Breckland: the significance of litter and of moles. *Journal of Applied Ecology,* **11,** 1157–1171.

Watts, W. A. (1980). The Late Quaternary vegetation history of the southern United States. *Annual Review of Ecology and Systematics,* **11,** 387–409.

Watts, W. A. & Stuiver, M. (1980). Late Wisconsin climate of northern Florida and the origin of species-rich deciduous forest. *Science,* **210,** 325–327.

Webb, S. D. (1984). Ten million years of mammal extinctions in North America. *Quaternary Extinctions, a Prehistoric Revolution* (Ed. by P.S. Martin & R.G. Klein), pp. 189–210. University of Arizona Press, Tucson.

West, R. G. (1980). Pleistocene forest history in East Anglia. *New Phytologist,* **85,** 571–622.

White, P. S. (1979). Pattern, process and natural disturbance in vegetation. *The Botanical Review,* **45,** 229–299.

Whitmore, T. C. (1974). Change with time and the role of cyclones in tropical rain forest on Kolombangara, Solomon Islands. *Commonwealth Forestry Institute Paper* **46.** Commonwealth Forestry Institute, Oxford.

Whittaker, R. H. (1975). *Communities and Ecosystems,* 2nd edition. Macmillan, New York.

Whittaker, R. H. (1977). Evolution of species diversity in land communities. *Evolutionary Biology,* **10,** 1–67.

Whittington, S. L. & Dyke, B. (1984). Simulating overkill: experiments with the Mosimann and Martin model. *Quaternary Extinctions, a Prehistoric Revolution* (Ed. by P.S. Martin & R.G. Klein), pp. 451–465. University of Arizona Press, Tucson.

Wiens, J. A. (1984). On understanding a non-equilibrium world: myth and reality in community patterns and processes. *Ecological Communities; Conceptual Issues and the Evidence* (Ed. by D.R. Strong, D. Simberloff, L.G. Abele & A.B. Thistle), pp. 439–457. Princeton University Press, Princeton, New Jersey.

Williams, C. B. (1964). *Patterns in the Balance of Nature.* Academic Press, London.

Williams, E. D. (1978). *Botanical Composition of the Park Grass.* Rothamsted Experiment Station, Harpenden.

Wills, C. (1981). *Genetic Variability.* Clarendon Press, Oxford.

Wright, D. H. (1983). Species-energy theory: an extension of species-area theory. *Oikos,* **41,** 496–506.

Wright, H. E. (1984). Sensitivity and response time of natural systems to climatic change in the late Quaternary. *Quaternary Science Reviews,* **3,** 91–131.

Yarranton, G. A. & Morrison, R. G. (1974). Spatial dynamics of a primary succession: nucleation. *Journal of Ecology,* **62,** 417–428.

6 PREDATOR–PREY AND HOST–PATHOGEN INTERACTIONS

M. P. HASSELL AND R. M. ANDERSON
Department of Pure and Applied Biology, Imperial College, London SW7 2AZ, UK

INTRODUCTION

Parasitism, predation, herbivory and competition for resources play a crucial part in constraining the growth of populations, and in shaping the evolution of plant and animal communities. The study of the dynamic effects of these processes has provided a theoretical framework upon which much of modern ecology rests. The development of this theory, hand-in-hand with detailed observations and experiments on a variety of population interactions, is crucial for the progress of the subject and for its application to fields such as pest and disease management and the harvesting of renewable resources. Studies on population interactions are also increasingly being extended to multispecies systems to explore how dynamics can influence the composition and behaviour of communities of organisms.

Biologists are often mainly concerned with the recondite details that make each particular case study unique. Our aim in this paper is very different; we shall focus on the basic concepts that underpin the dynamic interaction between host and parasite, and predator and prey populations in general, drawing parallels and highlighting differences where possible. A common criticism of theoretical ecology is the simplicity of most mathematical models despite the known biological complexity. Whilst a wide range of factors may indeed affect the dynamics of populations, we are encouraged by the frequency with which a few factors tend to dominate the generation of observed patterns. The rationale behind the exploration of relatively simple population models is thus akin to that adopted by the laboratory-based scientist in experimental design, where one or more factors are held constant as others are varied.

Despite important papers such as Skellam (1951), Hutchinson (1951) and Huffaker (1958), much of the theory of population ecology has, until recently, been set in homogeneous environments with uniformly distributed populations. The subject has now largely veered away from this idealized uniform world, and heterogeneities, be they due to spatial distributions, temporal asynchronies or phenotypic variability between individuals, are widely regarded as important factors affecting population dynamics. In the light of this, we shall pay special attention throughout this paper to the ways in which different forms of heterogeneity can influence the dynamics of predator–prey and host–pathogen interactions.

We begin by outlining some general mathematical models for the dynamics of predator–prey and host–pathogen systems. These greatly help in identifying the major processes that determine the dynamics of interacting populations. We then use these as 'building blocks' for examining the dynamics of mixed species interactions involving different types of predators and pathogens acting together. It is striking how unexpected dynamic properties can emerge, even on moving from a two- to a three-species system. Throughout the paper, much detail is omitted and the interested reader is encouraged to consult the source references. Some of the effects we describe have parallels with other consumer–resource interactions. These lie outside the scope of the present paper, but have been well reviewed elsewhere (e.g. Crawley 1983; Strong, Lawton & Southwood 1984).

BASIC THEORETICAL FRAMEWORKS

1 Predator–Prey Interactions

Most of the predator–prey models developed over the past few decades stem either from the original models of Lotka (1925) and Volterra (1926), or from those of Nicholson (1933) and Nicholson & Bailey (1935). Both types of model share a number of assumptions. Each is a closed system involving coupled interactions; neither involves explicit age structure; predation in each case is a linear function of prey density, implying insatiable predators with no handling time; and both models assume that the prey eaten are directly converted into new predators. The chief difference lies in the Lotka–Volterra models being framed in continuous time, while the Nicholson–Bailey models are based on difference equations giving discrete, non-overlapping generations.

The prominent position held by these models in the history of population ecology is perhaps surprising in the light of their dynamic properties. The neutral stability of the Lotka–Volterra model is generally regarded as 'pathological' in requiring the system to keep a 'memory' of its initial conditions, while the expanding oscillations from the Nicholson– Bailey model (see below) have only occasionally been seen in the laboratory (e.g. Gause 1934; DeBach & Smith 1941; Burnett 1958;

Huffaker 1958). Neither of these models therefore, at least in their original form, help to explain the persistence of predator–prey interactions under natural conditions. Predation could, of course, be a process which at best adds nothing to stability, in which case natural interactions are only persisting by grace of some other regulating factors. But it is more likely that these early models are deficient and that predator–prey interactions can persist without invoking other, additional density-dependent processes. Faith in this latter view has fuelled much of the more recent work on the dynamics of predator–prey systems.

The diversity of predators, and the different ways in which they interact with their prey, make it impossible to develop a single theoretical framework appropriate for studying the dynamics of all types of interaction. Different models are needed for different situations — for parasitoids (see below) rather than predators, for generalists rather than specialists, for interactions with discrete rather than continuous generations, and so on. In addition, since predators occur in all major groups from protozoans to mammals, the need to consider explicitly such complicating factors as the age structure and sex ratios of the interacting populations will depend on the taxonomic groups involved. The modern descendents of the original Lotka–Volterra and Nicholson–Bailey models have been used to examine the dynamic importance of many aspects of predation that were lacking from their progenitors such as a resource-limited prey population, interference between searching predators, a refuge for some of the prey, shifting predator sex ratios, patchily distributed prey populations, more complex life cycles of both populations, competition with other predator species, and a choice of alternative prey species.

Rather than review all these developments superficially, we shall focus for the main part on the dynamics of one class of relatively simple predator–prey models where the prey populations have discrete generations and the predators are specialists that are synchronized and coupled with their prey. Our aim is to highlight some general and important aspects of predation that shape the dynamics of the populations, with particular emphasis on the pervasive effects of different types of heterogeneity.

Much of this work on the dynamics of predation has involved a particular kind of predator — the insect parasitoids that develop as larvae on the tissues of other arthropods (usually insects), which they ultimately kill. Adult female parasitoids usually hunt actively for hosts and deposit one or more eggs on, in or near host individuals. The parasitoid larvae complete their development either within the host (as endoparasitoids) or on the host (as ectoparasitoids), so that the act of finding hosts by a female also defines her reproduction (see below). The sheer abundance and diversity of parasitoids make them an important category of predator. They differ from other predators principally in not having free-living, trophic juvenile stages and in only requiring one host individual to complete development.

Essential ingredients of any predator–prey model are descriptions of the predator's functional and numerical responses (Solomon 1949; Holling 1959a, b). The functional response defines the changes in the *per capita* predation rate as prey density increases, while the numercial response defines the changes in predator density as prey density increases. The latter is most easily defined for insect parasitoids where the total number of hosts attacked by the parasitoid population translates directly to parasitoid progeny entering the next generation. This is the assumption made in most of the simple predator–prey models. For all other predators, however, an explicit expression for the numerical response is needed to determine the density of predators searching for prey at any one time (see 'Generalist Predators' below).

Coupled host–parasitoid interactions

Predator–prey models have, for the most part, involved predator populations that are coupled to their prey in a closed system. Specialist natural enemies of this kind are much more common amongst insect parasitoids than amongst other predators, and are the subject of this section.

Parasitoids can regulate their hosts at levels well below the carrying capacity of the environment for that host, as shown by the laboratory example in Fig. 6.1(a). This need not, of course, necessarily be the case, as shown in Fig. 6.1(b), involving the same species, but under different conditions (see below). Situations such as these prompt the obvious but important questions of (i) what determines the average (equilibrium) population levels and (ii) what factors promote the persistence of the interacting populations?

A framework commonly used for examining these questions in the context of host–parasitoid interactions is given by:

$$N_{t+1} = Fg(fN_t)\,N_t f(N_t, P_t) \qquad (1a)$$
$$P_{t+1} = cN_t[1 - f(N_t, P_t)]. \qquad (1b)$$

Here N and P are the host and parasitoid populations, respectively, within successive generations, t and $t + 1$; $Fg\,(fN_t)$ is the *per capita* net

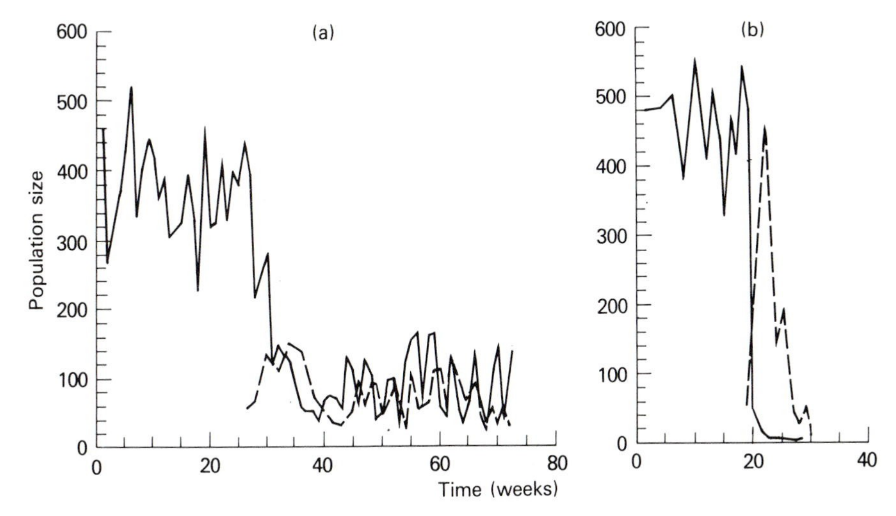

FIG. 6.1 Population dynamics of the bruchid beetle, *Callosobruchus chinensis* (————) feeding on black-eyed beans, and its pteromalid parasitoid, *Anisopteromalus calandrae* (— — — — —), in a laboratory system. (a) A 'patchy' environment with fifty beans, each in an individual container with restricted access to both hosts and parasitoids. (b) A 'non-patchy' environment with the fifty beans uniformly distributed on the floor of the arena. In both cases the parasitoids were introduced to the arena once the host population was fluctuating around its carrying capacity (V.A. Taylor & M.P. Hassell, unpublished).

rate of increase of the host population (which is density dependent when the fuction $g < 1$); c is the average number of adult female parasitoids emerging from each parasitized host (c therefore includes the average number of eggs laid per host parasitized, the survival of these progeny, and their sex ratio); and the function f defines the fraction of hosts that are not parasitized.

Comparable models framed as differential equations, and thus appropriate to populations with completely overlapping generations, have also been explored (e.g. Leslie & Gower 1960; Pielou 1969; Murdoch & Oaten 1975). We shall however, restrict ourselves here to interactions in which generations are effectively discrete and non-overlapping, as commonly occurs in temperate regions. The discrete generation framework also serves as a contrast to the host–pathogen models discussed below which are mainly framed in continuous time.

A feature of discrete-generation interactions is the different dynamics that can occur, depending upon the sequence of mortalities and reproduction in the host's life cycle (Wang & Gutierrez 1980; May *et al.* 1981; Hassell & May 1986). Equations (1*a*,*b*) describe the case where parasitism acts first, followed by the density dependence defined by g (see May *et al.* 1981 for a discussion of alternatives). In effect, therefore, the model represents a minimally complicated age-structured host population with pre- and post-parasitism stages. A fuller treatment of age-structured interactions is discussed below.

Within the framework of equations (1*a*, b), the degree to which the parasitoid population can reduce the average host population level (leaving aside whether or not this is a stable equilibrium) can be defined by the ratio q, of average host abundances with and without the parasitoid (i.e. $q = K/N^*$, where K is the carrying capacity for the host population in the absence of the parasitoid and N^* is the parasitoid-maintained equilibrium population). The magnitude of this depression depends upon the balance between:

(i) the host's net rate of increase [$Fg(fN_t)$ in equation (1*a*), and

(ii) the various factors affecting overall parasitoid performance contained within the function $f(N_t, P_t)$ and the term c. These include the *per capita* searching efficiency and maximum attack rate of adult females, the spatial distribution of parasitism in relation to that of the host (see below), and the sex ratio and survival of parasitoid progeny.

The original and most familiar version of the model in equations (1*a,b*) is that of Nicholson (1933) and Nicholson & Bailey (1935), where $c = g(N_t) = 1$ and f is given by

$$f = \exp(-aP_t). \tag{2}$$

Here a is the *per capita* searching efficiency of the parasitoids that sets the proportion of hosts encountered per parasitoid per unit time. The model thus assumes: (i) that each host is equally subject to attack (random search); (ii) that the parasitoids have a linear functional response; (iii) that each host parasitized produces one female progeny for the next generations (i.e. $c = 1$); and (iv) that the host population suffers no additional density dependent mortality due, for instance, to resource limitation [i.e. $g(f(N_t) = 1)$]. The model predicts expanding oscillations of host and parasitoid populations around an unstable equilibrium. The inclusion of a finite handling time, and thus a type II functional response, only makes this instability more acute (Hassell & May 1973).

Since such unstable interactions have only been observed from a few simple laboratory experiments [e.g. Fig. 6.1(b)], there has been much interest in factors that could be important in promoting the persistence of predator–prey interactions. These include sigmoid functional responses (Murdoch & Oaten 1975; Nunney 1980), mutual interference between searching predators (Hassell & Varley 1969; Hassell & May 1973), density-dependent sex ratios (Hassell, Waage & May 1983; Comins & Wellings 1985) and heterogeneity in the distribution of predation amongst the prey population (see below).

Of these, heterogeneity in levels of predation within the prey population has attracted the most attention, and is widely thought to be important to the dynamics of predator–prey interactions in general. It can arise from temporal asynchronies between prey and predators (Griffiths 1969; Hassell 1969), and from phenotypic variability in the susceptibility of individual prey to predation (Hassell & Anderson 1984). But it is most commonly discussed in the context of a patchy environment in which prey and predators are distributed.

There have been two approaches to considering the effects of such spatial patchiness.

(i) Models have been developed, often in continuous time, where patches can pass from being empty, to being colonized by prey, to being found by predators, to becoming empty again following the extinction of prey by the predators. In such systems asynchronies in the state of the different patches clearly have the potential to promote the persistence of the system as a whole (e.g. Maynard Smith 1974; Roff 1974; Hilborn 1975; Hastings 1977; Zeigler 1977; Crowley 1979; Chesson 1981).

(ii) Alternatively, there have been models in discrete time with explicit patches and predators, or more usually parasitoids, roaming over these, often with an emphasis on the details of the foraging strategy adopted by the natural enemy (e.g. Hassell & May 1973, 1974; Hassell 1978; Chesson & Murdoch 1986; Comins & Hassell 1987).

We now turn to models of this second kind, with the aim of emphasizing just how pervasive is the stabilizing effect of predation in a patchy environment.

Let us consider in the first place a habitat containing n patches amongst which an univoltine insect population is distributed. The adult insects in a given generation are the dispersing stage and each female lays her complement of eggs on the plants. The resulting larvae are hosts for a parasitoid whose population is coupled to that of the host. This now leads to the model discussed in detail in Hassell & May (1973, 1974), where $f(N_t, P_t)$ in equations (1*a,b*) is given by

$$f(N_t) = \sum_{i=1}^{n}\left[\alpha_i \exp\left(-\frac{a\beta_i P_t}{1+aT_h\alpha_i N_t}\right)\right]. \tag{3}$$

Here, N and P are the host and parasitoid populations respectively, in generations t and $t+1$, α_i is the fraction of hosts in the i th patch, β_i is the corresponding fraction for the parasitoids (such that $\Sigma\,\alpha_i = \Sigma\,\beta_i = 1$), a is the *per capita* searching efficiency and T_h is the handling time setting a maximum attack rate to the functional response. Thus the surviving hosts and parasitoids are redistributed amongst the n patches in each generation according to (α_i) and (β_i), and parasitism of hosts within a patch is defined by the Nicholson–Bailey term within equation (3). Any stability in the model, therefore, cannot stem from the within-patch functional response and must depend in some way on the spatial distribution of the populations. In short, stability is enhanced by a highly clumped host population distributed over many patches, by marked aggregation of parasitism amongst the host population, and by a low host rate of increase.

Central to this model is the distribution of parasitism in relation to that of the hosts. Originally, Hassell & May (1973) assumed an arbitrary host distribution α_i and searching parasitoids responding to this according to

$$\beta_i = c\alpha_i^{\mu}, \tag{4}$$

where μ is the aggregation index and c is a normalization constant. Thus, as μ increases from zero (an even distribution of parasitoids in all patches), the parasitoids increasingly aggregate in the patches of highest host density. If T_h is a small fraction of total time ($T_h << 1$), such aggregation will result in direct density-dependent spatial patterns of parasitism. However, as T_h increases, and thus the maximum attack rate within patches falls, there is an increased likelihood of inverse density dependence, even if the adult parasitoids are aggregating where host density is high (Hassell 1982; Lessells 1985). Both direct and inverse spatial patterns of parasitism have frequently been found in the field (Lessells 1985; Stiling 1987), and both promote stability within the framework of equations (1*a,b*) and (3) (Hassell 1984). Which has the greater effect, depends upon details of the host's distribution.

At the limit of $\mu = \infty$ all the parasitoids aggregate in the single patch of highest host density, leaving the remainder as complete refuges. A variety of models with such explicit prey refuges has been explored (Bailey, Nicholson & Williams 1962; Hassell & May 1973; Maynard Smith 1974; Hassell 1978). Here we consider the example where a fixed proportion $(1 - \Omega)$ of the host population in each generation is protected within a refuge, giving

$$N_{t+1} = f(1 - \Omega)\, N_t + F\, \Omega\, N_t f(N_t, P_t) \tag{5a}$$
$$P_{t+1} = \Omega\, N_t \{1 - f(N_t, P_t)\}. \tag{5b}$$

Assuming f given by equation (2), stability will be promoted provided

$$\Omega > 1 - \frac{1}{F} \tag{6}$$

leading either to a stable equilibrium or, if there are too few prey escaping predation, to stable limit cycles. If this condition is not met (i.e. too few susceptibles), the host population grows exponentially (until influenced by some other, unspecified constraints). The region of stability is broadened if there are a fixed *number* rather than proportion of prey within refuges in each generation. Refuge models of this general type apply to any situation where some host individuals are protected from parasitism. It may be, for example, that some hosts are immune from attack by virtue of their behaviour, physiology or relative timing (Hassell & Anderson 1984).

The models of spatial patchiness and foraging parasitoids described above have been further developed in two directions; either by moving towards greater detail and realism, or by striving for models of intermediate complexity that capture the main dynamic properties of their more complex cousins. More detailed models, for example, have included increasingly realistic distributions of both host and parasitoid populations (Hassell & May 1974; Comins & Hassell 1979). The constraint of complete host and parasitoid mixing and redistribution in each generation has also been removed, to permit interactions where some individuals tend to remain within the patch from which they originated, while the remainder behave as previously and disperse freely over the habitat (Hassell & May 1988; Reeve 1988). In this case there is now a continuum from complete to no mixing of both hosts and parasitoids. Let us assume for simplicity that dispersal from a patch is not density-dependent, and free from risks. The populations of hosts and adult parasitoids in the i th patch, N_i and P_i, respectively, are now given by

$$N_i(t+1) = F\left[S_i(1 - x_i) + \alpha_i\left\{\sum_{j=1}^{n} S_j x_j\right\}\right] \tag{7a}$$

$$P_i(t+1) = N_{ai}(1 - y_i) + \beta_i\left\{\sum_{j=1}^{n} N_{aj} y_j\right\}. \tag{7b}$$

Here x_i and y_i are the fractions of hosts and parasitoids, respectively, leaving the patch from which they emerged to enter a 'pool' for subsequent dispersal, and N_{ai} is the number of hosts parasitized and S_i is the number surviving from parasitism in the i th patch, derived from equation (2).

Interestingly, the heterogeneity in the distribution of parasitism now arises in part from the foraging behaviour of the parasitoids and in part from differential reproductive build-up of parasitoids within patches. This can introduce rich dynamics, which we illustrate here for one particular case where there is complete host mixing, but no parasitoid mixing. In addition, rather than assume that a patch of high host density in one generation remains so in the next, we assume that the location of the high host densities varies amongst the available patches from generation to generation. Parasitoids that disperse little, or not at all, are thus faced with a 'moving target' of high host density which they can only exploit heavily if by chance the same patch has a high host density for enough generations for the numerical response of the parasitoid to be effective. The examples in Fig. 6.2 show the populations fluctuating erratically, but persisting for long periods.

Alternatively, minimally complicated models can be developed that aim only to capture some of the dynamic properties of the more complicated ones (May 1978; Perry 1987). In May's (1978) model, for example, the distribution of parasitoid attacks amongst the host population follows the negative binomial distribution, rather than the Poisson as in equation (2). Thus the function f in equations (1a,b) is now given by the zero term of the negative binomial:

$$f = \left[1 + \frac{aP_t}{k(1 + aT_h N_t)}\right]^{-k} \tag{8}$$

where T_h is the handling time and k is the parameter expressing the degree of contagion in the distribution of parasitoid attacks (random when $k \rightarrow \infty$ and increasingly clumped as $k \rightarrow 0$). The model is stable if $k < 1$.

The precise relationship of this phenomenological model to more explicit models of patchily distributed hosts and aggregating parasitoids has been discussed by Chesson & Murdoch (1986). The negative binomial distribution of parasitism amongst hosts arises most readily when the frequency of adult parasitoids per patch follows a gamma distribution and is independent of the host distribution. In reality, the detailed outcome of a natural enemy's searching behaviour in response to a patchily distributed host population cannot usually be characterized so simply and more detailed models will be required (Hassell & May 1974; Chesson & Murdoch 1986; Perry & Taylor 1986; Karieva & Odell 1987). Equation (8) remains, however, a simple, albeit crude means of representing the non-random distributions of parasitism that can arise in many different ways within a host population.

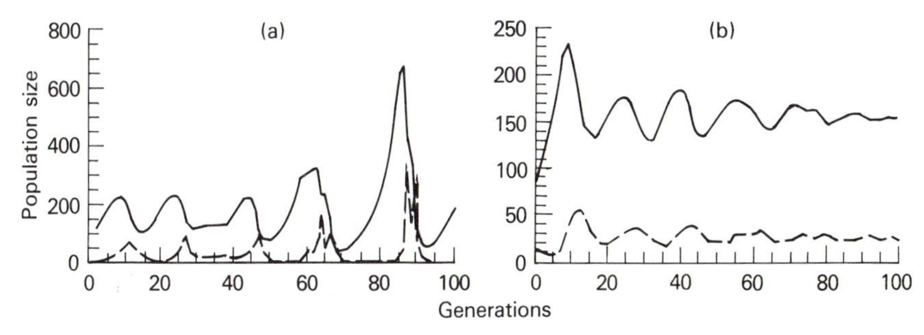

FIG. 6.2 Numerical examples from the model in equations (7*a,b*) with parasitism defined from equation (2). The host population (————) is completely redistributed over all patches in each generation, but the parasitoids (— — — — —) show little or no mixing. (a) All parasitoids remain in the patches from which they emerged. $x_i=1$, $y_i=0$, $n=5$, $a=0.01$ $F=2$ and $\alpha_i=0.5$; 0.1; 0.1; 0.1; 0.1. (b) Same parameters as (a) except $y_i=0.2$. (See Hassell & May 1988 for further details.)

An alternative, simple means of capturing the contribution of heterogeneity to stability is to examine the so-called 'apparent interference' or 'pseudointerference' relationship that is always associated with some hosts in the population being more susceptible to parasitism than others (Free, Beddington & Lawton 1977; Hassell 1978). These relationships between searching efficiency and parasitoid density arise because of the increasingly heavy exploitation of suceptible hosts as parasitoid density increases. As a result, the *per capita* searching efficiency measured over *all* hosts necessarily declines, and the steeper this decline, the greater is the contribution of the heterogeneity to stability. Looking for apparent interference relationships in this way is thus a convenient way of determining the dynamic contribution of *any* form of heterogeneity.

The coupled difference equation models discussed in this section are appropriate to many predator–prey systems, particularly those involving insect parasitoids. But even amongst these there are many natural situations where a proper understanding of the dynamics requires quite different kinds of population model. We illustrate this with two examples; first by including age-structure and permitting unequal host and parasitoid generation times, and second by uncoupling the predator population, as is appropriate for generalist predators whose population is unaffected by fluctuations in any one of their prey species.

Age-structured host–parasitoid interactions

The dynamics of age-structured host–parasitoid interactions have been explored by Auslander, Oster & Huffaker (1974), Godfray & Hassell

(1987, 1989), Murdoch *et al.* (1987) and Bellows & Hassell (1988), using either simulation models or systems of delayed differential equations. In this section, we illustrate briefly how age-structured host–parasitoid models can generate interesting dynamics that are absent from their simpler counterparts. In particular, we shall seek possible dynamic explanations for the patterns of fluctuations shown by the examples in Figs 6.3. and 6.4(a)

The example in Fig. 6.3 shows obvious cycles in host abundance of approximately one generation interval, despite the relatively aseasonal, tropical environment in which they occur. How this can result from the internal dynamics of a continuous host–parasitoid interaction (i.e. without environmental cues) has been discussed by Godfray & Hassell (1987, 1989). They envisaged a host population divided into pre-parasitism, susceptible, post-parasitism and reproductive adult stages, and a parasitoid with only two stages: immatures and searching adults.

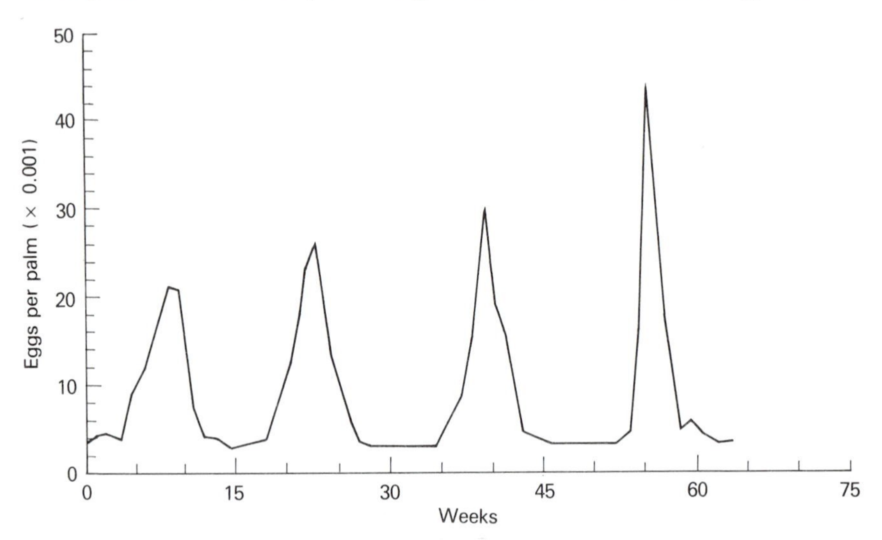

FIG. 6.3 An example of cycles of roughly one generation period in a tropical leaf-mining hispine beetle (*Coelaenomenodera minuta*), which is attacked by three species of enlophid parasitic wasps, on oil palm in West Africa. (From Godfray & Hassell 1989, after Mariau & Morin 1972.)

These four host and two parasitoid stages can be described by four equations giving the rate of change of numbers of susceptible hosts, adult hosts, adult parasitoids and the rate of change of survival through the susceptible stage. Further details and the properties of the model are fully described by Godfray & Hassell (1989). Of particular interest is the way that the internal dynamics of the interaction can result either in stable cycles of approximately one-generation period, or continuous generations with completely overlapping stages. Discrete generations are most likely when parasitoid generation times are either half, or one and a half, times those of their host. Parasitoids with very short, equal or double the generation times of their hosts are more likely to cause continuous generations. Interestingly, parasitoids with generation times approximately half those of their hosts are commonly found in tropical regions; they may, therefore, be an important cause of the discrete generation patterns recorded from hosts in the tropics (e.g. Fig. 6.3).

The example of fluctuations in host and parasitoid numbers shown in Fig. 6.4(a) comes from a laboratory host–parasitoid system (Utida 1950). One obvious interpretation of these irregular patterns of abundance is that they arise from some environmental stochasticity in the system. Alternatively, they could result deterministically from the internal dynamics of the interaction as illustrated by Bellows & Hassell (1988). Their simulation model is directly based on detailed studies of a laboratory system very similar to that of Utida (1950), using the same host species (*Callosobruchus maculatus;* Bruchidae) and a closely related parasitoid species (*Anisopteromalus puparum;* Pteromalidae). Their model is also age-structured with overlapping stages and unequal generation times. The predicted population dynamics [Fig. 6.4(b)] using parameters independently estimated in the laboratory, show not only a persisting host–parasitoid interaction, but also irregular, possibly 'chaotic' (May 1974; May & Oster 1976), fluctuations similar to those observed by Utida. Age-structured population dynamics clearly reveal interesting properties not apparent from much simpler, unstructured models.

Generalist predators

Forest lepidopteran populations are often relatively scarce for long-periods, interrupted by occasional spectacular outbreaks in abundance, quickly followed by an equally dramatic 'crash' in population size (Schwerdtfeger 1935). These characteristic dynamics are thought to result from generalist predators maintaining the prey population at a lower equilibrium from which it occasionally 'escapes'. Food limitation then causes the ensuing decline of the population to previous low levels (Southwood & Comins 1976; Hassell & May 1986; Hanski 1987). A good example of this is the spruce budworm, *Choristoneura fumiferana*, in Canada (Holling 1973; Ludwig, Jones & Holling 1978). The data suggest that, up to a certain budworm density, bird predation is density-dependent and sufficient to create a locally stable lower equilibrium. Above a threshold density, the population 'escapes' and rises towards a higher equilibrium determined by the availability of resources.

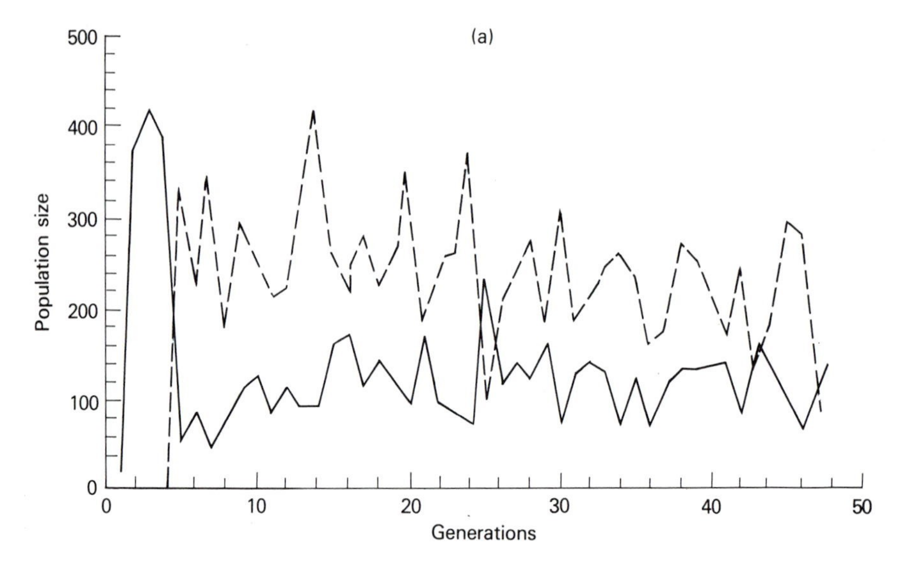

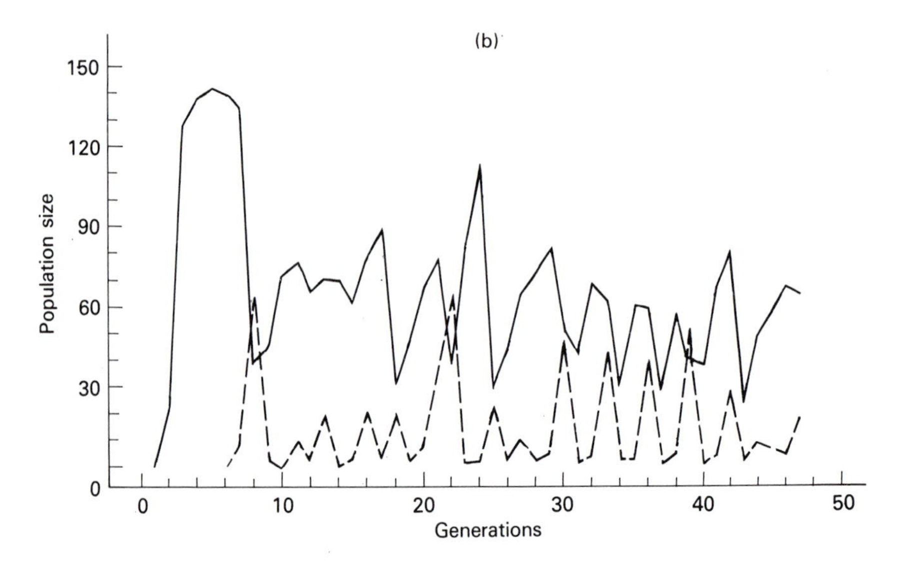

FIG. 6.4 The population dynamics of *Callosobruchus chinensis* (————) and its parasitoids (a) Observed interaction with the pteromalid parasitoid, *Anisopteromalus calandrae* (— — — — —) from the laboratory system of Utida (1950). (b) Predicted dynamics for the interaction with the pteromalid parasitoid, *Lariophagus distinguendus*, from the simulation model of Bellows & Hassell (1988).

This study has been the inspiration of several studies seeking similar interpretations. Unfortunately, the data purporting to show the necessary density-dependent predation have sometimes been less than convincing (e.g. Kiritani 1977; Rabinovich 1984), and examples of genuine alternative equilibrium states remain elusive. If the data are frail, the theory is certainly robust (May 1977). A number of explicit models with plausible parameters can produce multiple equilibrium points. These range from models for plant–herbivore interactions (Noy-Meir 1975), harvested populations (Clark 1976; Beddington & May 1977), competing species (Hassell & Comins 1975), host–pathogen interactions (Anderson & May 1979) and predator–prey systems (Southwood & Comins 1976; Hassell & May 1986).

In this section we present a simple framework for generalist predators, illustrating how they can maintain a prey population at a locally stable lower equilibrium. What scant field evidence exists, suggests that generalist predators often show relatively simple numerical responses to the density of a particular prey [Fig. 6.5(a)], such that they can be adquately described by the expression:

$$G_t = h\left\{1 - exp\left(-\frac{N_t}{b}\right)\right\}. \tag{9}$$

Here G_t are the number of predators searching in generation t for N_t prey, h is the saturation number of predators and b determines the typical prey density at which this maximum is approached (Southwood & Comins 1976). These responses can arise from the predators concentrating their search elsewhere or for different prey species when that prey is scarce, and increasingly tending to 'switch' from other diets or localities as the prey density rises (Murdoch 1969; Royama 1970).

The combination of such numerical responses with a type II functional response, gives predation that is directly density dependent, at least over a range of prey densities, as shown by the example in Fig. 6.5(b). A population model for the prey may now be written as

$$N_{t+1} = F\,N_t\,g(N_t), \tag{10}$$

where

$$g(N_t) = \exp -\left[\frac{aG_t}{1 + a\,T_H N_t}\right] \tag{11}$$

and G_t is given from equation (9). The map of population densities in successive generations is shown in Fig. 6.6. This shows that a generalist-maintained equilibrium (N^*) may occur if predation is sufficiently severe

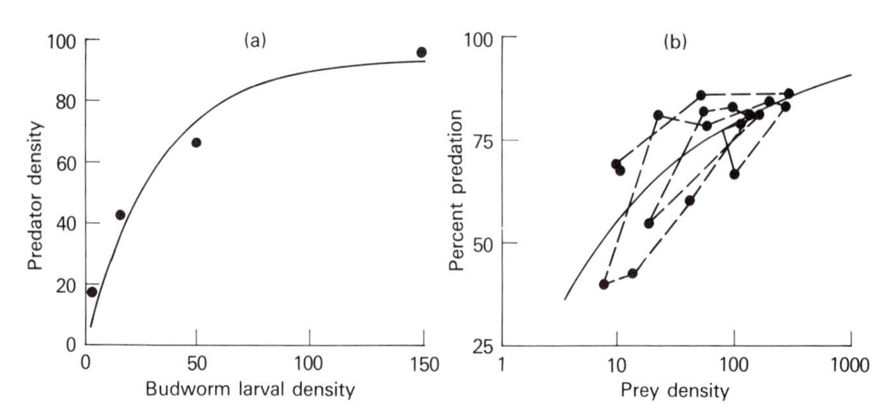

FIG. 6.5 Responses of generalist predators, (a) Numerical response of the bay-breasted warbler (*Dendroica fusca*) (nesting pairs per 100 acres) to third instar larvae of the spruce budworm (*Choristoneura fumiferana*) (numbers per 10 ft^2 of foliage) [From Hassell & May (1986); data from Mook (1963)]. (b) Density-dependent predation of pupae of the winter moth (*Operophtera brumata*). The fitted curve is from the relationship (k = 0.35 log density) expressed in terms of the k-values (Varley & Gradwell 1960) for pupal predation. The dotted line serially joins the points for eighteen generations, and the anticlockwise spiral emphasizes the slight time lag in the predators' response to prey density. (Data from Varley & Gradwell 1968.)

(defined by ah) in relation to the prey rate of increase, F. If $T_h = 0$, an equilibrium, if it occurs, is globally stable, but with $T_h > 0$ the prey population, if sufficiently large, eventually increases unchecked. Thus the prey population is only regulated provided it remains below some threshold value ($N < N_T$ in Fig. 6.6).

In short, by being uncoupled from their prey, generalists can cause direct density-dependent mortality which may be sufficient to regulate a host population. There are, therefore, none of the time delays inherent in coupled predator–prey interactions, and hence no tendency to produce typical host–parasitoid or prey–predator oscillations. Such generalists are likely to abound in the real world. However, in many cases, although feeding on a range of species, the abundance of a given host or prey may have some effect on the predator's reproductive success. Fig. 6.5(b) shows such an example where, superimposed on the overall density dependent predation relationship, is a clear indication of a time lag introduced by prey density affecting predator reproduction. Such examples fall into an intermediate category between the generalists discussed here and the specialists of the previous section.

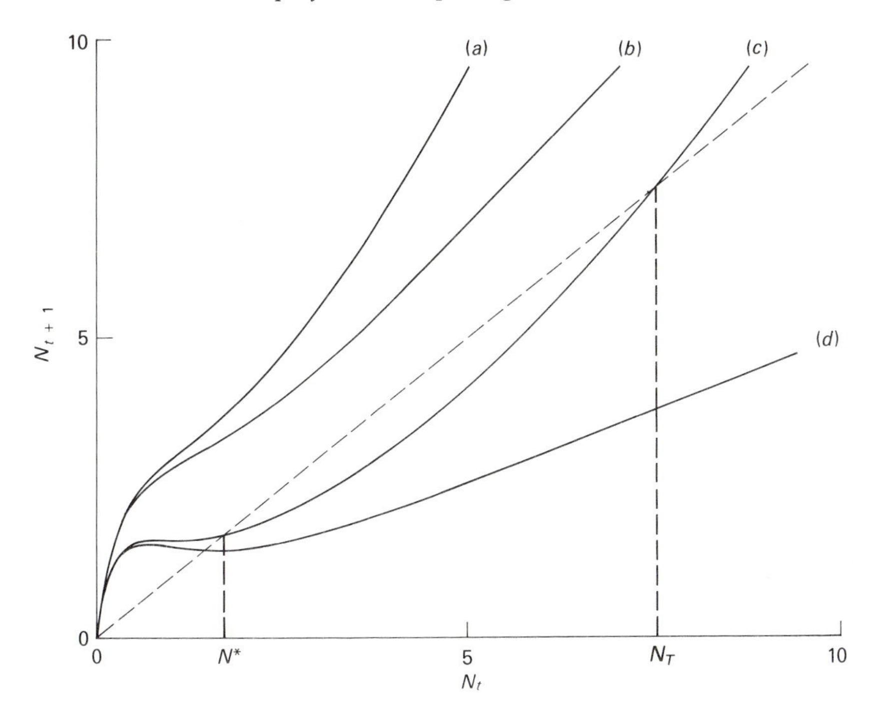

FIG. 6.6 Map of prey populations [scaled by dividing by b from equation (9)] in successive generations, t and $t+1$, obtained from the model in equations (10) and (11). Equilibria occur when the curves intersect the dotted 45° line. (a) $ah = 3$, $s = abT_h = 0.04$; (b) $ah = 3$, $s = 0$; (c) $ah = 2$, $s = 0.04$. A locally stable equilibrium occurs at N^* and an unstable, 'release' point at N_T. (d) $ah = 2$, $s = 0$. (From Hassell & May 1986.)

BASIC THEORETICAL FRAMEWORKS

2 Host–parasite interactions

Parasites, broadly defined to include viruses, bacteria, protozoans, helminths and arthropods, are increasingly thought to play a part analogous, or at least complementary, to that of predators, herbivores or resource limitation in influencing the population biology of plants and animals (Haldane 1949; Anderson 1979a; Anderson & May 1979; May & Anderson 1979; Michod & Levin 1988). The term 'parasite' encompasses a great diversity of infectious disease agents and life cycle structures that may be broadly classified as either microparasites or macroparasites.

This division is one of practical convenience, and in part derives from the unit of observation employed in ecological studies.

Microparasites (viruses, bacteria and Protozoa) are characterized by small size (relative to the host), short generation times, extremely high rates of reproduction within the host (thereby directly increasing population size) and a tendency to induce a degree of immunity to reinfection in those hosts that survive the initial onslaught. The duration of infection is typically short in relation to the expected lifespan of the host. There are, of course, exceptions, of which the human immunodeficiency virus (HIV), the etiological agent of AIDS, is a remarkable example. This virus persists for the life of the infected host since, once established, viral RNA is incorporated in the genome of host cells. Due to the small size of microparasites and the practical problems associated with detecting them and measuring their abundance within the host, the infected host serves as the basic unit of ecological study.

Macroparasites (helminths and arthropods) tend to have much longer generation times than microparasites, and multiplication rarely occurs within the definitive host. The immune responses and pathology that they elicit generally depend on the number, or burden, of parasites present in a given host. Macroparasitic infections tend to be of a persistent nature, with hosts being continually reinfected as the immunity elicited by infection is rarely fully protective. Due to the comparative ease (in contrast to microparasites) with which macroparasites can be counted within or on their host, the individual parasite normally forms the basic unit of ecological study.

Both types of organism may complete their life cycles by passing from one host to the next either directly or indirectly in one or more intermediate host or vector species. Transmission may be by contact between hosts or by free-living transmission stages of the parasite. The process is termed 'vertical transmission' when the infectious agent is conveyed by a parent to its unborn offspring, in contrast to 'horizontal transmission' involving contact between hosts, between host and vector, or with free-living infective stages.

In the following sections, we describe some simple models for directly transmitted micro- and macroparasites, using these to highlight the major processes that determine the dynamics of the system.

Microparasites

Models of the dynamics of host–microparasite interactions are typically compartmental in structure, where the host population is divided into a series of classes containing, for example, susceptible (= uninfected),

infected and immune individuals numbering $X(t)$, $Y(t)$ and $Z(t)$, respectively, at time t. The infected host is thus the unit of study and the total host population, $N(t)$ is simply, $X(t)+Y(t)+Z(t)$, at time t (Anderson & May 1979, 1981). The simplest set of differential equations that mimic temporal changes in $X(t)$, $Y(t)$ *and* $Z(t)$, are based on the assumption that in the absence of infection [$Y(t)=0$], the total population grows exponentially at a *per capita* rate r where $r = a - b$ and a and b are the *per capita* birth and death rates, respectively:

$$\frac{dX}{dt} = a(X + Y + Z) - bX - \beta XY + \gamma Z \tag{12a}$$

$$\frac{dY}{dt} = \beta XY - (\alpha + b + \sigma)Y \tag{12b}$$

$$\frac{dZ}{dt} = \sigma Y - (b + \gamma)Z. \tag{12c}$$

The equation for the total population is given by

$$\frac{dN}{dt} = rN - \alpha Y. \tag{12d}$$

This model assumes that the infectious agent is horizontally transmitted by direct contact (i.e. no vertical transmission) at a net rate proportional to the product of the density of susceptibles, times the density of infecteds (= infectious), scaled by a transmission coefficient β which measures the likelihood of contact and of transmission resulting from contact (βXY). The transmission coefficient is analogous to the parameter for searching efficiency in models of predator–prey and host–parasitoid interactions described above. Infected hosts are either killed by the disease at a rate α (without influencing host reproduction), or recover at a rate σ. Hosts are thus infected for an average duration of $1/(\alpha + b + \sigma)$ during which a fraction $(\alpha + b)/(\alpha + b + \sigma)$ dies, leaving a fraction $\sigma/(\alpha + b + \sigma)$ of those infected to recover. On recovery, hosts join an immune class and immunity is lost at a rate γ where $1/(b + \gamma)$ is the average duration of immunity to reinfection. Loss of immunity results in an individual rejoining the susceptible class X.

Equation (12b) shows that the introduction of a single infected individual into a susceptible population will only result in an increase in the number of infecteds, and thus the establishment of the infectious agent, provided $\beta X > (\gamma + b + \sigma)$. More generally, this can be written as

$$R_0 = \frac{\beta X}{\alpha + b + \sigma} > 1, \tag{13}$$

where R_0 is the basic reproductive rate of infection (equivalent to Fisher's (1930) net reproductive rate). The quantity R_0 defines the transmission potential of the infectious disease agent, and measures the average number of secondary cases of infection (the unit of study) generated by one primary case in a susceptible population of hosts. Alternatively we could express equation (13) in the form

$$X_T > \frac{\alpha + b + \sigma}{\beta}, \tag{14}$$

where X_T denotes the critical host density required to maintain R_0 above unity, the necessary condition for the infection to establish and persist within the host population. Note that the quantity R_0 is dimensionless since it is defined per generation time of the infection (Anderson 1981).

Equation (12*b*) introduces two related concepts that are central to an understanding of the population biology of host–parasite associations. The first is the definition of a basic reproductive rate, R_0, which measures the generation of secondary cases of infection in terms of a few parameters that denote the typical course of infection in an individual host and the spread between hosts. The second is the requirement that the density of susceptible hosts exceeds some critical value for parasite persistence and spread.

More complex life cycles result in more complex expressions for the basic reproductive rate and the threshold density of susceptibles. In brief, factors such as the involvement of vectors, vertical transmission and transmission via sexual contact between hosts tend to reduce the magnitude of the threshold criteria and hence enhance the likelihood of parasite persistence in low density host populations. Similary, R_0 will be increased by low pathogenicity linked with a long period during which infected hosts are infectious to others, and a high transmission efficiency (β large) (Anderson 1979b; Anderson & May 1979).

Regulation of host population growth. Directly transmitted infections are able to regulate host populations provided the mortality rate induced by infection, α is greater than the natural intrinsic growth rate of the host population r, weighted by a factor that takes account of host recovery and the duration of immunity. For the simple model defined by equations (12*a*) to (12*c*) the condition for regulation is

$$\alpha > r\left(1 + \frac{\sigma}{b + \gamma}\right). \tag{15}$$

This criterion is less easily satisfied if recovery from infection is rapid (i.e. σ large such that the average infectious period is short), and if immunity is long-lasting following recovery (i.e. γ small). Regulation is relatively easier if the infection reduces the reproductive capacity of infected hosts (i.e. the term aY in equation (12a) is replaced by afY where $0<f<1$). In the limit of no reproduction by infected hosts, regulation is always assured.

In all cases, regulation requires that the basic reproductive rate of the parasite exceeds unity. However, if $R_0 > 1$ but condition (15) is not satisfied, the host population will grow exponentially, but at a reduced rate p ($p < r$), defined in the case of equations (12a) to (12c) as

$$p \cong r - \alpha\left(1 + \frac{\sigma}{b + \gamma}\right). \qquad (16)$$

In these circumstances, the proportion of infected hosts (the prevalence of infection, Y/N) will settle to a constant value in the exponentially growing population (Anderson 1979a; Anderson, May & McLean 1988).

The simple models discussed above indicate that whether or not a microparasite is able to regulate host population growth and/or significantly depress the population growth rate (p) depends on a number of biological properties of the interaction. The most important are the reproductive rate of the host in the absence of infection, r, the pathogenicity of the parasite, α, the rate of recovery from infection, σ, and the duration of acquired immunity, $1/\gamma$. For short-lived host species, such as insects with high reproductive potentials, regulation will only result if the parasite is highly pathogenic such that few hosts recover from infection. Acquired immunity in such species is usually absent, so that recovery from infection is the major determinant of the regulatory impact of the infectious agent. In the case of vertebrate species however, with sophisticated immunological defences, the duration or efficacy of acquired immunity is typically the primary determinant of the impact of the pathogen on population size and growth.

For long-lived species with low population growth rates, the evolution of an effective immune system that enables infected individuals to recover and acquire resistance to reinfection, is central to survival in the face of continual exposure to a wide range of infectious diseases. The human species provides a good example of this point. Consider, for instance, the impact of a viral infection such as smallpox in a human population with a 4% growth rate per annum ($r = 0.04$) in the absence of the infection (typical of some developing countries such as Kenya). If we also assume that the average duration of infection $[1/(\alpha + b + \sigma)]$ is 2 weeks, that 50% of infecteds die from the disease induced by the virus, and that life expectancy ($1/b$) in the absence of infection is 50 years, then the magnitudes of γ and σ are 12.98 and 13.0, year^{-1}, respectively. The criterion defined for regulation in equation (15) is now not satisfied unless the duration of immunity is very short. This point is made more explicitly in Fig. 6.7 where the boundary between regulation and exponential growth of the population is plotted for various combinations of the percentage of infecteds who die and the average duration of acquired immunity in those who recover. For diseases of low to moderate pathogenicity, the duration of immunity following recovery must be short if the infection is to regulate human population growth. Retrospective analyses of this kind suggest that diseases such as smallpox and plague were able significantly to depress growth rates but were unable to regulate population size at an equilibrium level. In contrast however, the emerging pattern of the spread of HIV-1, the etiological agent of AIDS, suggests that this new disease may be able to turn human population growth rates negative in developing countries on time scales of a few

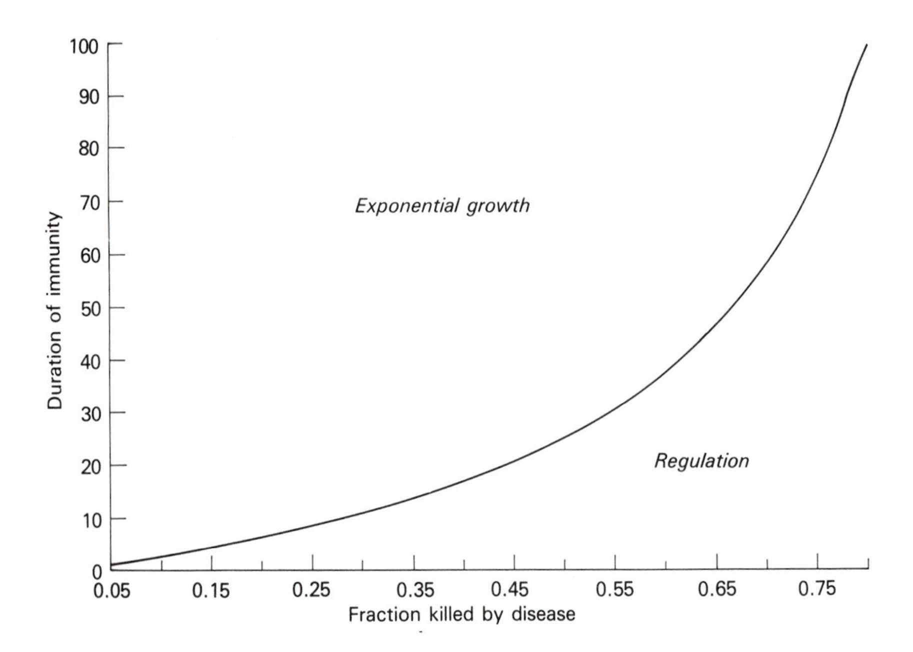

FIG. 6.7 Boundary between regulation of the host population to a stable equilibrium, and exponential growth, in the parameter space created by the duration of acquired immunity and the fraction of infecteds killed by the infectious disease. As predicted by the model, the boundary is defined in equations (12*a*) to (12*d*) with parameter values, $r = 0.04$ year^{-1}, $b = 0.02$ year^{-1}, $1/(\alpha+b+\sigma) = 2$ weeks, $1/y$ (= duration of immunity) and f variable. Note that the shorter the duration of immunity the more easily the pathogen can regulate host population growth.

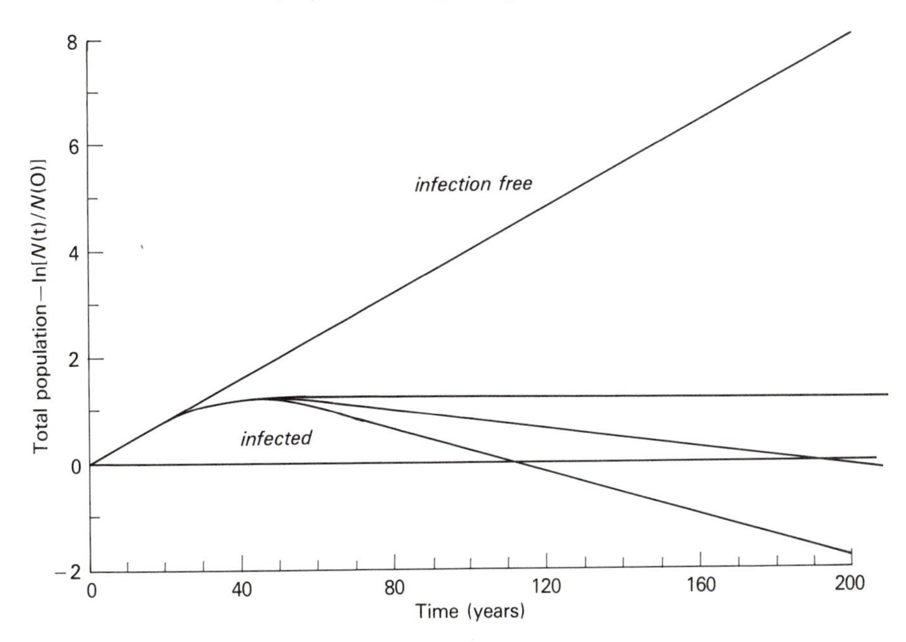

FIG. 6.8 Trajectories with time of human population growth in the presence and absence of the disease AIDS as predicted by a model that combines epidemiological and demographic processes (see Anderson, May & McLean 1988). The top line denotes 4% growth in an uninfected population. The remaining trajectories, from bottom to top, denote predictions where the fraction of babies born to infected mothers who die from HIV infection, is set at 0.3, 0.5 and 0.7, respectively. Note the prediction that AIDS is capable of turning positive human population growth rates to negative ones over a time scale of a few decades.

decades (Fig. 6.8). This stems from (i) the ability of HIV to transmit both horizontally via sexual contact and vertically from mother to unborn offspring, (ii) the high mortality associated with infection, and (iii) the apparently long period over which infected persons are asymptomatic but infectious to their sexual partners (Anderson, May & McLean 1988; May, Anderson & McLean 1988).

Macroparasites

In the case of macroparasites, the basic unit of study can be the individual parasite rather than the infected host, since direct counts of parasite burdens per host are often possible. Models can thus be based on a detailed description of the distribution of parasite numbers within the host population. The distribution of macroparasites is also important in that the mortality they cause, and the effectiveness of resistance to reinfection, typically depend on the burden of parasites harboured, and not simply on whether or not the host is infected (Anderson & May 1978; May & Anderson 1978).

The simplest models of directly transmitted macroparasites describe rates of change with respect to time in the number of hosts, $N(t)$, parasites, $P(t)$, and the infective stages, $W(t)$, that are responsible for transmission between hosts. For example:

$$\frac{dN}{dt} = (a - b)N - \alpha P \tag{17a}$$

$$\frac{dP}{dt} = \beta NW - (b + \mu)P - \alpha N E\{i^2\} \tag{17b}$$

$$\frac{dW}{dt} = \lambda P - (d + \beta N)W. \tag{17c}$$

Here a and b are the birth and death rates of the host and α is the pathogenicity of the parasite assuming that the death rate of a host infected with i parasites is αi. The parameter β denotes the transmission coefficient for contacts between hosts and infective stages and μ is the *per capita* death rate of the parasite. The term $\alpha NE\{i^2\}$ gives the net parasite losses resulting from parasite-induced host deaths (at a *per capita* rate αi). Finally, $E\{i^2\}$ is the mean-square number of parasites per host, the precise value of which depends on the form of the probability distribution of parasite numbers per host. Specifically,

$$E\{i^2\} = \sigma^2 + M^2, \tag{18}$$

where σ^2 and M are respectively, the variance and the mean of the distribution. Even if the precise form of the distribution is unknown, the mean-square term can be defined empirically from the relationship (e.g. a power function) between the variance and the mean (Taylor 1961).

Typically, parasites are highly aggregated within their host populations such that most hosts harbour few parasites and a few harbour many. The negative binomial probability distribution has proved to be a good empirical model of observed patterns (Crofton 1971; Anderson 1976). This distribution is defined by two parameters, the mean M and a parameter k which varies inversely with the degree of parasite aggregation (Pielou 1969). The mean square term of this distribution is given by

$$E\{i^2\} = M^2\left(\frac{k+1}{k}\right) + M. \tag{19}$$

For the Poisson distribution the aggregation parameter k tends to infinity such that the term $(k+1)/k$ in equation (18) becomes unity. More complex models of macroparasite transmission and distribution show that the simple phenomenological model of equations (17*a*) to (17*c*), in which the distribution is assumed fixed and unaffected by changes in host and parasite densities, is a good approximation of the more realistic situation in which the form of the distribution is itself a dynamic variable (Hadler & Dietz 1984).

In equation (17*c*) the parameters λ and d denote the *per capita* rates of production and mortality, respectively, of the infective stages, W. In practice, the life expectancy of these stages ($1/d$) is usually short compared with that of the host and adult parasites. It is therefore possible to uncouple the system and reduce the model to two equations. For the negative binomial assumption, this gives

$$\frac{dN}{dt} = rN - \alpha p \tag{20a}$$

$$\frac{dP}{dt} = \left[\frac{\lambda NP}{H_0 + N}\right] - (\mu + b + \alpha)P - \left[\frac{\alpha(k+1)P^2}{kN}\right], \tag{20b}$$

where $r = a - b$ and $H_0 = d/\beta$.

The basic reproductive rate, R_0, and the critical density for parasite establishment and persistence, N_T, from this model are defined by

$$R_0 = \frac{\lambda \beta N}{(\mu + b + \alpha)(d + \beta N)} \tag{21}$$

$$N_T > \frac{(\mu + b + \alpha)(d + \beta N)}{\beta \lambda}. \tag{22}$$

For macroparasites, R_0 should be interpreted as the average number of offspring produced by one reproductively mature parasite that themselves attain reproductive maturity in the host population in the absence of density-dependent constraints on parasite population growth. Note that R_0 is enhanced by high transmission efficiency (β large), large host populations (N large) and low rates of parasite and host mortality.

Regulation of host population growth. Like microparasites, macroparasites are able to regulate host population growth within certain parameter constraints. For the model defined above, the condition for regulation is

$$\lambda - (\mu + b + \alpha) > \frac{r(k+1)}{k}. \tag{23}$$

In other words, the parasite's effective rate of reproduction $\lambda - (\mu + b + \alpha)$ must be greater than the host population growth rate r, weighted by a factor $(k+1)/k$ to allow for the aggregation of the parasites within the host population. If equation (23) is not satisfied, the host population grows exponentially but at a reduced rate p, defined by

$$p = r - [\lambda - (\mu + b + \alpha)]\left(\frac{k}{(k+1)}\right). \tag{24}$$

As always, the parasite is maintained within the population provided $R_0 > 1$.

The similarities between the predictions of the microparasite and macroparasite models are striking. The concepts of reproductive success (R_0), critical densities of hosts for parasite establishment and persistence (N_T) and regulation determined by specific parameter combinations, emerge from both types of models. Both of the basic frameworks can be modified to incorporate much greater biological detail and increased life cycle complexity. The concepts outlined above, however, remain of central importance.

Dynamic behaviour

In considering the ability of infectious diseases to regulate their host populations, we have so far focused on the parameter combinations that permit locally stable equilibria. More generally, microparasites and macroparasites can also induce oscillations in host abundance, and in this respect mirror predator–prey interactions.

These cyclic changes in host and pathogen abundance are more likely to arise within microparasite–host associations. Several factors determine the likelihood that oscillations will occur, such as: (i) a short duration of infection relative to host life expectancy; (ii) long latent periods of infection when hosts are infected but not yet infectious; (iii) lasting immunity on recovery; (iv) high pathogenicity relative to the host's natural intrinsic growth rate, (v) long-lived infective stages for transmission between hosts; (vi) discrete non-overlapping generations of hosts; (vii) seasonal changes in transmission; and (viii) resource limitation constraining host population growth at high densities (Anderson 1979b; Anderson & May 1979, 1981, 1982; May 1982). Examples include the widely-observed oscillations in the incidences of childhood viral and bacterial infections in human populations, such as the 2-year cycle of measles prior to mass vaccination (Anderson, Grenfell & May 1984; Anderson & May 1985a), cycles in abundance of lepidopteran insects which may be driven by their granulosis and nucleopolyhedrosis viruses (Anderson & May 1980) and recurrent rabies epidemics in European Red Fox (*Vulpes vulpes*) populations that occur every 3–5 years (Anderson *et al.* 1981). Factors that reduce the likelihood of recurrent epidemics include a component of vertical transmission, long infectious periods relative to life expectancy and low pathogenicity.

Macroparasite–host associations are typically much more stable than microparasitic ones due to the persistent nature of the infections and their inability to induce lasting immunity to reinfection. However, in some circumstances they too may induce recurrent epidemics. These are promoted by the presence of long-lived, free-living infective stages coupled with high pathogenicity, and pathogenicity also linked to host density via the nutritional status of the hosts (May & Anderson 1978; Anderson 1979b, Hudson, Dobson & Newborn 1985). The conditions for oscillations to occur from the macroparasite model of equations (17*a*) to (17*c*) with the parasites distributed in a negative binomial pattern, are that $R_0 > 1$, that equation (23) is satisfied and

$$\frac{\omega}{d} < \frac{1}{k}, \tag{25}$$

where $\psi = \mu + \alpha + (a - b)(k + 1)/k$ (May & Anderson 1978). If condition (25) is met, both host and parasite populations oscillate in stable limit cycles. A study by Hudson, Dobson & Newborn (1985) of the red grouse (*Lagopus L. scoticus*) and a gut nematode parasite, *Trichostrongylus tenuis*, suggests that the observed cycles in grouse abundance in certain habitats (moors) may result from the impact of the nematode parasite. They also provide empirical evidence to support the prediction of equation (25) that cycles are less likely to arise when the parasites are highly aggregated within the bird-host population (k small).

Heterogeneity

Heterogeneity in host exposure, susceptibility and resistance to infection is a pervasive force in determining pattern and dynamics in host–parasite associations. In this section we highlight a few areas of particular interest where heterogeneity has a profound influence on the ecology and epidemiology of infectious diseases.

Distribution of infection and disease. The observed aggregation of macroparasites within their host population can result from many factors—genetic, behavioural, social, spatial, temporal and nutritional —either acting alone, or concomitantly (Anderson 1982). This is well illustrated by the simple macroparsite model defined by equations (17*a*) to (17*c*). The model contains a single density-dependent term, $\alpha HE\{i^2\}$, representing the action of parasite-induced host mortalities on the mortality of the total parasite population. The distribution of parasites per host was assumed to be negative binomial with aggregation parameter k (Fig. 6.9), which resulted in a stable equilibrium if $R_0 > 1$, provided equation (23) was satisfied. However, had the parasite distribution been random (Poisson) or more regularly dispersed (positive binomial), an unstable equilibrium would have occurred (Anderson & May 1978). Thus, as in the predator–prey interactions above, aggregation is acting as a stabilizing process. The underlying mechanism stems from the combination of parasite aggregation and the positive relationship between host death rate and parasite burden. This results in the death of a few hosts (those with heavy burdens) and in turn to the loss of a large number of parasites. The parasite is thus prevented from having too great an impact on the host population, which is analogous to the apparent interference effect in predator–prey systems discussed on p. 157.

More broadly, it is interesting to consider what the major processes are that generate aggregation. Host genetics and spatial distribution are

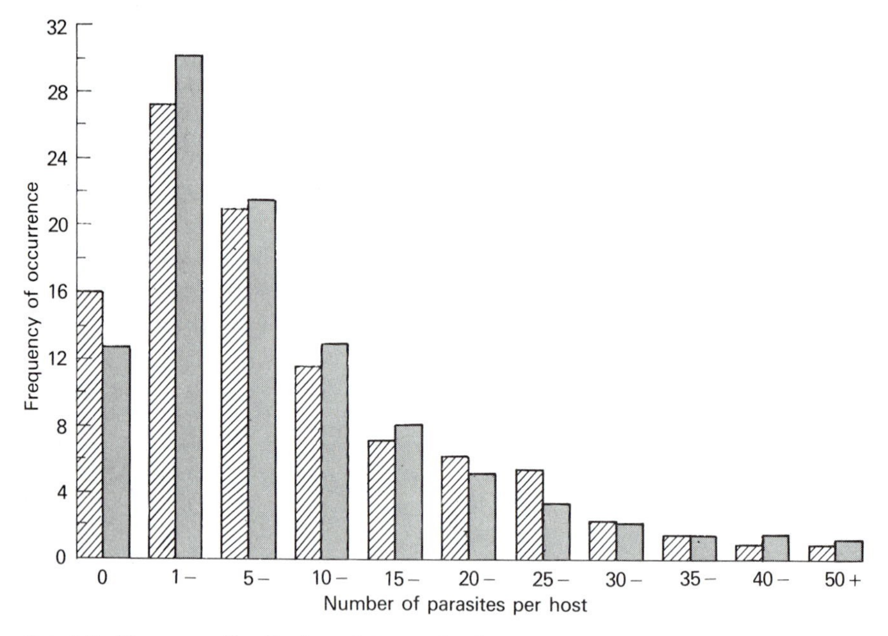

FIG. 6.9 Frequency distribution of *Ascaris lumbricoides*, the gut nematode of humans, in a sample of patients from a fishing village in southern India (data from Elkins, Haswell-Elkins & Anderson 1986). The observed frequencies and the predictions of the best fit negative binomial distribution ($k = 0.805$) (cross-hatched bars), are recorded.

both likely to be important in natural habitats. For instance, studies on helminth parasites of humans suggest that the 'wormy' people with high parasite burdens are predisposed to this state (Schad & Anderson 1985; Anderson 1986*a,b*). Following chemotherapeutic treatment to expel existing worm loads, those with heavy worm burdens re-acquire heavy infection following a period of further exposure (Elkins, Haswell-Elkins & Anderson 1986). Predisposition is thought to arise from a combination of genetic, spatial and behavioural factors (Bundy 1988).

Laboratory studies of helminth infection in rodents show a clear genetic component in susceptibility/resistance to infection. A simple illustration of how genetic background or spatial factors can influence aggregation is provided by the macroparasite model of equations (17*a*) to (17*c*). Suppose a proportion p of the host population is refractory to infection due to genotype or spatial location. If the parasites are randomly distributed in the susceptible fraction $(1 - p)$ of the population, then the variance V, of the parasite's distribution in the total population is

$$V = M + M^2\left(\frac{p}{1-p}\right). \tag{26}$$

In other words the variance exceeds the mean (M) and the parasites are aggregated in their distribution (Anderson 1982). This example thus parallels the predator–prey refuge model on p. 154. Similar, but more complex, examples of the generation of aggregation by the genetic control of host immunological defences in vertebrate–macroparasite associations are described in Anderson (1986a). In the case of human infections, for example, genetic variability in the ability to acquire immunity as a consequence of past exposure to infection can generate complex patterns of change in the average parasite burden per person with host age. Convex changes in worm load where average burdens change from low levels in infants to high levels in teenagers and back to low levels in adults may arise, not from age-related changes in exposure to infection, but from genetic heterogeneity in the ability to acquire immunity (Anderson & May 1985b).

Examples of the stabilizing effect of heterogeneity can also be found amongst microparasite–host associations. In the case of many bacterial infections of humans, such as typhoid and tuberculosis, certain individuals may be infected and highly infectious for long periods of time, but show no overt symptoms of disease. Genetic factors are thought to play an important role in determining whether or not the infection persists in a given individual. The symptomless, persistent carriers of infection can play an extremely important role in the endemic maintenance of an infection within a community, even if they constitute only a small fraction of those infected. This is apparent from the definition of the basic reproductive rate R_0 in equation (24) where the average duration of infection is defined as $1/(a+b+\sigma)$. Long durations of infection can result in high reproductive potential. More generally, if a fraction, f, of the susceptible population passes into the carrier state once infected, the basic reproductive rate for carriers and non-carriers combined is

$$R_0 = R_{01}(1 - f) + R_{02}f. \tag{27}$$

Here R_{01} is the basic reproductive rate in the absence of carriers and R_{02} is the rate if all the population are carriers. Clearly, the overall R_0 may be maintained above unity, even when non-carriers constitute the majority of the population and their R_{01} is less than unity, provided $R_{02} f >> 1$. A good example of the importance of carriers is provided by the persistence of bovine tuberculosis (*Mycobacterium bovis*) in populations of the European badger (*Meles meles*) (Anderson & Trewhella 1985). Apart from enhancing persistence, the presence of carriers also tends to damp oscillations arising in microparasite–host interactions. This, therefore, reduces the likelihood of stochastic 'fade out' of the infection during the troughs in the density of susceptible hosts that occur when the amplitude of recurrent epidemic cycles is large.

Host behaviour. Some of the clearest examples of the significance of heterogeneity generated by variability in host behaviour, occur in the transmission of childhood viral and bacterial infections, and sexually transmitted diseases.

In the case of childhood infections, the rate at which individuals acquire infection [the term βY in equation (12b)], often referred to as the force of infection, is age and time dependent. This force of infection in age class a at time t is defined as

$$\lambda(a,t) = \int_0^{\infty} \beta(a',a)\, Y(a',t)\, da', \tag{28}$$

where $Y(a',t)$ is the number of infectious individuals at age a' at time t. The transmission coefficient $\beta(a',a)$ records the rate at which contacts occur between susceptible and infectious persons of various ages, and the rate at which such contacts result in new cases of infection. The *per capita* rate, $\lambda(a,t)$ is therefore dependent on differing degrees of contact with infectious persons in their own and all other age classes. If heterogeneity in contact between age classes due to behavioural factors is assumed to be the dominant influence on transmission, then the transmission function, $\beta(a', a)$ defines 'who mixes with whom' (Anderson & May 1984). An illustration of observed age-dependent changes in the force of infection with the measles virus of humans is shown in Fig. 6.10. The rate changes from a low level in infants to a high level in children and teenagers and back to a low level in adults. This pattern is thought to reflect behavioural differences in the frequency of intimate contact between age classes. The significance of such age-dependent heterogeneity lies in the increased or decreased tendency (compared with homogeneous mixing) for the prevalence of infection to oscillate in recurrent epidemic cycles. This depends on the precise structure of the 'who mixes with whom' function and the magnitudes of the age-dependent forces of infection (Anderson & May 1985b).

Heterogeneity in mixing within the host population is perhaps of greatest significance in the study of the transmission dynamics of sexually transmitted infections such as gonorrhea or the human immunodeficiency virus (HIV-1) (Anderson *et al.* 1986; May & Anderson 1987). Human sexual behaviour varies greatly both between individuals in a

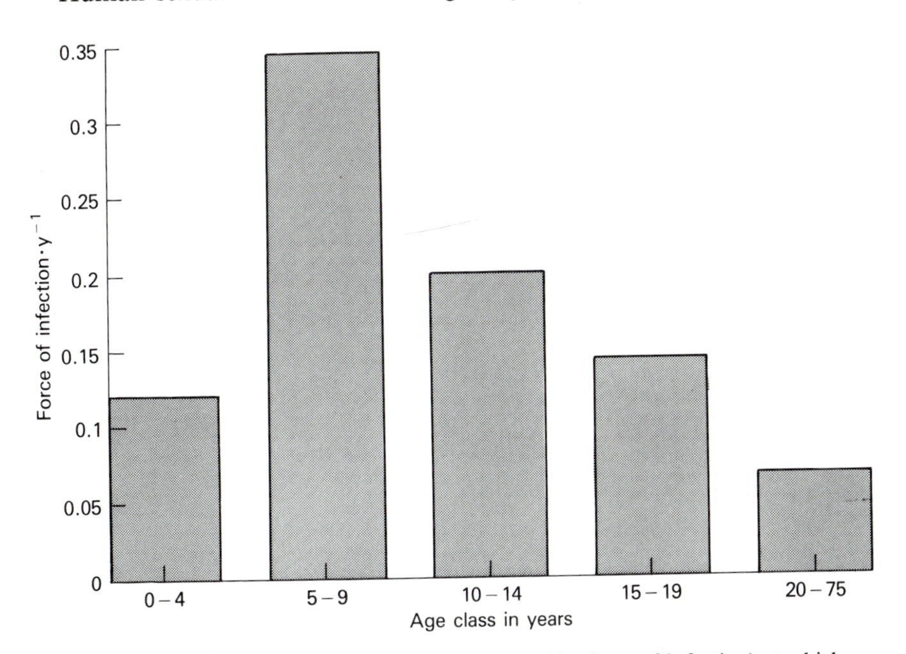

FIG. 6.10 Changes with age in the *per capita* rate (the force of infection) at which susceptible individuals acquire measles virus infection in developed countries (USA, England and Wales). [Data from Anderson & May (1985b).]

given community and between communities or populations. One quantitative measure of such variability is the rate per unit time at which individuals acquire new sexual partners. Figure 6.11 records two distributions of this rate, one from a male homosexual population and the other from a heterosexual population of males and females. The variances of both these distributions exceeds their respective means (Anderson 1988)–most individuals have few partners and a few have many partners.

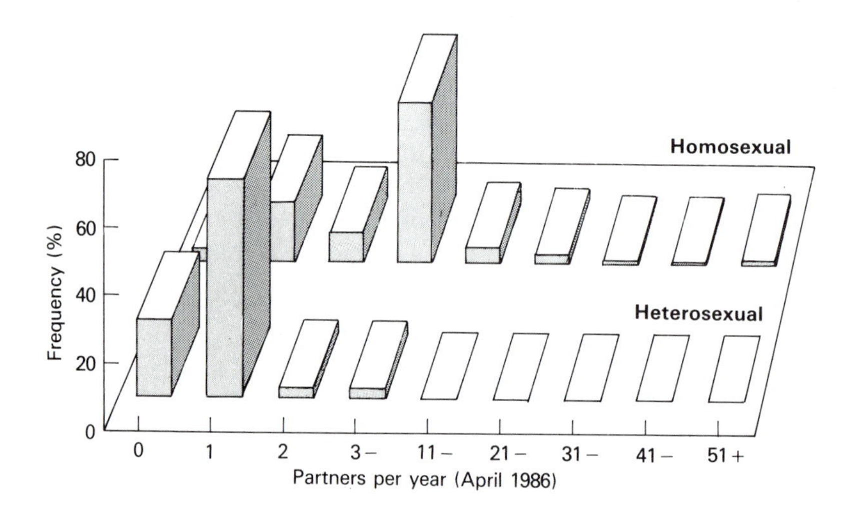

FIG. 6.11 Frequency distribution of the number of claimed sexual partners per year in samples of male homosexuals and female and male heterosexuals interviewed in England in April 1986 (see Anderson 1988). Note the marked heterogeneity apparent in each distribution.

The importance of this heterogeneity to the spread and persistence of a sexually transmitted infection, such as HIV, is best illustrated by reference to the parameters that determine the magnitude of the basic reproductive rate, R_0, defined in this case as

$$R_0 = \beta c D, \qquad (29)$$

where β defines the transmission probability per partner contact, c is the effective average number of partners per unit of time and D is the average duration of infectiousness of an infected person. Assuming homogeneous mixing, the effective average c is simply the mean number of different sexual partners per unit of time, m. With heterogeneous mixing, where few individuals have many partners and many have few, c is approximately defined by

$$c = m + \frac{\sigma^2}{m} \tag{30}$$

where σ^2 is the variance of the distribution of partner changes (Anderson *et al.* 1986; May & Anderson 1987, 1988; Anderson & May 1988). Current estimates (admittedly rather crude) for β and D for HIV infection are 0.1 and 8 years, respectively. With a mean rate of partner change of 1 per year, the assumption of homogeneous mixing would result in the estimate of R_0 being less than unity, and the infection would thus be unable to spread and persist. However if $m = 1$, but the variance is a factor 10 greater, due to a small proportion of highly sexually active individuals, the magnitude of R_0 is above unity. The conclusion from this over-simplified example is that a small fraction of highly active individuals can maintain the infection within the general population, despite the fact that average behaviour is insufficient to maintain transmission.

Parasite genetics. Genetic variability between parasites, like heterogeneity in the host population, can have important effects on the dynamics of host–parasite interactions. In general, there has been rather little attention paid to combining models of population dynamics and genetics (Roughgarden 1979; May & Anderson 1983; Beck 1984). It is clear, however, that non-linearities in the dynamics also generate frequency and density-dependent selective pressures in models of gene frequency changes. Such dependencies can result in more than one stable state in a given system, where initial conditions determine the population and gene frequency trajectories through time, together with complex oscillatory changes in the population densities and gene frequencies (Beck 1984).

An example that combines epidemiological and genetic processes in models of host–pathogen interactions is given by May & Anderson (1983). They considered a host population with discrete, non-overlapping generations, and focused on a single locus with two alleles, A and a, which determines host susceptibility to infection by a given strain of parasite. For a diploid host, the relationship between the frequency of allele A in generation $t(P_t)$ and the frequency of P_{t+1} in the next generation is given by

$$P_{t+1} = P_t (P_t W_{AA} + q_t W_{Aa}) W_t . \tag{31}$$

Here q_t is the frequency of allele a, and W_{AA} and W_{Aa} are the fitnesses of the host genotypes AA and Aa, respectively. The average host fitness in generation t, W_t, is

$$\bar{W}_t = P_t^2 W_{AA} + 2P_t q_t W_{Aa} + q_t^2 W_{aa}. \tag{32}$$

The fitness of each genotype depends both upon the relative frequencies of the various genotypes and on the total number of hosts. By making specific assumptions about the form of the fitness function, we can explore the effects of frequency-dependent and density-dependent selection upon the evolution of the association. A lucid account of analogous studies of predator–prey systems, where the fitness functions are usually constructed in an *ad hoc* way, is given by Roughgarden (1979). For host–pathogen interactions, however, it is possible to derive these functions based on conventional epidemiological considerations. For example, the single mass-action model of microparasite transmission first described by Kermack & McKendrick (1927) [(see equations (12*a*) to (12*c*)] yields a simple result for the fraction of a host population infected I, from the course of a single epidemic through a given host generation. For host genotype i, the fraction infected of type i, I_i, is given by

$$1 - I_t = \exp\left(-\frac{I_i N_i}{N_{T,i}}\right). \tag{33}$$

Here N_i is the initial population density of hosts of genotype i, and $_{T,i}$ is the familiar threshold density of susceptibles for establishment of the infection in the population of genotype i (Gillespie 1975; May & Anderson 1983). Note that equation (33) is derived from a simple epidemiological model of changes in the densities of infected and susceptible hosts over the course of single epidemic. It provides a measure of the fitness of a given host genotype under the selective pressure imposed by one or more strains of pathogen. The changes in host density from one generation to the next are given by

$$N_{t+1} = \sum_i N_{t,i} W_{t,i} = N_t \bar{W}_{t,} \tag{34}$$

where $N_{t,i}$ is the density of genotype i at time t and W_t is the average fitness [see equation (32)].

Equations (31) to (34) provide a combined population genetic and dynamic model incorporating frequency- and density-dependent selection. As such, the model describes the changes in host and pathogen abundance and in the frequency of the different genetic types (of host or pathogen) within the overall population. Models of this type can exhibit very complex (sometimes chaotic) patterns of fluctuation resulting from the non-linearities introduced both by the frequency and density-dependence.

MULTISPECIES INTERACTIONS

The emphasis in population ecology has largely centred on the dynamics of single- or two-species systems, divorced from the more complex webs of multispecies interactions of which they are an integral part. Increasingly, however, this emphasis is shifting towards studies on more complex systems with the ultimate aim of determining how population dynamics can influence the structure of simple communities (e.g. Pimm & Lawton 1977, 1978; Lawton & Pimm 1978; Pimm 1982). To achieve this end, it is clearly important in the first place to determine how the dynamics of two interacting species are influenced by the additional linkages typically found with other species in the food web. With this in mind, the dynamics of a wide range of different three-species systems have been examined. These include a natural enemy species (predator or pathogen) attacking competing prey or host species (e.g. Roughgarden & Feldman 1975; Comins & Hassell 1976; Anderson & May 1986); competing natural enemy species sharing a common prey or host species (e.g. May & Hassell 1981; Kakehashi, Suzuki & Iwasa 1984; Anderson & May 1986; Hochberg, Hassell & May 1989); a prey species attacked by both generalist and specialist natural enemies (Hassell & May 1985), and various interactions of three trophic levels (Beddington & Hammond 1977; May & Hassell 1981; Anderson & May 1988). Interestingly, in several of these systems the dynamics are not just the expected blend of the component two-species interactions. Rather, the additional non-linearities introduced by the third species lead to quite unexpected dynamic properties. For instance, Hassell & May (1986) showed how the interaction of a generalist predator and specialist parasitoid attacking a common prey can lead to complex alternative stable states, a quite unexpected property on the basis of the dynamics of the different two-species interactions.

We now concentrate on a further two examples to illustrate how the dynamic possibilities broaden on moving from two- to three-species interactions.

A pathogen–predator–prey interaction

In this section the basic Lotka–Volterra predator–prey model with resource limited prey [a continuous time analogue of equations (1*a*,*b*)] is modified to include an invading pathogen which attacks the predator, giving

$$\frac{dN}{dt} = rN - cN[X + (1 - f)] - \gamma N^2 \tag{35a}$$

$$\frac{dX}{dt} = \alpha NX - bX - \beta XY \tag{35b}$$

$$\frac{dY}{dt} = \beta XY - (b + d)Y \tag{35c}$$

$$\frac{dP}{dt} = \alpha NX - bP - dY . \tag{35d}$$

Here X and Y define the densities of susceptible and infected predators, respectively, while P is the total density of predators ($P = X + Y$). The rate of change in the size of the prey population, N, with respect to time is determined by the intrinsic growth rate, r, the severity of intraspecific competition for resources, γ, and the rate of prey capture by susceptible, c, and infected, $c(1 - f)$, predators. Infected predators are assumed to be less efficient at capturing prey items than their uninfected counterparts (Anderson & May 1988). The predators increase at a rate αN dependent on prey captures, die at a rate b, and become infected at a rate βY. Infected predators are assumed to be unable to reproduce and have an additional death rate, d, over and above that acting on uninfected animals.

This simple three-species model has a range of possible dynamic behaviour. In the absence of the predator and pathogen, the prey population settles to a stable predator-free equilibrium, K, where $K = r/\gamma$. In the presence of the predator, but with the pathogen absent, there is a predator–prey equilibrium (P^*, N^*) where

$$N^* = b/\alpha, \; P^* = \left[\frac{r\alpha - \gamma b}{\alpha c}\right]. \tag{36}$$

The predator is only able to invade and persist provided that the predator-free equilibrium K, exceeds a critical density N_T (see below).

The pathogen is able to invade the predator–prey interaction and persist provided that the density of susceptible predators necessary for pathogen persistence, X_T (where $X_T = (d + b)/\beta$), is less than P^* [given in equation (36)]. In short, the coexistence of all three species requires

$$\frac{r\alpha - \gamma b}{\alpha c} > \frac{d + p}{\beta}. \tag{37}$$

If this condition is satisfied, the equilibrium still need not be locally stable, in which case limit cycles will occur. Thus, the invasion of a pathogen can transform a stable predator–prey interaction into one showing oscillatory fluctuations.

Parasitoids and pathogens attacking a common host

This example differs from that above in specifying insect hosts and parasitoids, rather than predators in general, and in the pathogen now competing with the parasitoids for the shared host species [see Hochberg, Hassell & May (1989) for full details]. The hosts and parasitoids are assumed to have discrete, synchronized generations, and the pathogen (virus, bacterium or protozoan) to be directly transmitted between hosts by a free-living infective stage released into the environment on the death of the infected host. Defining the densities of hosts (infected and uninfected) at time t as N_t, of infective stages as W_t, and of parasitoids as P_t and the fraction of the host population killed by the pathogen in a single generation as I_t, leads to the model

$$N_{t+1} = FN_t(1 - I_t)f(P_t) \tag{38a}$$

$$W_{t+1} = \frac{g\lambda}{\beta}\ln\left[\frac{1}{1 - I_t}\right] \tag{38b}$$

$$P_{t+1} = cN_t(1 - \phi I_t)[1 - f(P_{t)}] \tag{38c}$$

$$1 - I_t = \exp\left[-\frac{1}{\mathrm{N}_T}[(N_t\, I_t\,\{f(P_t) + \phi(1 - f(P_t))\}) + \left\{\frac{1}{\theta - q}W_t\right\}\right]. \tag{38d}$$

Here F, c and $f(Pt)$ are as defined in equations (1*a*,*b*) and (8). The parameter β is the usual transmission coefficient for the rate of infection of susceptible hosts, λ is the rate at which the free-living infective stages of the pathogen are transferred in a 'soil' reservoir of long-lived infective states [e.g. viral particles of the nucleopolyhedrosis viruses of lepidopteran insects (Anderson & May 1981)] and ϕ measures the dominance of the pathogen over the parasitoid when they co-occur within the same host individual. The immature parasitoid outcompetes the pathogen when $\phi = 0$ and vice versa when $\phi = 1$. Finally, q denotes the average number of infective stages required to infect a susceptible host (for viral particles on a leaf surface it is the number consumed by the insect host), g is the fraction of the reservoir of infective stages that becomes accessible to the next generation of hosts and N_T is the critical density of susceptible hosts required to initiate an epidemic of the pathogen in the absence of long-lived stages and for $\phi = 1$ [i.e. X_T of equation (14)].

The model is complex and analysed in detail in Hochberg, Hassell & May (1989). It combines the continuous production of infective stages and the continuous mortality of infected hosts within a single generation, with the discrete, generation-to-generation processes linked with host reproduction and attack by parasitoids. A range of conditions defines whether the pathogen or the parasitoid is able to invade an interaction of the other two species, or whether the three species are unable to coexist together. Parasitoid invasion requires that their dominance (ϕ) within co-infected hosts is high, and that the product of their survival and searching efficiency is also high relative to pathogen survival and transmission efficiency. More formally, it requires

$$acN_T(1 - r) > \left[\frac{F - 1}{\ln(F)\{F - \phi(F - 1)\}}\right]. \tag{39}$$

Here a is the *per capita* searching efficiency of the parasitoids and $r = g\lambda/(\lambda + \mu)$ where μ is the death rate of the infective stages of the pathogens.

For the pathogen to invade an established host–parasitoid interaction requires

$$[acN_T(1 - r)]^{-1} > \left[\frac{F - 1}{k(F^{1/k} - 1)\{1 + \phi(F - 1)\}}\right], \tag{40}$$

where k is the aggregation parameter for the distribution of parasitoid attacks on hosts. If both inequalities (39) and (40) hold, then either natural enemy can invade the host population. Interestingly, the addition of the third coexisting species may, or may not, further depress host abundance below the level attained in the absence of either enemy, which sounds a warning note for the use of such mixed natural enemy complexes in biological control programmes.

As in the other three-species examples discussed above, this model also highlights complexities that can arise on moving from two- to three-species interactions. Thus, the addition of the third species may either stabilize the existing two-species interaction, or it may induce oscillatory, or even chaotic, fluctuations depending on the precise combination of parameter values (Fig. 6.12).

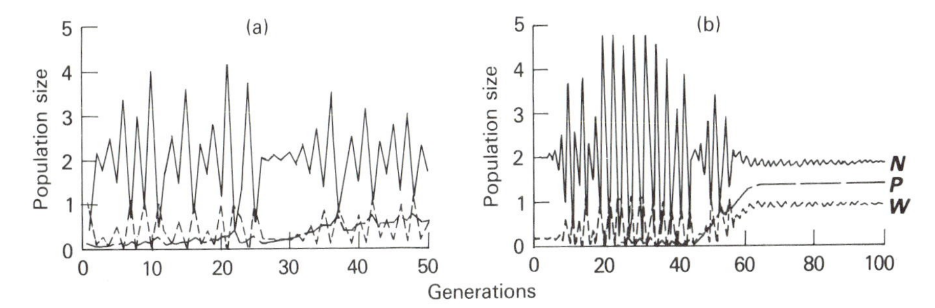

FIG. 6.12 Changes in host (N), pathogen (W) and parasitoid (P) densities with time governed by equations (38*a,d*). (a) Chaotic dynamics resulting from both single-species equilibria being unstable. $F = 5$, $g = 0$, $\lambda = 0.01$, $\mu = 0.01$, $\beta = 0.05$, $a = 1$, $k = \infty$, $\theta = 0.4$, $\phi = 0.5$, $d = 0$. (b) Constant equilibrium trajectories resulting from the introduction of the parasitoid in the twenty-fifth generation even though both single-species equilibria are unstable. Parameters as in (a) except $a = 1.1$, $\phi = 0.45$. (From Hochberg, Hassell & May 1989.)

DISCUSSION

Predators, parasitoids and parasites differ in a number of obvious ways summarized in Table 6.1. Despite such contrasts, there are also close parallels to be drawn between the different interactions and some of their fundamental parameters. Thus the probability of a predator or parasitoid locating a prey or host individual is given by its *per capita* searching efficiency [a in equation (2)], which is analogous to the transmission coefficient [β in equation (12)] that determines in part the likelihood of contact between pathogen and susceptible host.

The kinds of dynamic patterns generated by predators, parasitoids, and parasites are also similar. They all tend to depress the abundance of their prey or hosts, to an extent that depends on the degree to which they reduce survival and/or reproductive rates. The reduced populations can then persist around a stable equilibrium or may show cyclic behaviour. In general, neither predators, parasitoids nor pathogens can depress their prey or host populations to the point of eradication. Thus, for pathogens there is a critical density of susceptible hosts above which the disease can become established in the host population, persist and spread. This occurs when $R_0 > 1$, where R_0 is the basic reproductive rate [see

TABLE 6.1. Comparisons between some characteristics of the life histories of microparasites, macroparasites, parasitoids and predators (From May 1982)

Life-history characteristic or other property	Microparasite	Macroparasite	Parasitoid	Predator
Ratio of average lifespan to that of host or prey	$<<1$ (very small)	<1 (fairly small, $1-10^{-2}$)	~ 1 (usually about equal)	>1 (usually live longer)
Ratio of body sizes	Much smaller than hosts	Smaller than hosts	Mature stages often of similar size to host	Usually larger than their prey
Intrinsic growth rate of population	Much faster than hosts	Faster than hosts	Comparable, but usually slightly slower than hosts	Usually slower than prey
Interaction with individual hosts, as observed in natural populations	One host usually supports a number of populations of different species	One host usually supports from a few to many individuals of different species	One host can support one or several individuals of one (or rarely two) species	One prey item can feed one or a few individuals of the same predator species, but many individual prey are required during a predator's lifespan.
Effect of the above interaction on the host individual	Mildly to fairly deleterious	Variable: not usually too virulent in definitive host; can be very virulent in intermediate host	Eventually fatal	Usually fatal immediately
Ratio between numbers of species, at the population level	Many species of parasites recorded from each member of the host population	Many species of parasites recorded from each population of hosts	Most host species support several parasitoid species, both specialist and generalist (but only a proportion of hosts are actually attacked)	Individual species of predators tend to use more than one prey species.
Degree of overlap of the ranges of the two species	Occur as diffuse foci throughout host's range	Occur as diffuse foci throughout host's range	Usually present throughout host's range	Range is usually greater than that of prey
Genotypes per host or prey	Single or multiple	Multiple	Single or multiple sibships	Single

equations (13) and (21)], measuring the production of secondary cases of infection in terms of the parameters defining the typical course of infection in an individual host, and the spread between hosts.

Similar threshold conditions may be derived from models of host–parasitoid or predator–prey interactions. Suppose that a host population in the absence of parasitoids fluctuates about an average population level, N, (given from equations (1a,b) with $Fg(N) = f(N_t,P_t) = c = 1$). Parasitoids [with f defined in equation (2)] will now be unable to invade and persist if N_t is below a threshold defined by $N_t = 1/ac$, where a is the *per capita* searching efficiency and c is the average number of adult female parasitoids emerging from each parasitized host (May & Hassell 1988). This is another way of saying that each female parasitoid must produce at least one female progeny during her lifetime for the population to persist, and is thus directly analogous to the criterion, $R_0 > 1$. The equivalent threshold criterion from the Lotka–Volterra predator–prey model is d/b, where d is the predator death rate and b expresses the rate of prey capture and the translation of these captures into new predator births.

A central theme in our discussions of both predation and parasitism has been the role of various kinds of heterogeneity in promoting the persistence of the interacting populations. The heterogeneity can arise in many ways, but in all cases can either be expressed in terms of the different probabilities of prey or host individuals being attacked by a predator or pathogen, or the probability distribution of parasites or parasitoids per host. Such variability between individuals is of central importance in promoting the persistence of populations, whether one considers the dynamics of single species (e.g. DeJong 1979; Lomnicki 1988), of competing species in a patchy environment (e.g. Atkinson & Shorrocks 1981; DeJong 1981: Hanski 1981, 1983; Hassell & May 1985; Ives & May 1985; Comins & Hassell 1987), of plant–herbivore interactions (e.g. Crawley 1983; Strong, Lawton & Southwood 1984) or of the natural enemies considered in this paper.

Although the predator–prey interactions discussed here have been, for the main part, set in a patchy environment where the probability of predation varies from patch to patch, the prey could equally well differ in their susceptibility due to behavioural, physiological or other phenotypic differences between individuals. Likewise, different probabilities of infection leading to aggregated distributions of macroparasite burdens per host can arise from heterogeneity in the genetic, behavioural, social, spatial, temporal and nutritional status of individuals in the host population. In all these cases, however, the effect on dynamics depends not on the details of the generating processes, but on the end-product in terms of the probability distribution of attacks on prey or the degree of aggregation of parasites per host.

In exploring the dynamics of host–parasite interaction, the negative binomial distribution provides a good empirical description of parasite burdens per host, and has proved a convenient tool in the analysis of population models. There has been little need, therefore, to seek more elaborate descriptions for the aggregated distribution of infective stages. This is less true in host–parasitoid, and other predator–prey models where one can usually only work with the zero and one-minus-zero terms of the frequency distribution (the probabilities of a prey surviving or being attacked). Consequently, the empirical justification for using the negative binomial distribution is based on the very few cases where it provides a good description of the observed distribution of parasitoid larvae per host (e.g. May 1978; Hassell 1980). Alternative measures of the effect of heterogeneity on predator–prey interactions are thus required that are both easy to apply in the field and whose properties can be readily explored in population models.

Assessing dynamics from field studies

Despite the well-developed theory on the dynamics of all kinds of species interactions, there remain relatively few natural systems where the dynamics are unequivocally understood. Some of the best cases come from the medical literature. For example, the 2-year cycles of measles in human communities that have been recorded over several decades are predicted from a simple Lotka–Volterra model for the infection (Anderson & May 1982). At least one reason for the paucity of such examples is the lack of consensus about what should be measured in the field and over what period. There is, for instance, no clear protocol on how best to detect the regulatory role of the various kinds of heterogeneities discussed above.

One school of thought (Dempster 1975, 1983; Dempster & Pollard 1986) holds that to be important to the dynamics, the net regulatory effect of heterogeneity must be detectable from life-table procedures (e.g. key-factor analysis) using data on average population densities per generation. Such life-table methods have certainly been successful in identifying several density-dependent factors acting on natural populations (e.g. Blank, Southwood & Cross 1967; Varley & Gradwell 1968; Krebs 1970; Southern 1970; Sinclair 1973). There have also, however, been instances where such analyses reveal no regulation (e.g. Richards & Waloff 1961; Southwood & Reader 1976). In these cases, it remains an open question whether regulatory processes were indeed absent, or just not detected by the methods used. The problem is an important one for the design of field population studies (Hassell, Southwood & Reader 1987).

In an ideal, deterministic world, there is no real problem. Regulatory processes stemming from heterogeneity translate directly to density-dependent relationships as a function of average or total population size. Unfortunately, in the real world such determinism rarely applies and variability in some of the parameters, such as those governing the degree of clumping of the host population, or the 'quality' of different patches, can easily result in the density dependent 'signal' being obscured by the 'noise' (Hassell 1987). The detection of regulation in such cases requires a more flexible approach to field studies. Long-term population studies combined with experiments and observations remain crucial, but more attention needs to be paid to variability between individuals, whether this is due to spatial pattern, behaviour or genetic differences. Techniques to do this are now available (e.g. DNA 'fingerprinting'), and will make quantifying the variability within populations much easier. With such developments we can look forward to field studies playing a much greater part in stimulating and driving the direction of ecological theory.

ACKNOWLEDGMENTS

We are very grateful to Charles Godfray, Michael Hochberg and Vicky Taylor for their helpful comments on the manuscript.

REFERENCES

Anderson, R. M. (1976). Dynamic aspects of parasite population ecology. *Ecological Aspects of Parasitology* (Ed. by C.R. Kennedy), pp. 431–462 North Holland Publishing Company, Amsterdam.

Anderson, R. M. (1979a). The influence of parasite infection on the dynamics of host population growth. *Population Dynamics* (Ed. by R.M. Anderson, B.D. Turner & L.R. Taylor), pp. 245–281. Blackwell Scientific Publications, Oxford.

Anderson, R. M. (1979b). The persistence of direct life cycle infectious diseases within populations of hosts. *Some Mathematical Questions in Biology*, Vol. **12** (Ed. by S.A. Levin), pp. 1–67. American Mathematical Society, Providence, Rhode Island.

Anderson, R. M. (1981). Population ecology of infectious disease agents. *Theoretical Ecology: Principles and Applications* (Ed. by R.M. May), pp. 318–355. Blackwell Scientific Publications, Oxford.

Anderson, R. M. (1982). Parasite dispersion patterns; generative mechanisms and dynamic consequences. *Aspects of Parasitology* (Ed. by E. Meerovitch), pp. 1–40. McGill University, Montreal.

Anderson, R.M. (1986a). Genetic variability and resistance to parasitic invasions: Population implications for invertebrate host species. *Immune Mechanisms in Invertebrate Vectors* (Ed. by A.M. Lackie). Symposium of the Zoological Society, London, Oxford University Press, Oxford.

Anderson, R. M. (1986b). The population dynamics and epidemiology of intestinal nematode infections. *Transactions of the Royal Society of Tropical Medicine & Hygiene,* **80,** 686–696.

Anderson, R. M. (1988). The epidemiology of HIV infection: variable incubation plus infectious periods and heterogeneity in sexual activity. *Journal of the Royal Statistical Society, London A,* **151**, 66–93.

Anderson, R. M., Grenfell, B. T. & May, R. M. (1984). Oscillatory fluctuation in the incidence of infectious disease and the impact of vaccination: time series analysis. *Journal of Hygiene Cambridge*, **93**, 587–608.

Anderson, R. M., Jackson, H. C., May, R. M. & Smith A. D. M. (1981). Population dynamics of fox rabies in Europe. *Nature*, **288,** 765–771.

Anderson, R. M. & May, R. M. (1978). Regulation and stability of host-parasite population interactions: I. Regulatory processes. *Journal of Animal Ecology*, **47,** 219–247.

Anderson, R. M. & May, R. M. (1979). Population biology of infectious diseases: Part I.*Nature*, **280,** 361–367.

Anderson, R. M. & May, R. M. (1980). Infectious diseases and population cycles of forest insects. *Science*, **210,** 658–661.

Anderson, R. M. & May, R.M. (1981). The population dynamics of microparasites and their invertebrate hosts. *Philosophical Transactions of the Royal Society, London. B*, **291,** 451–524.

Anderson, R. M. & May, R. M. (1982). Directly transmitted infectious diseases: control by vaccination. *Science*, **215,** 1053–1060.

Anderson, R. M. & May, R. M. (1984). Spatial, temporal and genetic heterogeneity in host populations and the design of immunization programmes. *IMA Journal of Mathematics Applied to Medicine and Biology*, **1,** 233–266.

Anderson, R. M. & May, R. M. (1985a). Vaccination and herd immunity to infectious disease. *Nature*, **318,** 323–329.

Anderson, R. M. & May, R. M. (1985b). Age-related changes in the rate of disease transmission: implications for the design of vaccination programmes. *Journal of Hygiene, Cambridge*, **94,** 365–436.

Anderson, R. M. & May, R. M. (1986). The invasion, persistence and spread of infectious diseases within animal and plant communities. *Philosophical Transactions of the Royal Society, London, B,* **314,** 533–570.

Anderson, R. M. & May, R. M. (1988). Epidemiological parameters of HIV transmission. *Nature*, **333,** 514–518.

Anderson, R. M., May, R. M. & McLean, A. R. (1988). Possible demographic consequences of AIDS in developing countries. *Nature*, **332,** 228–234.

Anderson, R. M., May, R. M., Medley, G. F. & Johnson, H. (1986). A preliminary study of the transmission dynamics of the human immunodeficiency virus (HIV), the causative agent of AIDS. *IMA Journal of Mathematics Applied in Medicine and Biology*, **3,** 229–263.

Anderson, R. M. & Trewhella, W. (1985). Population dynamics of the badger (*Meles meles*) and the epidemiology of bovine tuberculosis (*Mycobacterium bovis*). *Philosophical Transactions of the Royal Society, London, B*, **310,** 327–381.

Atkinson, W. D. & Shorrocks, B. (1981). Competition on a divided and ephemeral resource: a simulation model. *Journal of Animal Ecology*, **50,** 461–471.

Auslander, D. M., Oster, G. F. & Huffaker, C. B. (1974). Dynamics of interacting populations. *Journal of the Franklin Institute*, **297**, 345–376.

Bailey, V. A. Nicholson, A. J. & Williams, E. J. (1962). Interaction between hosts and parasites when some host individuals are more difficult to find than others. *Journal of Theoretical Biology*, **3**, 1–18.

Beck, K. (1984). Coevolution: mathematical analysis of host–parasite interactions. *Journal of Mathematical Biology*, **19**, 63–78.

Beddington, J. R. & Hammond, P. S. (1977). On the dynamics of host–parasite–hyperparasite interactions. *Journal of Animal Ecology*, **46**, 811–821.

Beddington, J. R. & May, R. M. (1977). Harvesting natural populations in a randomly fluctuating environment. *Science*, **197**, 463–465.

Bellows, T. S. Jr. & Hassell, M. P. (1988). The dynamics of age-structured host–parasitoid interactions. *Journal of Animal Ecology*, **57**, 259–268.

Blank, T. H., Southwood, T. R. E. & Cross, D. J. (1967). The ecology of the partridge. 1. Outline population processes with particular reference to chick mortality and nest density. *Journal of Animal Ecology*, **37**, 549–556.

Bundy, D. A. P. (1988). The epidemiology of intestinal helminth infections. *Philosophical Transactions of the Royal Society, London, B*, **321**, 405–420.

Burnett, T. (1958). A model of host–parasite interaction. *Proceedings of the 10th International Congress of Entomology*, **2**, 679–686.

Chesson, P. L. (1981). Models for spatially distributed populations: the effect of within-patch variability. *Theoretical Population Biology*, **19**, 288–325.

Chesson, P. L. & Murdoch, W. W. (1986). Aggregation of risk: relationships among host–parasitoid models. *American Naturalists*, **127**, 696–715.

Clark, C. W. (1976). *Mathematical Bioeconomics.* Wiley, New York.

Comins, H. N. & Hassell, M. P. (1976). Predation in multi-prey communities. *Journal of Theoretical Biology*, **62**, 93–114.

Comins, H. N. & Hassell, M. P. (1979). The dynamics of optimally foraging predators and parasitoids. *Journal of Animal Ecology*, **48**, 335–351.

Comins, H. N. & Hassell, M. P. (1987). The dynamics of predation and competition in patchy environments. *Theoretical Population Ecology*, **31**, 393–421.

Comins, H. N. & Wellings, P. W. (1985). Density-related parasitoid sex-ratios: influence of host–parasitoid dynamics. *Journal of Animal Ecology*, **54**, 583–594.

Crawley, M. J. (1983). *Herbivory. The Dynamics of Animal–Plant Interactions.* Blackwell Scientific Publications, Oxford.

Crofton, H. D. (1971). A quantitative approach to parasitism. *Parasitology*, **62**, 179–194.

Crowley, P. C (1979). Predator-mediated coexistence: an equilibrium interpretation. *Journal of Theoretical Biology*, **80**, 129–144.

DeBach, P. & Smith, H. S. (1941). The effect of host density on the rate of reproduction of entomophagous parasites. *Journal of Economic Entomology*, **34**, 741–745.

De Jong, G. (1979). The influence of the distribution of juveniles over patches of food on the dynamics of a population. *Netherlands Journal of Zoology*, **29**, 33–51.

De Jong, G. (1981). The influence of dispersal pattern on the evolution of fecundity. *Netherlands Journal of Zoology*, **32**, 1–30.

Dempster, J. P. (1975). *Animal Population Ecology.* Academic Press, London, New York.

Dempster, J. P. (1983). The natural control of populations of butterflies and moths. *Biological Reviews*, **58**, 461–481.

Dempster, J. P. & Pollard, E. (1986). Spatial heterogeneity, stochasticity and the detection of density dependence in animal populations. *Oikos*, **46**, 413–416.

Elkins, D. B., Haswell-Elkins, M. & Anderson, R. M. (1986). The epidemiology and control of intestinal helminths in the Pulicat Lake region of southern India. I. Study design and pre- and post-treatment observations on *Ascaris lumbricoides* infection. *Transactions of the Royal Society of Tropical Medicine and Hygiene*, **80,** 774–792.

Fisher, R. A. (1930). *The Genetical Theory of Natural Selection.* Clarendon Press. Oxford.

Free, C. A., Beddington, J. R. & Lawton, J. H. (1977). On the inadequacy of simple models of mutual interference for parasitism and predation. *Journal of Animal Ecology*, **46,** 543–554.

Gause, G. F. (1934). *The Struggle for Existence.* Williams & Wilkins, Baltimore. Reprinted 1964 by Hafner, New York.

Gillespie, J. H. (1975). Natural selection for resistance to epidemics. *Ecology*, **56,** 493–495.

Godfray, H. C. J. & Hassell, M. P. (1987). Natural enemies can cause discrete generations in tropical insects. *Nature, London*, **327,** 144–147.

Godfray, H. C. J. & Hassell, M. P. (1989). Discrete and continuous insect populations in tropical environments. *Journal of Animal Ecology*, **58,** 153–174.

Griffiths, K. L. (1969). Development and diapause in *Pleolophus basizonus* (Hymenoptera: Icheumonidae). *Canadian Entomologist*, **101,** 907–914.

Hadler & Dietz, K. (1984). Population dynamics of killing parasites which reproduce in the host. *Journal of Mathematical Biology*, **21,** 45–66.

Haldane, J. B. S. (1949). Disease and evolution. *Symposium sui fattori ecologici e genetici della speciazone negli animali, Ricerca Scentifica*, **19,** (Suppl.) 3–11.

Hanski, I. (1981). Coexistence of competitors in patchy environment with and without predation. *Oikos*, **37,** 306–312.

Hanski, I. (1983). Coexistence of competitors in patchy environments. *Ecology*, **64,** 493–500.

Hanski, I. (1987). Pine sawfly population dynamics: patterns, processes, problems. *Oikos*, **50,** 327–335.

Hassell, M. P. (1969). A study of the mortality factors acting upon *Cyzenis albicans* (Fall.), a tachinid parasite of the winter moth (*Operophtera brumata* (L.)). *Journal of Animal Ecology*, **38,** 329–340.

Hassell, M. P. (1978). *The Dynamics of Arthropod Predator-prey Systems.* Princeton University Press, Princeton.

Hassell, M. P. (1980). Foraging strategies, population models and biological control: a case study. *Journal of Animal Ecology*, **49,** 603–628.

Hassell, M. P. (1982). Pattern of parasitism by insect parasitoids in patchy environments. *Ecological Entomology*, **7,** 365–377.

Hassell, M. P. (1984). Parasitism in patchy environments: inverse density dependence can be stabilizing. *IMA Journal of Mathematics Applied in Medicine and Biology*, **1,** 123–133.

Hassell, M. P. (1987). Detecting regulation in patchily distributed animal populations. *Journal of Animal Ecology*, **56,** 705–713.

Hassell, M. P. & Anderson, R. M. (1984). Host susceptibility as a component in host–parasitoid systems. *Journal of Animal Ecology*, **53,** 611–621.

Hassell, M. P. & Comins, H. N. (1975). Discrete time models for two-species competition. *Theoretical Population Biology*, **9,** 202–221.

Hassell, M. P. & May, R. M. (1973). Stability in insect host–parasite models. *Journal of Animal Ecology*, **42,** 693–726.

Hassell, M. P. & May, R. M. (1974). Aggregation in predators and insect parasites and its effect on stability. *Journal of Animal Ecology*, **43,** 567–594.

Hassell, M. P. & May, R. M. (1985). From individual behaviour to population dynamics. *Behavioural Ecology* (Ed. by R.M. Sibly & R.H. Smith) pp. 3–32. British Ecological Society Symposium No. 25. Blackwell Scientific Publications, Oxford.

Hassell, M. P. & May, R. M. (1986). Generalist and specialist natural enemies in insect predator–prey interactions. *Journal of Animal Ecology*, **55**, 923–940.

Hassell, M. P. & May, R. M. (1988). Spatial heterogeneity and the dynamics of parasitoid–host systems. *Annales Zoologici Fennici*, **25**, 55–61.

Hassell, M. P., Southwood, T. R. E. & Reader, R. M. (1987). The dynamics of the viburnum whitefly (*Aleurotrachelus jelinekii*): A case study of population regulation. *Journal of Animal Ecology*, **56**, *283–300.*

Hassell, M. P. & Varley, G. C. (1969). New inductive population model for insect parasites and its bearing on biological control. *Nature, London*, **223**, 1133–1137.

Hassell, M. P., Waage, J. K. & May, R. M. (1983). Variable parasitoid sex ratios and their effect on host-parasitoid dynamics. *Journal of Animal Ecology*, **52**, 889–904.

Hastings, A. (1977). Spatial heterogeneity and the stability of predator–prey systems. *Theoretical Population Biology* **12**, 37–48.

Hilborn, R. (1975). The effect of spatial heterogeneity on the persistence of predator–prey interactions. *Theoretical Population Biology*, **8**, 346–355.

Hochberg, M. E., Hassell, M. P. & May, R. M. (1989). The dynamics of host–parasitoid–pathogen interactions. *American Naturalist* (in press).

Holling, C. S. (1959a). The components of predation as revealed by a study of small mammal predation of the European pine sawfly. *Canadian Entomologist*, **91**, 293–320.

Holling, C. S. (1959b). Some characteristics of simple types of predation and parasitism. *Canadian Entomologist*, **91**, 385–398.

Hollings, C. S. (1973). Resilience and stability of ecological systems. *Annual Review of Ecology and Systematics*, **4**, 1–23.

Hudson, P. J., Dobson, A. P. & Newborn, D. (1985). Cyclic and non-cyclic populations of red grouse: a role for parasitism. *Ecology and Genetics of Host-parasite Interactions* (Ed. by D. Rollinson & R.M. Anderson), pp. 77–90. Academic Press, London.

Huffaker, C.B. (1958). Experimental studies on predation: dispersion factors and predator–prey oscillations. *Hilgardia*, **27**, 343–383.

Hutchinson, G. E. (1951). Copepodology for the ornithologist. *Ecology*, **32**, 571–577.

Ives, A. R. & May, R. M. (1985). Competition within and between species in a patchy environment: relations between microscopic and macroscopic models. *Journal of Theoretical Biology*, **115**, 65–92.

Kakehashi, N., Suzuki, Y. & Iwasa, Y. (1984). Niche overlap of parasitoids in host–parasitoid systems: its consequence to single versus multiple introduction controversy in biological control. *Journal of Applied Ecology*, **21**, 115–131.

Kareiva, P. & Odell, G. M. (1987). Swarms of predators exhibit 'preytaxis' if individual predators use area restricted search. *American Naturalist*, **130**, 233–270.

Kermack, W. O. & McKendrick, A. G. (1927). A contribution to the mathematical theory of epidemics. *Proceedings of the Royal Society of London, A*, **115**, 700–721.

Kiritani, K. (1977). A systems approach to pest management of the green rice leafhopper. *Proceedings of a Conference on Pest Management, 25–29 October, 1976,* (Ed. by G.A. Norton & C.S. Holling), pp. 229–252. International Institute for Applied Systems Analysis, Laxenburg, Austria.

Krebs, J. R. (1970). Regulation of numbers of the Great Tit (Aves: Passeriformes). *Journal of Zoology*, **162**, 317–333.

Lawton, J. H. & Pimm, S. L. (1978). Population dynamics and the length of food chains. *Nature*, **272**, 189–190.

Leslie, P. H. & Gower, J. C. (1960). The properties of a stochastic model for the predator–prey type of interaction between two species. *Biometrika*, **47**, 219–224.

Lessells, C. M. (1985). Parasitoid foraging: should parasitism be density dependent? *Journal of Animal Ecology*, **54**, 27–41.

Lomnicki, A. (1988). *Population Ecology of Individuals.* Monographs in Population **25,** Princeton University Press, Princeton.

Lotka, A. J. (1925). *Elements of Physical Biology.* Williams and Wilkins, Baltimore (Reissued as *Elements of Mathematical Biology* by Dover, 1956).

Ludwig, D., Jones, D. D. & Holling, C. S. (1978). Qualitative analysis of insect outbreak systems: the spruce budworm and forest. *Journal of Animal Ecology*, **47,** 315–332.

Mariau, D. & Morin, J. P. (1972). La biologie de *Coelaenomedera elaeidis* IV. La dynamique de populations du ravageur et de ses parasites. *Oleagineaux*, **27,** 469–474.

May, R. M. (1974). Biological populations with non-overlapping generations: stable points, stable cycles and chaos. *Science*, **186,** 646–647.

May, R. M. (1977). Thresholds and breakpoints in ecosystems with a multiplicity of stable states. *Nature*, **269,** 471–477.

May, R. M. (1978). Host–parasitoid systems in patchy environments: a phenomenological model. *Journal of Animal Ecology*, **47,** 833–843.

May, R. M. (1982). Introduction. *Population Biology of Infectious Diseases* (Ed. by R.M. Anderson & R.M. May), pp. 1–12. Springer-Verlag, Berlin & New York.

May, R. M. & Anderson, R. M. (1978). Regulation and stability of host–parasite population interactions. II. Destabilizing processes. *Journal of Animal Ecology*, **47,** 249–267.

May, R. M. & Anderson, R. M. (1979). Population biology of infectious diseases. Part II. *Nature*, **280,** 455–461.

May, R. M. & Anderson, R. M. (1983). Epidemiology and genetics in the coevolution of parasites and hosts. *Proceedings of the Royal Society B*, **219,** 281–313.

May, R. M. & Anderson, R. M. (1987). The transmission dynamics of HIV infection. *Nature*, **326,** 137–142.

May, R. M. & Anderson, R. M. (1988). The transmission dynamics of human immunodeficiency virus (HIV). *Philosophical Transactions of the Royal Society, London B*, **321,** 565–607.

May, R. M., Anderson, R. M. & McLean, A. R. (1988). Possible demographic consequences of HIV/AIDS epidemics: I. Assuming HIV infection always leads to AIDS. *Mathematical Biosciences*, (in press).

May, R. M. & Hassell, M. P. (1981). The dynamics of multiparasitoid–host interactions. *American Naturalist*, **117,** 234–261.

May, R. M. & Hassell, M. P (1988). Population dynamics and biological control. *Philosophical Transactions of the Royal Society, London B*, **318,** 129–169.

May, R. M., Hassell, M. P., Anderson, R. M. & Tonkyn, D. W. (1981). Density dependence in host–parasitoid models. *Journal of Animal Ecology*, **50,** 855–865.

May, R. M. & Oster, G. F. (1976). Bifurcations and dynamic complexity in simple ecological models. *American Naturalist*, **110,** 573–600.

Maynard Smith, J. (1974). *Models in Ecology.* Cambridge University Press, Cambridge.

Michod, R. E. & Levin, B. R. (eds) (1988). *The Evolution of Sex: and Examination of Current Ideas.* Blackwell Scientific Publications, Oxford.

Mook, L. J. (1963). Birds and the spruce budworm. *The Dynamics of Epidemic Spruce Budworm Populations* (Ed. by R. F. Morris), pp. 268–271. Memoirs of the Entomological Society of Canada No. 31.

Murdoch, W. W. (1969). Switching in general predators: experiments on predator and stability of prey populations. *Ecological Monographs*, **39,** 335–354.

Murdoch, W. W., Nisbet, R. M., Blythe, S. P., Gurney, W. S. & Reeve, J. D. (1987). An invulnerable age class and stability in delay-differential parasitoid–host models. *American Naturalist*, **129,** 263–282.

Murdoch, W. W. & Oaten, A. (1975). Predation and population stability. *Advances in Ecological Research*, **9,** 1–131.

Nicholson, A. J. (1933). The balance of animal populations. *Journal of Animal Ecology*, **2,** 131–178.

Nicholson, A. J. & Bailey, V. A. (1935). The balance of animal populations. Part 1. *Proceedings of the Zoological Society of London*, **1935,** 551–598.

Noy-Meir, I. (1975). Stability of grazing systems: An application of predator–prey graphs. *Journal of Ecology*, **63,** 459–481.

Nunney, L. (1980). The influence of the type 3 (sigmoid) functional response upon the stability of predator–prey difference models. *Theoretical Population Biology*, **18,** 257–278.

Perry, J. N. (1987). Host–parasitoid models of intermediate complexity. *American Naturalist*, **130,** 955–957.

Perry, J. N. & Taylor, L. R. (1986). Stability of real interacting populations in space and time: implications, alternatives and the negative binomial k_c. *Journal of Animal Ecology*, **55,** 1053–1068.

Pielou, E. C. (1969). *An Introduction to Mathematical Ecology*. John Wiley, New York.

Pimm, S.L. (1982). *Food Webs*. Chapman & Hall, London.

Pimm, S. L. & Lawton, J. H. (1977). Numbers of trophic levels in ecological communities. *Nature*, **268,** 329–331.

Pimm, S. L. & Lawton J. H. (1978). On feeding on more than one trophic level. *Nature*, **275,** 542–544.

Rabinovich, J. E. (1984). Chagas' Disease: Modelling transmission and control. *Pest and Pathogen Control Strategic, Tactical and Policy Models* (Ed. by G.R. Conway), John Wiley & Sons, New York.

Reeve, J. D. (1988). Environmental variability, migration, and persistence in host-parasitoid systems. *American Naturalist*, **132**, 810–836.

Richards, O. W. & Waloff, N. (1961). A study of a natural population of *Phytodecta olivacea* (Forster) (Coleoptera: Chrysomeloidea). *Philosophical Transactions of the Royal Society, B,* **244,** 205–257.

Roff, D. A. (1974). Spatial heterogeneity and the persistence of populations. *Oecologia*, **15,** 245–258.

Roughgarden, J. (1979). *Theory of Population Genetics and Evolutionary Ecology. An Introduction.* Macmillan, New York.

Roughgarden, J. & Feldman, M. (1975). Species packing and predation pressure. *Ecology*, **56,** 489–492.

Royama, T. (1970). A comparative study of models of predation and parasitism. *Researches in Population Ecology*, **1,** 1–91.

Schad, G. A. & Anderson, R. M. (1985) Predisposition to hookworm infection in man. *Science*, **228,** 1537–1540.

Schwerdtfeger, F. (1935). Studien uber den Massenwechsel einiger Forstschadlinge. *Zeitschrift fur Forst-v. Jaqdwesen*, **67,** 15–38, 85–104, 449–482 & 513–540.

Sinclair, A. R. E. (1973). Regulation, and population models for a tropical ruminant. *East African Wildlife Journal,* **11**, 307–316.

Skellam, J. G. (1951). Random dispersal in theoretical populations. *Biometrika*, **38,** 196–218.

Solomon, M. E. (1949). The natural control of animal populations. *Journal of Animal Ecology*, **18,** 1–35.

Southern, H. N. (1970). The natural control of a population of tawny owls (*Strix aluco*). *Journal of Zoology, London*, **162,** 197–285.

Southwood, T. R. E. & Comins, H. N. (1976). A synoptic population model. *Journal of Animal Ecology*, **45,** 949–965.

Southwood, T. R. E. & Reader, P. M. (1976). Population census data and key factor analysis for the viburnum whitefly, *Aleurotrachelus jelinekii* (Frauenf.) on three bushes. *Journal of Animal Ecology*, **45**, 313–325.

Stiling, P. D. (1987). The frequency of density dependence in insect host–parasitoid systems. *Ecology*, **68**, 844–856.

Strong, D., Lawton, J. H. & Southwood, T. R. E. (1984). *Insects on Plants. Community Patterns and Mechanisms.* Blackwell Scientific Publications, Oxford.

Taylor, L. R. (1961). Aggregation, variance and the mean. *Nature*, **189**, 732–735.

Utida, S. (1950). On the equilibrium state of the interacting population of an insect and its parasite. *Ecology*, **31**, 165–175.

Varley, G. C. & Gradwell, G. R. (1960). Key factors in population studies. *Journal of Animal Ecology*, **29**, 399–401.

Varley, G. C. & Gradwell, G. R. (1968). Population models for the winter moth. *Insect abundance* (Ed. by T.R.E. Southwood), pp. 132–142. Symposium of the Royal Entomological Society of London No. **4**. Blackwell Scientific Publications, Oxford.

Volterra, V. (1926). Variazioni e fluttuazioni del numero d'individui in specie animali conviventi. *Memorie della R. Accademia Nazionale dei Lincei*, **2**, 31–113 [Translation in Chapman, R.N. (1931) *Animal Ecology*, pp. 409–448. McGraw-Hill, New York.]

Wang, Y. H. & Gutierrez, A. P. (1980). An assessment of the use of stability analyses in population ecology. *Journal of Animal Ecology*, **49**, 435–452.

Zeigler, B. P. (1977). Persistence and patchiness of predator–prey systems induced by discrete event population exchange mechanisms. *Journal of Theoretical Biology*, **67**, 687–713.

7 POPULATION REGULATION IN ANIMALS

A. R. E. SINCLAIR
The Ecology Group, Department of Zoology, University of British Columbia, Vancouver V6T 2A9, Canada.

'. . . and yet in the long run the forces are so nicely balanced, that the face of nature remains uniform for long periods of time . . .'

CHARLES DARWIN,
The Origin of Species, 1859
Chapter 3

INTRODUCTION

In 1798 the Rev. T.R. Malthus, in his essay on population growth, pointed out that whereas animal populations can increase geometrically, their food supplies usually increase arithmetically; and consequently animal populations always increase up to the limit of their resources. Mortality from lack of food through starvation and from disease then acts to prevent further increase. It was not until Charles Darwin read this essay that he understood the significance of intraspecific competition for resources for his theory of Natural Selection. The Malthusian hypothesis has been at the centre of all subsequent debate on what regulates populations. Before reviewing the history of this debate it is necessary to explain the meaning of the terms commonly used.

LIMITATION, REGULATION AND PERSISTENCE

I shall use the term 'population' to mean a group of co-existing individuals which interbreed if they are sexually reproductive. Some populations frequently go extinct (e.g. bird populations on islands, Diamond 1969) and are then re-established through founding immigrants. Many other populations have remained extant for long periods and show no signs of becoming extinct. It is this 'persistence' of populations which has led some of the earlier population ecologists to suggest that there is a demographic mechanism which at lower density prevents extinction and at high density provides an upper limit to population numbers in concordance with the Malthusian principle.

Populations have inputs of births and immigrants (production) and outputs of deaths and emigrants (losses). Figure 7.1 expresses the processes of production and loss as a percentage of population size N. For simplicity, production P is held constant with density. In temporal sequence a constant mortality (m_1), say at an early age, reduces this production [Fig. 7.1(a)]. There follows a mortality (m_2) which increases linearly as population increases. This percentage increase in mortality (or

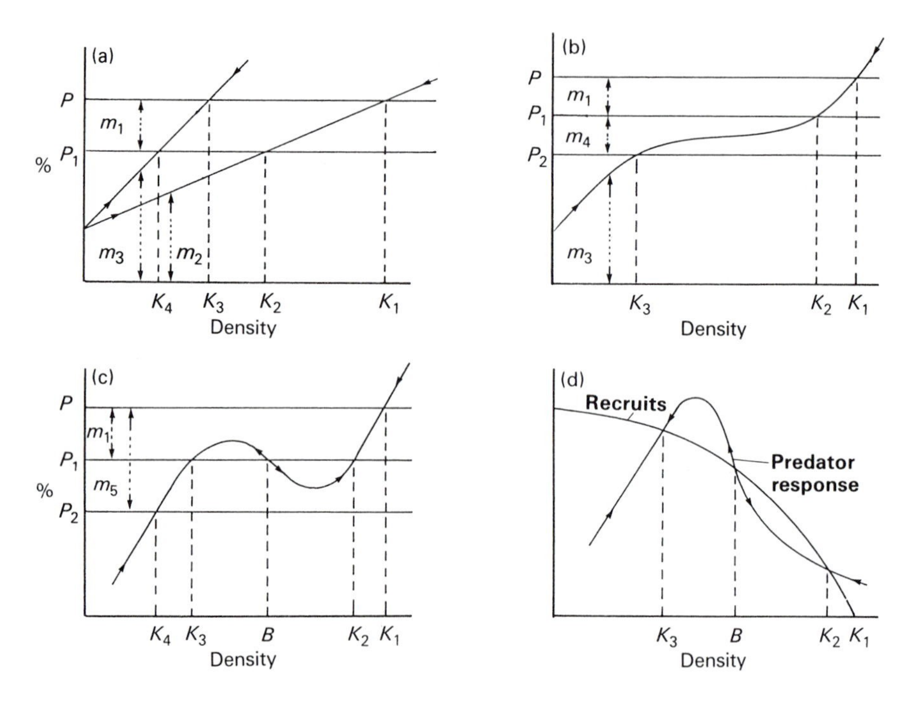

FIG. 7.1 Model of density-dependent and density-independent processes,
(a) Production (P), is held constant over all densities. In sequence a density-independent mortality, DI, (m_1) reduces production (P_1). There follows a density-dependent mortality, DD (m_2 or m_3). The intersect of production with density-dependent mortality determines the equilibrium densities ($K_1 - K_4$) [developed from Krebs (1985), after Enright (1976).]
(b) Density-dependent mortality (m_3) is curvilinear. Both m_1 and m_4 are density independent and reduce production by the same amount. A constant DI reduces the equilibrium to a smaller degree (K_1 to K_2) when DD is stronger, than when DD is weaker (K_2 to K_3).
(c) There are two ranges of density with density-dependent mortality and an intermediate range when there is inverse density dependence. A density-independent mortality, m_1, results in two stable equilibria, K_2, K_3 and an unstable equilibrium at boundary, B. A larger DI (m_5) results in only one lower equilibrium K_4.
(d) Production reduced by density-dependent mortality through lack of resources (as mimicked by the logistic equation) produces the (recruits) curve with equilibrium K_1. The predator total-response curve intersects at stable equilibria K_2, K_3 and at boundary point B.

equivalent percentage drop in production) with population increase is termed *density dependent* and the causes of this relationship are called *density-dependent factors.* Conversely, processes which show no relation to population size (being either constant or random, such as m_1), are termed *density independent* and their causes *density-independent factors.* In addition the reverse relationship to *density dependence* is termed *inverse density dependence*. Where m_2 intersects P_1 there is a stable equilibrium point K_2 because losses balance production remaining after the initial mortalities. Any temporary perturbation of the population above K_2 induces the mortality m_2 to be higher than P_1 and the population returns to K_2. If for some reason the density-independent mortality (m_1) is not imposed then the stable point moves to a higher population size K_1. If the intensity of the density-dependent mortality is strengthened (m_3) the slope increases and the population decreases toK_4.

Several points emerge from Fig. 7.1(a):

(i) The position of the equilibrium point is set by both density-dependent and density-independent factors. The process which sets the equilibrium point is termed *limitation* and the factors which cause the changes in production or loss are *limiting factors.* It is obvious that *any* factor causing a change in production or loss is a limiting factor. Therefore, it is a trivial question to ask whether a particular mortality factor limits a population. The more interesting questions ask: (i) to what extent; and (ii) how does a particular factor change the equilibrium position?

(ii) The process whereby a population returns to its equilibrium is termed *regulation* and the factors causing this are *regulating factors* (e.g. Nicholson 1933; Murdoch 1970). Such factors must have a density-dependent effect.

(iii) A change in the density-independent mortality causes a greater change in equilibrium (K_1 to K_2) when the density-dependent mortality is weak (lower slope) than when the latter mortality is strong (K_3 to K_4).

(iv) A change in the strength (slope) of the density-dependent process alters the equilibrium set point to a greater extent when the mortality acts earlier in an animal's life and before much of the density-independent mortality (K_1 to K_3) than when density dependence occurs later (K_2 to K_4).

Figure 7.1(a) is a simple model in which it is assumed that production is constant; the conclusions do not alter if production is also density dependent. Furthermore, in Fig 7.1(a) I have assumed that density dependence is linear over all values of N. There is no *a priori* justification for this and density dependence could be curvilinear or even take the complex form of mortality (m_3) shown in Fig. 7.1(b). Here there are two regions of N with strong density dependence and a middle region where density dependence is weak. It follows from the points discussed above that in this model, the same degree of density-independent mortality (m_1 or m_4) will have greater effect on altering the equilibrium point if it is in the middle region of N (K_2 to K_3) than at either end (e.g. K_1 to K_2).

Figure 7.1(c) develops the sequence of models so that the population now experiences an inverse density-dependent mortality at intermediate values of N. Imposition of a density-independent mortality m_1, results in there being two stable equilibria K_2 and K_3 and an unstable boundary point B. If the density-independent mortality is larger (m_5) then only one, lower stable point K_4 exists. This kind of scenario could arise if a prey population was regulated at low densities by predators and at high densities by intraspecific competition for resources. In fact Fig. 7.1(c) does not differ from the familiar Rosenzweig–MacArthur models (e.g. Ricklefs 1979, p. 618) for prey recruitment and predator total response curves [Fig. 7.1(d)].

Figure 7.1 illustrates the difference between limitation and regulation. It is clear from the literature that these two terms as well as the term 'control' have been used interchangeably and one has to infer, if possible, from the context, what meaning is intended. Often one finds density-dependent regulation which is a tautology, and density-independent regulation which is a contradiction — it should be density-independent persistence. I shall confine the use of control to the context of biological control where one wants to depress a pest population to some arbitrary but economically suitable low level (Milne 1957).

HISTORY OF CONCEPTS

The balance of nature

Nicholson (1933) revolutionized thinking about population limitation when he introduced the concept of regulation through density-dependent factors. This idea was stimulated by the observation that populations rarely went extinct. His premise was that persistence would occur only if some form of governor was acting on the population. Thus, in his own words (Nicholson 1933) 'Populations must be in a state of balance with their environments' (p. 176); 'For the production of balance, it is essential that a controlling factor should act more severely against an average individual when the density of animals is high, and less severely when the density is low' (p. 135); '[that is] it is essential that the actions of a controlling factor should be governed by the density of the population controlled, and competition seems to be the only factor that can be governed in this way . . ., generally competition between animals when seeking the things they require for existence, or competition between natural enemies that hunt for them' (p. 176).

Factors which are governed in their action by the density of the population were more aptly described as density dependent by Smith (1935), and although Nicholson persisted in using his own more cumbersome terminology, Smith's terms have now become the common usage. Smith developed his terms from two economic entomologists Howard & Fiske (1911).

Nicholson reiterated his ideas but they remained essentially unchanged (Nicholson, 1954a, b, 1958; Nicholson & Bailey 1935) and were accepted more or less by Varley (1947), Elton (1949), Solomon (1949) and Lack (1954, 1955). The main points are: (i) that the environment is composed of density-dependent and density independent factors; (ii) that regulating factors must be density-dependent, the chief of which is intraspecific competition for resources, but others such as predators and parasites can also be density dependent; (iii) there can be more than one density-dependent factor some of which act together whilst some can compensate for each other; (iv) density dependence may not operate all the time — there are times of increase and times when density-independent factors cause a decrease without compensating effects from the density-dependent factors (Nicholson 1958, p.168); (v) scramble (exploitation) competition can cause wastage or spoilage of resources with the result that populations overshoot the equilibrium and cycles occur, whereas contest (interference) competition causes less wastage and greater dampening of cycles; and (vi) parasite–host interactions in insects are unstable and can cause local extinction because parasites act in a delayed density-dependent way (Varley 1947). However, since in nature populations live in a patchy environment, patches can be recolonized by emigrants from other patches. In particular, interference competition (e.g. territoriality) promotes the production of emigrants which act to regulate the home population while recolonizing other areas (Nicholson 1958, p. 169).

This theory came under heavy criticism in the 1950s. Some of the main criticisms were: (i) the logical necessity for density dependence is a mathematical result of the premise that there is an equilibrium. If this premise is false then the rest of the theory collapses and it is unlikely to be applicable in nature (Andrewartha 1958; Milne 1958); (ii) Andrewartha & Birch (1954, p.19) stated '. . . that density-independent factors do not exist . . . [so] there is no need to attach special importance to density-dependent factors . . .'; and (iii) Milne (1957, p.203) considered that

'. . . numbers cannot stop falling unless the density independent environment is or becomes favourable for that purpose. The ultimate control of fall in numbers rests, therefore, with the density independent factors'.

The first of these criticisms is fair enough. The critics claimed that since there is no obvious balance in nature, there is no need to dream up a mechanism for it. By the same argument, however, there is no reason why regulation should not actually exist. The test to distinguish between the two hypotheses is to measure density dependence in the field and predict the population equilibrium. Varley (1947) was the first to attempt this in his studies of parasitism of the knapweed gallfly, but Milne (1957, p. 199) countered that the data did not support the process of density dependence. By 1958 therefore, 25 years after the theory of population regulation had been proposed, there was little if any incontrovertible evidence in nature in its favour. Indeed, Andrewartha (1958) considered it was untestable. Both of the other two criticisms above reflect a misunderstanding of how regulating factors work.

Weather and population persistence

Early theories (e.g. Uvarov 1931) proposed that weather alone limited animal populations. These were superseded by the theory of Thompson (1929, 1939) who considered that '. . . populations are not self-governing systems.' They are limited by a variety of factors; occasionally all factors favour a population which can then 'outbreak' by rapid increase. For the most part, however, one or other factor keeps the population down. Thompson makes no reference to density dependence except for the enigmatic remark (1939) that '. . . the Universe is a density-dependent factor'. Thompson, himself regarded his theory as philosophical, even metaphysical. Milne (1957) agreed that it was '. . . certainly subtle and elusive'. One is left with little idea of how to test such a theory and the uncomfortable feeling that it may not be testable.

Andrewartha & Birch (1954), Birch (1958) and Andrewartha (1958, 1961) advanced the idea of environmental limiting factors, and as mentioned above, did not consider the environment could be divided into density-dependent and density-independent factors. In their words:

> The numbers of animals in a natural population may be limited in three ways: (a) by shortage of material resources, such as food, places in which to make nests, etc.; (b) by inaccessibility of these material resources relative to the animals' capacities for dispersal and searching; and (c) by shortage of time when the rate of increase, r, is positive. Of these three ways, the first is probably the least, and the last is probably the most, important in nature. Concerning (c), the fluctuations in the value of r may be caused by weather, predators, or any other component of environment which influences the rate of increase.

In essence (a) and (b) are the same — populations can be limited by resources but this rarely happens because other factors, notably weather, impose their effect before resources run out. Their idea is effectively the same as that of Thompson. Their work was conducted largely in Australia where both weather and populations fluctuate markedly; they were concerned with explaining these fluctuations in numbers. Two of their classic studies were on the grasshopper *Austroicetes cruciata* and the rose thrips *Thrips imaginis*. Populations are divided into sub-populations; those in the centre of the species' range are limited by many factors, including migration of dispersers. Populations on the edge of the range experience harsh climate and these populations may go extinct, later to be recolonized. Thus Birch (1958, p. 206) wrote 'We have maintained that there is a chance of extinction for the grasshopper in any year in any one place . . . but the chance of extinction in all places in any one year is infinitesimally small under present climatic conditions.'

It is evident that Andrewartha and Birch were concerned with limitation and limiting factors, and their conclusions concerning these are quite correct. Their disagreement with Nicholson was unnecessary because they were arguing at cross purposes — Nicholson was concerned primarily with regulation. Indeed, Nicholson's view of limitation (see above) does not differ substantially from that of Andrewartha & Birch. Another distinction was that Nicholson was concerned with populations in moderate and stable environments near their equilibrium, whereas Andrewartha and Birch examined those not near their equilibrium and in fluctuating environments.

However, the most contentious aspect of Andrewartha & Birch's theory is that populations could persist without regulation. It is generally recognized that such populations would show a random walk through time and would eventually go extinct; hence Elton's (1955) review of their book stated, 'I cannot believe that the natural communities of the world will be found to have evolved solely by guess and by chaos.'

Milne (1957, 1958) reviewed the various theories, and finding none to his liking, came up with a modified version of both Nicholson's and Andrewartha & Birch's ideas. He claimed that first, intraspecific competition is the only perfectly density-dependent factor and thus operates only at high density, but rarely; second, populations are usually held at low levels by a combination of density-independent and imperfect density-dependent factors such as predators and parasites; third, since density-dependent factors cannot stop a population declining to extinction, density-independent factors must do so. Milne considered the idea of imperfect density dependence as his contribution, but in fact

Nicholson had already suggested it. Milne's third point reflects a confusion between the actions of limiting and regulating factors and is illogical (Solomon 1959).

This covers the main theories and some of the important criticisms up to the late 1950s. Peculiarly, they centred on insects in Australia. Nicholson's view had developed from theory, was more complete and more robust, but apart from some early work by Reynoldson (see Reynoldson 1966 for review), there was little hard field evidence in its favour. Andrewartha & Birch based their ideas on extensive field measurements but the theoretical development was incomplete and unsatisfactory.

Extrinsic factors, intrinsic factors and group selection

By 1960 Nicholson's views were the more popular and there was a burst of field studies to test them. In so doing new debate arose concerning the cause and mechanism of regulation. Debate shifted to Britain and centred largely on bird studies.

Lack (1954, 1966), as the leading spokesman for the Nicholson view, proposed that extrinsic factors — factors external to the organism such as resources, predators and parasites — acted to regulate populations. Summarizing the specific mechanism Lack (1966, p.290) argued that '. . . in birds the reproductive rate of each species, evolved through natural selection, is that which results in the greatest number of surviving offspring per pair, and that the population density is regulated by density dependent mortality, in most species by food shortage outside of the breeding season'. He conceded that other species, particularly phytophagous insects, could be regulated by predators. He based his ideas on the long term studies of great tits (*Parus major*) at Wytham Wood (Perrins 1965) and other studies, some of which are collated in his 1966 book. The two essential elements of his idea are first, that the density dependence occurs outside the breeding season, and second, that behavioural attributes of the animals do not result in regulation. In particular, territoriality does not regulate numbers but merely determines who is able to breed.

This view was in stark contrast to a substantially new theory that was developing. It proposed that intrinsic factors,those characteristics of the individual such as behaviour, physiology and genes, could act to regulate the population at a level below that imposed by the food supply. This is the 'self-regulation hypothesis'. Chitty (1960), in an important early paper, suggested that the decline phase of vole (*Microtus agrestis*) population cycles occurred as a consequences of intraspecific aggression during periods of high density. This aggression caused voles to become susceptible to a variety of secondary environmental factors, mortality ensued and the population declined. Chitty (1967) developed this idea into the 'Genetic hypothesis', (Krebs 1978) in which he suggested that the population is genetically polymorphic for aggression. A non-aggressive high reproducing morph is favoured by natural selection at low and increasing densities, but at high densities a large, aggressive but low reproducing morph is favoured. Because of both low reproduction and aggression the population then declines. Although Chitty considered that aggression is selected for ultimately through competition for food, proximately it is the degree of social interaction to which the individual now responds as the population no longer experiences food limitation. Recently, Chitty (1987) has modified the mechanism of social interaction but maintains the concept of a genetic polymorphism.

A major conceptual problem with the genetic hypothesis is how a population becomes emancipated from its food supply. If aggressive behaviour has evolved through natural selection by intraspecific competition, then the effects of food limitation must still be present to maintain the behaviour in the population (i.e. the population remains food limited). If food limitation does not occur, then it would appear that a mutant individual which could breed at high density would be selected for and the population would increase until it is again limited by food. So far this problem has not been resolved. However the hypothesis has stimulated many studies to test the behavioural (Myers & Krebs 1971; Krebs *et al.* 1973; Lidicker 1975; Tamarin 1980) and genetic aspects (Krebs & Myers 1974; Tamarin 1978). Two significant developments were C.J. Krebs' experiments demonstrating the 'fence effect' — the radical change in population when dispersal is inhibited (Krebs *et al.* 1973) and Lidicker's (1975) proposal of regulation through density-dependent dispersal. I will return to the behavioural aspects later, but the genetic evidence in favour of the idea remains inconclusive (McGovern & Tracy 1981). A recent test of heritability of Boonstra & Boag (1987) indicated little heritability.

While Chitty's ideas were gaining popularity, Christian suggested mechanisms for how aggression could affect the animal (Christian 1961; Christian & Davis 1964). Endocrine responses to increasing density affected reproduction primarily through: (i) aggression, (ii) stimulation of adrenal function and production of steroids; and (iii) a sensitivity of the juvenile hypothalamus. These, Christian (1980) suggests, are density dependent, although details of the mechanisms remain unknown. This work has been largely confined to mammals and to what extent it applies to other groups of animals is unclear. However the general idea of

Chitty's genetic theory of population regulation was readily applied to other classes of animals: Pimental (1968) used it to interpret changes in insect populations.

In 1962, Wynne-Edwards (1962, 1965, 1986) proposed his theory of regulation through behaviour specifically functioning to keep the population below the level determined by food supply. He stated (1965) 'It is readily apparent that, although food is generally the commodity that ultimately limits the carrying capacity of the habitat, the population dependent on it must not be allowed simply to increase until further growth in numbers is chronically held in check by general starvation. Famine is a catastrophe especially likely to damage the resource and lead to its permanent depletion'. Two necessary conditions follow from this proposition: (i) the mechanism of regulation operates to limit reproduction usually by behavioural means so that mortality from starvation is avoided (he concedes some mortality through stress could occur); and (ii) such regulatory mechanisms evolve through 'group selection'.

Group selection (or interdemic selection) requires that: (i) there is effectively no genetic interchange between groups or demes; and (ii) groups which overexploit their resources go extinct. Thus a gene that causes animals to limit reproduction so as to keep a population lower than the limit set by food would prevent demes from going extinct, so eventually only demes with this gene would survive.

This theory of Wynne-Edwards caused controversy. Its predictions were the opposite to those of Lack's (1966) theory. Wynne-Edwards considered reproduction was regulated through behaviour such as territoriality, dominance, aggression, inhibition of breeding and disperal, while Lack proposed regulation through mortality in the non-breeding season, with behaviour playing little or no part. Such a juxtaposition of ideas naturally stimulated an explosion of research in population dynamics on the one hand, and behavioural mechanisms of regulation and their evolution on the other (for example see the symposia edited by Watson 1970; Ebling & Stoddart 1978; Alexander & Tinkle 1981; Sibly & Smith 1985). This work produced much evidence for density dependence from population studies and a plethora of behavioural mechanisms which potentially could regulate populations, though regulation was rarely demonstrated (see below). Group selection and the evolution of 'self-regulation' has been severely criticized on theoretical grounds and it is now generally discounted (Maynard Smith 1964, 1976; Wiens 1966; Bell 1987). There are a few special cases where group selection might occur (Wilson 1983; Harvey 1985) but they differ from the type of group selection promoted by Wynne-Edwards. On the whole observations do not support the two conditions necessary for group selection to operate. Wynne-Edwards' hypothesis rests on the assumption that populations regulated by resources, damage those resources (i.e overexploit them) and so decline in numbers. There is no *a priori* reason why this should be so and there is now plenty of evidence to show the assumption is false.

REGULATION AND TESTING FOR DENSITY DEPENDENCE

One of the major problems in determining whether or not regulation occurs, has been the difficulty of distinguishing between non-regulated and regulated populations. The most robust evidence comes from perturbation experiments; populations are artifically reduced from or increased above their starting densities. If they return to the level of the unaltered control population, then this is evidence for regulation and such convergence is evidence for density dependence (Murdoch 1970). These experiments are rarely conducted.

The most common approach involves the statistical analysis of census data. The earlier approaches (Morris 1963; Solomon 1964) used a modified Ricker (1954) curve. A plot of log population numbers in one year (N_{t+1}) against log numbers in the year before (N_t) with a slope less than one, if it is not due to some change in weather, could be interpreted as showing density dependence. Morris (1963) provides the classic example with his spruce budworm plot (Fig. 7.2). However this approach has been criticized on the statistical grounds that it can show spurious density independence (Maelzer 1970; St. Amant 1970).

A more robust approach was that of Varley & Gradwell (1960, 1968). The mortality is expressed as log (N_t/N_s) (called the k-value) where N_t is the original population density and N_s is the density of survivors. The k-values are plotted against log N_t, and with appropriate statistical precautions, a positive slope indicates density-dependence. Thus

$$k\text{-value} = \log (N_t/N_s) = \log a + b \log N_t \qquad (1)$$

where the slope 'b' is a measure of the strength of density dependence. This model is linear and it does not always fit the data. Other models have been developed which in logarithmic form are curvilinear (e.g. May *et al.* 1974), and perhaps the best of these is that of Hassell (1975):

$$\log (N_t/N_s) = b \log (1 + a N_t). \qquad (2)$$

This model has the advantage that the k-value plotted against log N_t is

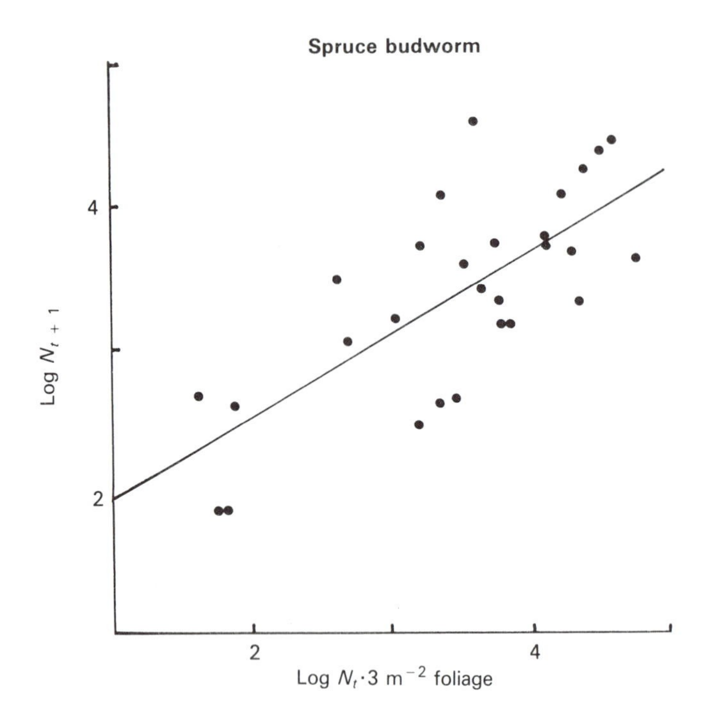

FIG. 7.2 Log density of spruce budworm (*Choristoneura fumiferana*) in each generation (t) plotted against log density in the following generation ($t + 1$). [After Morris (1963).]

nearly linear at high densities. It describes density-dependent relationships from field data reasonably well (Fig. 7.3).

The k-value approach of Varley & Gradwell (1968) and related models (Hassell 1975) have been criticized for being too conservative — they miss too much of the density dependence (Slade 1977; Vickery & Nudds 1984). The autoregression method of Bulmer (1975) is effective only when the population is stationary, while the method of Slade (1977) requires the population to be changing. Vickery & Nudds (1984) suggest another method using a simulation regression test. Recently Gaston & Lawton (1987) have tested these alternative methods using census data from natural populations known from independent evidence to be subject to density-dependent processes. All methods failed to detect density dependence reliably, irrespective of sample size and the dynamics of the population. They concluded that none of the methods was a useful test of density dependence in sequential censuses. These results strengthen the argument for returning to well-designed perturbation experiments.

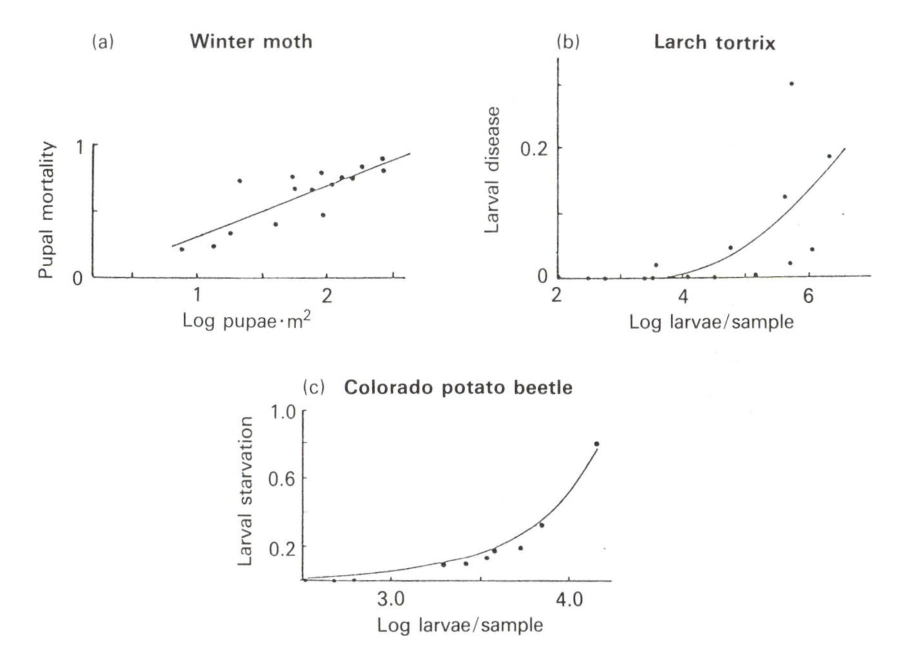

FIG. 7.3 Examples of density dependence in insects from field data, described by equation (2) in text. All are log (N_t/N_s) against log N_t.
(a) Winter moth (*Operophtera brumata*) pupal mortality. [After Varley & Gradwell in Hassell (1975).]
(b) Larch tortrix (*Zeiraphera diniana*) larval disease. [After Varley & Gradwell in Hassell (1975).]
(c) Colorado potato beetle (*Leptinotarsa decemlineata*) larval starvation. [After Harcourt in Hassell (1975).]

Despite all this, most workers are still using the Varley & Gradwell k-value approach. Dempster (1983) used this method to review a number of Lepidopteran populations. He concluded that in one third of the cases no density dependence could be found. Hassell (1985, 1987) challenged this by arguing that life-table studies using average census data may hide density dependence by predators reacting to spatial patchiness in host density or other forms of within-generation heterogeneity. Thus in viburnum whitefly (*Aleurotrachelus jelinekii*) inter-generation density dependence was not detected (Southwood & Reader 1976) because the important density dependence operates from leaf to leaf within a viburnum bush (Hassell, Southwood & Reader 1987). Jones *et al.* (1987) also found that density-dependent mortality of *Pieris rapae* was spatially and temporally patchy.

Murray (1982) has resurrected an old argument of Andrewartha & Birch (1954) claiming that in a territorial species the number of territories can determine population size without the action of density-dependent factors; and like the earlier arguments he has confused limitation with regulation. Territorial space limits the population but competition for that space provides the regulation and is a function of population size, as is explicitly demonstrated in his own model. One other problem is that he defines the density-dependent process strictly as that in Fig. 7.1(a), with uniform effect over all densities; and if it is not uniform then it is not density dependent. His conclusion follows from his definition, but the latter is arbitrary and not one that is generally accepted.

It is now more than 30 years since the 1957 Cold Spring Harbor Symposium on population studies when there was no known way to detect regulation in the field. The intervening period has shown a progressive sophistication in measuring density dependence through statistical means, positions are less rigid in the debates and we are more aware of the pitfalls in measurement. Nevertheless we are likely to be missing much of the density dependence in the data and some methods may be providing spurious answers (Royama 1977). We should proceed with caution.

EVIDENCE FOR REGULATION

Referring to the statistical problems of measuring density dependence, Murdoch (1970) commented 'It is in these circumstances that the beauty of the convergence experiment as a test for regulation becomes apparant'. Yet there are still very few examples of this type of manipulation. Because of logistical difficulties, most are incomplete or unbalanced. For example, a population of elk (*Cervus elephus*) in Yellowstone National Park, USA was reduced from around 15 000 in 1933 to 4000 animals in 1968 and then left alone (Fig. 7.4): It returned to 12 000 in a few years and is now levelling out at about 18 000 (D. B. Houston 1982 and personal communication). Regulation, it seems, was achieved through competition for winter food supplies. In a similar way, the wildebeest (*Connochaetes taurinus*) population in the Serengeti increased from a level (250 000) at which it was held by the Rinderpest disease in the 1950s to a new level of 1.3 million in 1977 at which it levelled out [Fig. 7.5(a)]. This example is less satisfactory than the elk example for we do not know what the pre-Rinderpest density was, but evidence on reproduction and condition does suggest Rinderpest held the population below the level of the food supply, a necessary condition to demonstrate regulation. There is however, direct evidence [Fig. 7.5(b)] that the

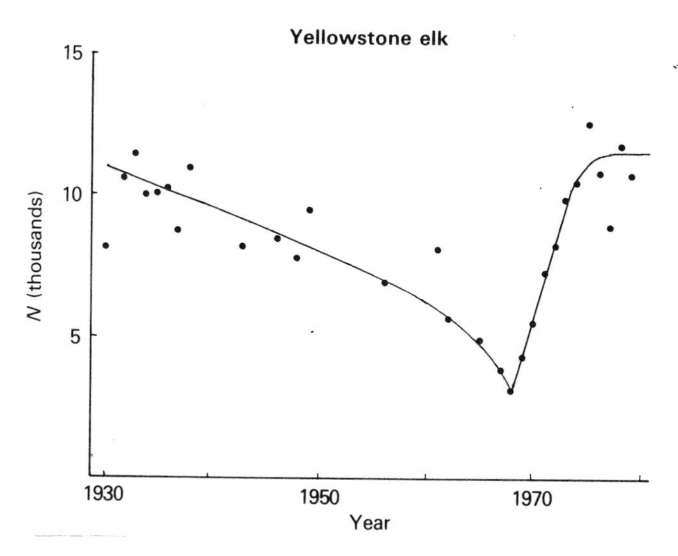

FIG. 7.4 Perturbation of the northern yellowstone elk (*Cervus elephus*) population by culling which stopped in 1968. The population has returned to approximately pre-culling levels. [After Houston (1982).]

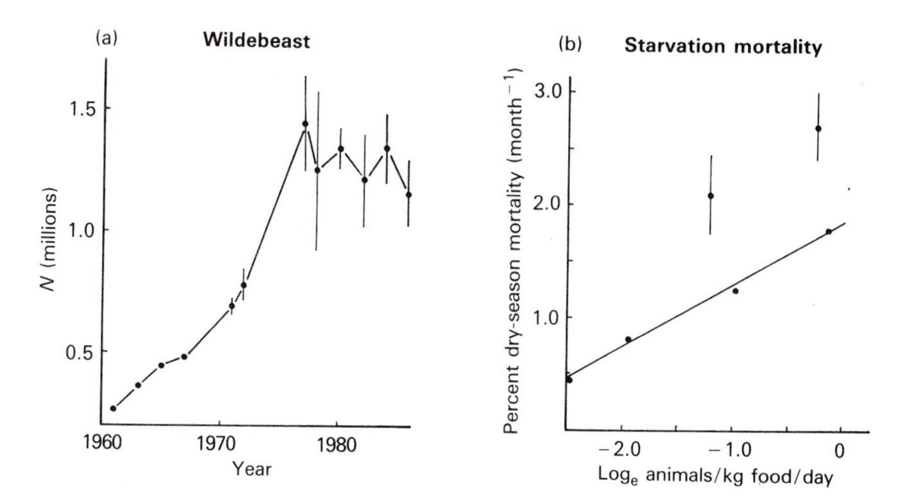

FIG. 7.5 (a) Increase and levelling off of the wildebeest (*Connochaetes taurinus*) population in Serengeti after the virus disease Rinderpest, died out in 1963. Vertical lines are one standard error.
(b) Density-dependent dry season mortality caused by lack of food as the population increased (partly from Sinclair, Dublin & Borner (1985) and from unpublished data). Regression line for the years 1968–72 was used to predict the slope of the line for the two points 1982–83. The slope remained the same, although absolute mortality was higher. Vertical lines are 95% confidence limits.

mechanism for regulation was through density-dependent dry season mortality through lack of food (Sinclair, Dublin & Borner 1985).

Another even less common approach to detecting regulation is a method discussed by Reddingius (1971). The existence of a relationship between an external factor and density, measured from separate populations not influenced by migration, indicates that regulation is taking place (Fig. 7.6), since in an environment exhibiting random variations, this effect on mortality and reproduction will also vary randomly, sometimes resulting in greater mortality than reproduction and sometimes the reverse. Although any one population can remain within relatively narrow limits without regulating factors playing a part, it is extremely improbable that several independent populations can do this and also show mean densities that are correlated with some environmental factor. Some examples include African finches (Schluter 1988), thrips (Klomp 1962) and African buffalo (Sinclair 1977).

By far the most abundant evidence for regulation comes from measures of density dependence. It should be recognized, however, that: (i) density dependence is often overlooked (Gaston & Lawton 1987); (ii) the identification of density dependence may not be sufficient to account for the observed regulation, for there may be other undetected density-dependent factors; and (iii) density-dependent effects may not invariably lead to dampening oscillations, as some factors may have effects leading to stable limit cycles or increasing fluctuations, as seen in some predator–prey models (May *et al.* (1974). This third caveat applies more to insect–parasitoid interactions than to vertebrate populations. On the whole, the detection of density dependence is a less satisfactory form of evidence for the regulation of populations.

Density dependence in insect populations has been reviewed by Solomon (1964), Podoler & Rogers (1975), Stubbs (1977), Dempster (1983) and Strong, Lawton & Southwood (1984) amongst others. Examples of such relationships are shown in Fig. 7.3. Dempster (1983) considered that as much as one third of Lepidoptera studies failed to detect density dependence, but as mentioned above, this proportion may be too high. Nevertheless, in thirty-one studies covering a wide array of insects, Strong, Lawton & Southwood (1984) record eleven with no identifiable density dependence, although subsequently in one, the viburnum whitefly, density dependence has been reported (Hassell *et al.* 1987). In general, the occurrence of density dependence is the rule and I suspect that as census techniques improve, more examples will be reported.

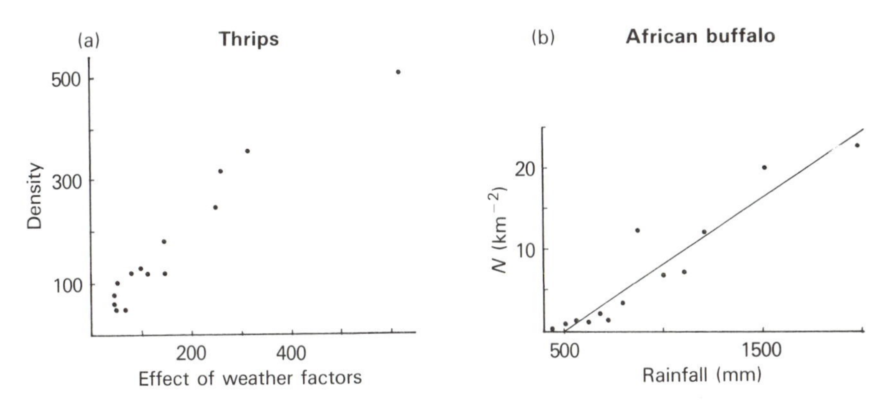

FIG. 7.6 (a) Geometric mean density per rose of thrips (*Thrips imaginis*) from separate populations plotted against an index of weather, as evidence for regulation. [Klomp (1962) after Andrewartha & Birch (1954).]
(b) Mean density of African buffalo (*Syncerus caffer*) from separate populations in East Africa as a function of rainfall, which determines the amount of grass food. [After Sinclair (1977).]

Amongst vertebrates, there are fewer data perhaps because of the long-term records that are necessary. For fish populations there are few attempts to measure density dependence directly, and the work of Elliott (1987) on trout stands out as he demonstrated density dependence both in first winter, and in summer survival [Fig. 7.7(a)]. On the other hand, Doherty (1983) found no evidence for density dependence in the territorial responses of damsel fish. Evidence of regulation in other fish populations, particularly marine fishes, and in marine invertebrates (molluscs, crustaceans), comes indirectly from Ricker-type stock-recruitment curves (Larkin 1973). This evidence is sufficiently strong to warrant mention. Ricker plots show reproductive offspring versus parent stock, the data describing a convex curve if regulation occurs. These curves can be shallow with weak regulation or domed with strong regulation but if the dome is too high, cyclicity or instability occurs. The literature contains many examples of stock-recruitment curves implying regulation (see reviews in Cushing 1981; Parrish 1973). Cushing & Harris (1973) found evidence of regulation in all thirty-one fish stocks they examined, ranging from relatively weak in Atlantic herring and Pacific salmon, intermediate in flatfish, to strong in gadoid fishes. Elasmobranchs show weak to intermediate degrees of regulation (Holden 1973). In molluscs there is evidence of regulation in cockles (*Cardium edule*) and perhaps in the Pacific razor clam (*Siliqua patula*), but for oysters,

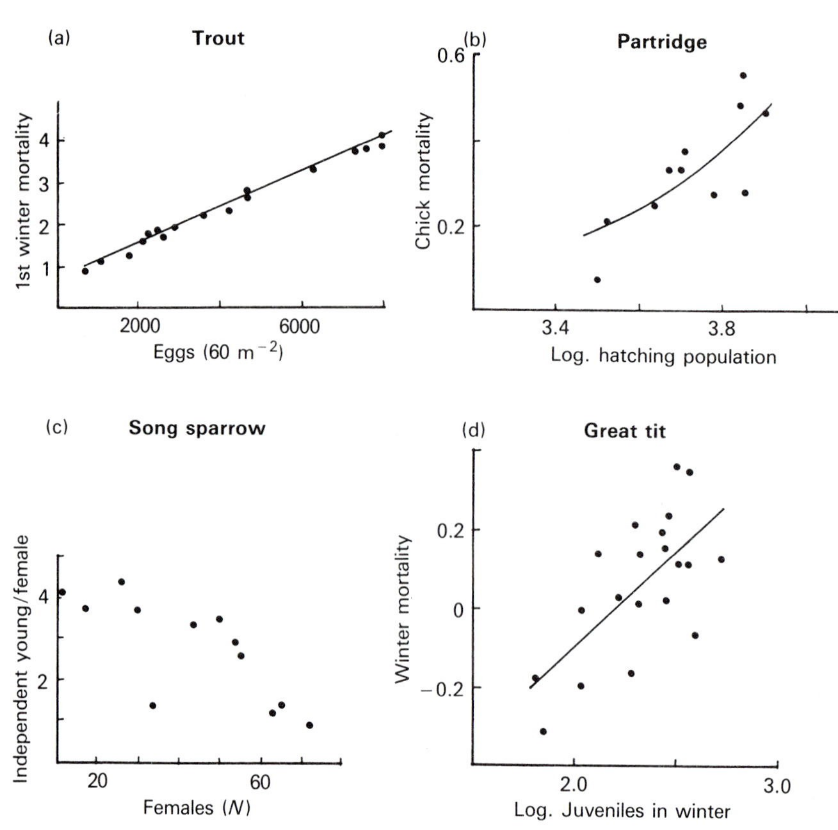

FIG. 7.7 Examples of density dependence in some vertebrate populations.
(a) Trout (*Salmo trutta*) first winter mortality. [After Elliott (1987).]
(b) Partridge (*Perdix perdix*) chick mortality as log hatching population/log population at 6 weeks on a Hampshire estate. [After Blank, Southwood & Cross (1967).]
(c) Song sparrow (*Melospiza melodia*, independent young per female in autumn on Mandarte Island, British Columbia. [After Arcese & Smith (1988).]
(d) *k*-value for winter mortality (log juveniles in winter/first year breeding population) of great tits (*Parus major*) against log juvenile density in winter at Wytham Wood, Oxford. [After McCleery & Perrins (1985).]

mussels and scallops the density-independent effects of temperature and other factors obscure any possible stock-recruitment relationship (Hancock 1973). For crustacea, stock-recruitment curves show evidence of regulation in some cases, such as crabs (Jamieson 1986), lobsters (Addison 1986; Caputi & Brown 1986; Fogarty & Idoine 1986) and prawns (Penn & Caputi 1985), but in many other cases, recruitment was highly influenced by occasional good year classes surviving because of environmental conditions and consequently regulation was not detected.

Finally, for aquatic and marine zooplankton there has been little attempt to find evidence of regulation. The reasons for this are probably the difficulty of measuring single populations for stock-recruitment or life-cycle census data, although Kerfoot, DeMott & DeAngelis (1985) have measured density dependence in *Daphnia* reproduction. Nevertheless, plankton populations are particularly suited to perturbation experiments and these would seem a profitable and exciting area for further research.

Before 1975, there were few studies on birds which showed density dependence, those of Perrins (1965), Krebs (1970) and Kluyver (1971) on great tits, Southern (1970) on tawny owls, and Blank, Southwood & Cross (1967) [Fig. 7.1(b)] on partridges stand out. In the 1980s there has been an increase in reports of density dependence, for example, in pied and collared flycatchers (Stenning, Harvey & Campbell 1988; Torok & Toth, 1988) and song sparrows (Arcese & Smith 1988) [Fig 7.7(c)]. Some of the early great tit data have now been re-examined with longer records. For the years 1947–1961 Perrins (1965) found that clutch size declined as the number of breeding pairs increased, but the data for the period 1962–1983 show no relationship (McCleery & Perrins 1985). Thus there are temporal changes in density dependence, a feature also noted by Stenning, Harvey & Campbell (1988) but which has not previously been considered important in population studies. The present great tit data show that winter mortality is now density dependent [Fig. 7.7(d)] whereas previously it was hardly detectable (Krebs 1970). The greater sample sizes are showing up new, or confirming previously less robust density-dependent relationships (see also Tinbergen, Van Balen & Van Eck 1985). Nevertheless, even with relatively long data records, 17 years for pied flycatchers (Stenning, Harvey & Campbell 1988) and 22 years for great tits in Holland (Tinbergen, Van Balen & Van Eck 1985), the density-dependent effects can be weak and hard to detect.

Regulation of mammal populations is better known for large mammals (see Fig. 7.8 and reviews by Fowler 1981, 1987; Fowler & Smith 1981) than for small mammals. With the latter group, work has been concerned with limiting factors and causes of cycles, rather than with a search for density dependence. Studies on large mammals are often confounded by human interference through commercial harvesting, yet such harvesting can provide the right conditions for a perturbation experiment. Fowler (1987) reports over 100 studies of separate populations of terrestrial and marine large mammals where density dependence has been detected. It is not yet possible to give an estimate of the prevalence of density dependence but I suspect it will be found in most cases. Fowler (1987) notes that density-dependent mortality is non-linear with density, the strongest effects occurring at high density [resembling the high densities of Fig. 7.1(b)].

Stubbs (1977) has suggested that animals living in temporary habitats (the *r* strategists) tend to suffer high, overcompensating and destabilizing density-dependent effects due to scramble competition; all the competitors use the limiting resource until it begins to run out and then only a few individuals survive. Here the density dependence is curvilinear. Animals in permanent habitats (*K* strategists) have exact or undercompensating mortalities due to interference competition; this allows some of the competitors to obtain sufficient resource to reproduce while others obtain very little and die. Fowler (1981) however, concluded that large mammals (*K* strategists) show most density-dependent change at high population levels, while insects, fish and similar *r* strategists show most density-dependent changes at low population levels. Fowler's conclusion may not contradict that of Stubbs because most of the *r*-type species that Fowler refers to could be regulated by predators rather than by competition (Strong, Lawton & Southwood 1984). Stubbs (1977) refers mostly to those *r* strategists which experience intraspecific competition. Nevertheless, the pattern of density dependence in its relation to life-history strategies is still vague. This is an area for further analysis.

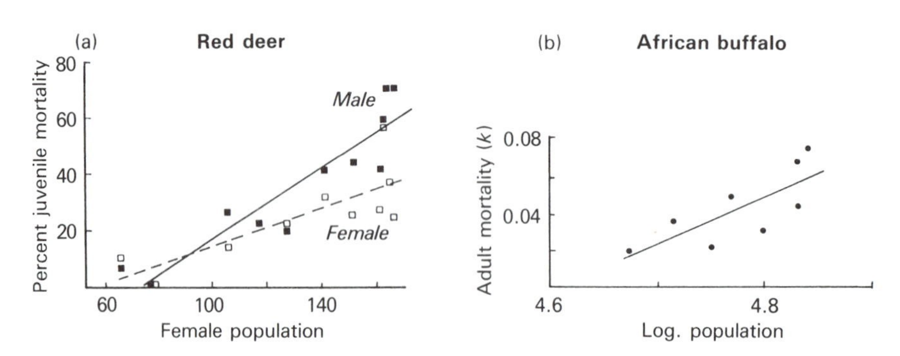

FIG. 7.8 Density dependence in large mammals.
(a) Mortality of female and male juvenile red deer (*Cervus elephus*, on Rhum. [After Clutton-Brock, Major & Guinness (1985).]
(b) *k*-value of adult mortality for African buffalo. [After Sinclair (1977).]

DENSITY DEPENDENÇE IN THE LIFE CYCLE

I have examined which stages in the life cycle are subject to regulatory influences, and I have compared these stages between animal groups. Data were obtained from the reviews of Cushing & Harris (1973), Stubbs (1977), Fowler (1981, 1987), Strong, Lawton & Southwood (1984),

McLaren & Smith (1985), and a number of subsequent reports (Table 7.1). I have divided the studies into insects, fish, birds, small mammals (largely rodents), large marine mammals (seals, whales) and large terrestial mammals, which include ungulates, carnivores and primates.

Life stages examined for insects are eggs, larvae (early juvenile), pupae (late juvenile) and adults. Out of the forty-seven case studies, most reported density dependence occurred at the larval stage. Fecundity or egg production was the next most common stage for density dependence and only a few reported regulation in adult stages. For marine fishes, Cushing (1981) found that the degreee of regulation acting on a population was directly related to the fecundity of the species; thus gadoids with high fecundity experienced strong regulation, while salmonids with relatively low fecundity showed weak regulation, and elasmobranchs with the lowest fecundity showed even weaker regulation (Holden 1973). Nevertheless, it was the early larval stages of the life cycle where density dependence occurred in the thirty one stocks examined by Cushing & Harris (1973), as it was for trout (*Salmo trutta*) (Elliott 1987) and Pacific sardine (Iles 1973). However, Le Cren (1987) found that fertility was density dependent in perch (*Perca fluviatilis*) and this was suggested for elasmobranchs (Holden 1973). In general, fisheries biologists assume that juvenile mortality is density dependent while adult mortality is density independent (Charnov 1986).

Similarly in marine crustaceans, density-dependent mortality was assumed to occur in juvenile prawns (Penn & Caputi 1985), crabs (Jamieson 1986) and western rock lobster (*Panulirus longipes*) (Chittleborough & Philips 1975). However, there may be density-dependent mortality of the larger adult lobsters in some species because they cannot find sufficient shelters (Addison 1986).

In contrast to insects and fish, 74% of the bird studies reported density-dependent mortality in the later juvenile stages which included independent young during their first summer and winter, and in two studies included adult mortality as well. There are some reports of clutch size and early juvenile (nestling, fledging) mortality being density dependent but these were less frequent. Overall 56% of studies reported density dependence in the breeding season (eggs to fledgings) and 61% in the non-breeding season (autumn recruitment and winter mortality). Thus the data are split and do not unequivocally support either of the hypotheses of Lack (1966) or Wynne-Edwards (1962). Since all of these studies are in temperate regions, these data may well be biased, as processes in the tropics could be acting on earlier life stages (e.g. nestling mortality through predation). This is an area for new work.

Few small mammal studies address the question of regulation. All

TABLE 7.1. Number (%) of reports of separate populations demonstrating density dependence at different life stages

Group	Fertility/ egg prodn N (%)	Early juvenile mortality N (%)	Late juvenile mortality N (%)	Adult mortality N (%)	Total no. populations
Insects[1]	14 (30)	19 (40)	13 (28)	6 (13)	47
Fish[2]	2 (6)	33 (94)	0	0	35
Birds[3]	5 (26)	6 (32)	14 (74)	4 (21)	19
Small[4] mammals	0	0	12 (92)	1 (8)	13
Large[5] marine mammals	34 (83)	10 (24)	0	1 (2)	41
Large[6] terrestrial mammals	49 (68)	35 (49)	1 (1)	12 (17)	72

[1] From Clark *et al.* 1967; Gage, Miller & Mook 1970; Solomon *et al.* 1976; Stubbs 1977; Kenmore *et al.* 1984; Strong, Lawton & Southwood 1984; Chambers, Wellings & Dixon 1985; Jones *et al.* 1987; Hill 1988; Roland 1988; Southwood & Reader 1988.
[2] From Cushing & Harris 1973; Holden 1973; Iles 1973; Elliott 1987; Le Cren 1987.
[3] From Blank, Southwood & Cross 1967; Southern 1970; Ryder 1975; Goss-Custard 1980; Zwickel 1980; Hannon, Sopuck & Zwickel 1982; Alatalo & Lundberg 1984; Ekman 1984; Hill 1984; Greig-Smith 1985; Hudson, Dobson & Newborn 1985; McCleery & Perrins 1985; Tinbergen *et al.* 1985; Newton & Marquiss 1986; Nilsson 1987; Porter & Coulson 1987; Arcese & Smith 1988; Stenning, Harvey & Campbell 1988; Torok & Toth 1988.
[4] From Healey 1967; Jannett 1978; Cockburn 1981; Hestbeck 1982, 1987; Boonstra & Rodd 1983; Galindo & Krebs 1987; Trostel *et al.* 1987.
[5] From Fowler 1981, 1987; Fowler & Smith 1981.
[6] From Fedigan 1983; Skogland 1985; Fowler 1987; Kruuk & Parish 1987.

eleven rodent studies concluded that there was density-dependent exclusion of late juveniles into the breeding population prior to breeding. In contrast, a study on snowshoe hares (*Lepus americanus*) (Trostel *et al.* 1987) has recorded adult mortality overwinter as the possible regulating stage. In large marine mammals, almost the opposite result is found; either female fertility (all whale species reported and some seals) or early juvenile mortality (suckling seal pups) were density dependent. Large terrestrial mammals also reflect the effects of density-dependent factors on fertility and juvenile mortality but a few studies record density dependence in winter (or dry season) mortality of late juveniles and adults.

On the face of it, there are large differences between groups in where regulation occurs during the life cycle. However, there are inherent biases in the way species are studied which distort the distribution. Thus in marine mammals, almost nothing is known about late juvenile and adult mortality because it takes place at sea. Similarly, little is known about fertility changes and nestling mortality of small animals because of the difficulty of measuring these life stages. For large terrestrial mammals, few studies have attempted to measure adult mortality directly so this stage is likely to be under-represented as it probably is in birds. Thus generalities must remain tentative and future research must concentrate on obtaining a more complete picture of mortality during the life cycle.

Stubbs (1977) concluded that most of the density-dependent mortality in the life cycle acts on the young stages for species living in temporary habitats, while it acts more on fecundity in those species living in permanent habitats. With the present data one can modify this. Species with very high reproductive rates (insects, fish) have early juvenile (eggs, larvae) density-dependent mortality; those with intermediate reproductive rates (birds, small mammals) have late juvenile and prebreeding regulation, while large mammals with low reproductive rates are at least partly regulated through changes in fertility. I suspect this is an incomplete picture and in large mammals at least, more cases of late juvenile density dependence will be found. As noted earlier from Fig. 7.1 the earlier in the life cycle that regulation takes place, the greater the sensitivity the population shows to changes in regulation and hence, one sees greater fluctuations in population size for species with early life-cycle regulation.

CAUSES OF DENSITY-DEPENDENT MORTALITY

A large number of studies have considered the causes of mortality that limit populations. As discussed above, this is a relatively uninformative exercise unless one also knows whether the mortality is density-dependent or not. The *k*-factor analysis of Varley & Gradwell (1968) can provide information on 'key factors', those mortalities which account for most of the total population change from year to year. These can be either density dependent or density independent (Podoler & Rogers 1975) and it is important to identify the causes of key-factor mortality. However, in studies of vertebrate populations even this analysis is rare, so the results of most studies are of limited usefulness. Many, but not all cases were from those referenced in Table 7.1. In addition, some reports describe causes without reference to life stage. Table 7.2 presents the frequency of case histories where the cause of density-dependent mortality was identified.

TABLE 7.2. Number (%) of reports of separate populations recording cause of density dependence. Population totals differ from those in Table 7.1 because some studies report causes but not the life stage for density dependence, and other studies the reverse. For insects, parasites include parasitoids

Group	Space *N* (%)	Food *N* (%)	Predators *N* (%)	Parasites *N* (%)	Disease *N* (%)	Social *N* (%)	Total no. popu-lations
Insects	0	23 (45)	20 (39)	19 (37)	5 (10)	4 (8)	51
Large marine mammals	0	6 (60)	4 (40)	0	0	0	10
Large terrestrial mammals	1 (1)	71 (99)	0	0	2 (3)	0	72
Small mammals	14 (67)	5 (24)	4 (19)	0	0	14 (67)	21
Birds	5 (33)	8 (53)	0	1 (6)	0	7 (47)	15

Sources as for Table 7.1 but in addition, for insects Faeth & Simberloff 1981; Jones 1987; for small mammals other sources in Erlinge *et al.* 1983; Hestbeck 1987; for marine mammals McLaren & Smith 1985; for large terrestrial mammals Sinclair 1977.

Insects

The insects provide an interesting paradox. On the one hand, Table 7.2 indicates that from fifty-one studies, 45% of the populations were at least partly regulated through competition for food causing either larval mortality or reduced fecundity; a conclusion also reached by Dempster (1983) for Lepidoptera. On the other hand, some insect ecologists (e.g. Myers 1987; Price 1987) point to the strong indirect evidence for predator–parasitoid regulation: (i) biological control experiments where introduced pests are regulated by their natural predators, or where weeds are not controlled by introduced insect herbivores because the latter are regulated by predators; (ii) the use of pesticides which release pests by killing their natural predators. For example, the brown planthopper (*Nilaparvata lugens*) is a pest of rice plants and is regulated by spiders causing density-dependent nymphal mortality. Application of insecticide killed the spiders and caused an 800-fold increase in brown planthopper densities (Kenmore *et al.* 1984); and (iii) insect antipredator defences can confer a 30% or greater survival advantage over individuals without the defence (Price 1987). The same life-table data used by Dempster (1983) were analysed by Strong, Lawton & Southwood (1984) and they reached the opposite conclusion, namely that predation was the most frequent cause of regulation. It would appear that life-table analyses are applied mostly to pest species whose populations are known to erupt and reach food limitation, thus they are a biased sample of insect populations and we know little about most insect species that do not noticeably affect their food supply. Moreover, this ignorance lies at the centre of the debate on whether predation or interspecific competition structures ecological communities.

The effect of diseases caused by viruses, bacteria and fungi has been largely overlooked in field studies, but their significance as a cause in the collapse of insect outbreaks (Anderson & May 1980) and in the production of insect population cycles is becoming more apparent (Ewald 1987; Myers 1988). The few cases involving 'social' causes of regulation cited emigration of adults due to crowding. It is clearly not a common form of regulation but it is seen in aphids, locusts and planthoppers.

Fish

Causes of density-dependent mortality in fish stocks are so little known that they have not been included in Table 7.2. It is generally considered that there is an interaction between food supplies and predators affecting early larval stages. High density leads to competition and lack of food which causes slow growth and hence prolongs the stage where fish are vulnerable to predators such as birds, other fish species or even adults of their own species (e.g. Jones 1973; Cushing 1981; Lasker 1985; Le Cren 1987). In essence fish show density-dependent growth.

Large mammals

In marine mammals the predominant cause of regulation has been assumed to be food supplies affecting fertility and growth of juveniles. Recently McLaren & Smith (1985) have suggested that predators on seals such as polar bears, arctic foxes, leopard seals and sharks, may be regulating the populations of several prey species.

For large terrestrial mammals, the overwhelming cause of regulation in Table 7.2 is food supply. Many of these studies involved direct measures of food (e.g. for wildebeest, Sinclair, Dublin & Borner 1985) but in other studies, food is inferred from effects on fertility. In contrast to this, there is a general perception amongst North American wildlife ecologists that large mammal herbivores are regulated by predators (Keith 1974; Bergerud, Wyett & Snider 1983; Bergerud 1988). Unfortunately, no studies demonstrate the density-dependent effects of predation, most contenting themselves with the observation that populations increase when predators are removed. This rather predictable result merely allows the trivial conclusion that predators are limiting their prey populations but it does not imply regulation. In Africa, there is some circumstantial evidence for regulation of prey populations by predators, for example resident antelopes in the Serengeti Park by predators in general (Sinclair 1985) and wildebeest by lions in the Kruger Park (Smuts 1978), but conclusive evidence is lacking. The absence of critical evidence for density-dependent effects by predators highlights the lack of appropriate questions and research programmes in wildlife ecology.

To demonstrate regulation of prey by predators from a predator removal experiment, it is not enough to show an increase in the prey population. Regulation is demonstrated only if: (i) the prey population increases when predators are removed; and (ii) the prey population does not return to its original density after predators have been allowed to return to their previous levels. This can be illustrated by reference to Fig. 7.1(c) where a prey population moves from K_3 (predator regulated) to K_1 then back to K_2 (predator limited but not regulated). In any other case (initial prey population at K_4 or K_2) the removal experiment produces the same qualitative behaviour and therefore, ambiguous results; in these cases k-factor analysis is needed in addition. So far none of these conditions has been met. Large mammal populations may well be regulated by predators but this has not yet been demonstrated.

Small mammals

In fourteen of the twenty-one small mammal studies (Table 7.2) the important cause of regulation was density-dependent exclusion of juveniles from breeding colonies, and presumably this exclusion was due to competition for space. Lidicker (1975) considered that in the microtine rodents which show population cycles, juveniles emigrate voluntarily in increasing populations before resources become limiting. However, the difficulties of measuring resources for rodents are great and it remains open whether or not emigrants are detecting early signs of food depletion and using this as a cue to leave (which would mean regulation by interference competition for food). Various food addition experiments (e.g. Taitt & Krebs 1981; Ims 1987) show that rodent populations are monitoring their food supplies, and Ostfeld (1985) considered that food supplies are the underlying cause of territoriality in microtines.

Predation is implicated in the regulation of populations of three species, field voles (*Microtus agrestis*), wood mice (*Apodemus sylvaticus*) and snowshoe hares (Keith *et al.* 1977; Erlinge *et al.* 1983; Trostel *et al.*

1987). In all cases, predators acted in a delayed density-dependent way (Fig. 7.9). For voles and mice in southern Sweden, the delayed action of predators (mainly foxes and buzzards) was within the annual cycle. Erlinge *et al.* (1983) argued that heavy predation acted to dampen the 4–5 year cycles which occur in northern Sweden. However, Kidd & Lewis (1987) have shown that these predators act, if anything, in an inverse density-dependent way and are thus destabilizing; they concluded that these predators are merely limiting and not regulating. Trostel *et al.* (1987) reported the delayed action of raptors (great horned owls, goshawks) and carnivores (lynx, coyotes) causing winter mortality on snowshoe hares, and suggested through a simulation model, that this process could produce the 10-year cycle of numbers.

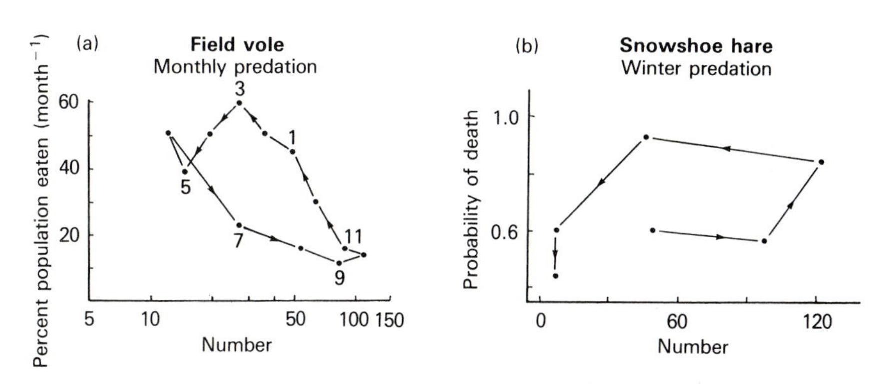

FIG. 7.9 Delayed density-dependent predation on small mammals, indicated by an anti-clockwise spiral when points are joined in temporal sequence.
(a) 'Monthly predation mortality of field voles (*Microtus agrestis*). [After Erlinge *et al.* (1983).]
(b) Annual predation mortality in winter on snowshoe hares (*Lepus americanus*). [After Trostel *et al.* (1987).]

In most studies of small mammals, predation has not been measured. Indirect evidence, for example from the provision of cover (Taitt *et al.* 1981), showed that predation by birds could be important in causing the decline phase in vole cycles and suggests more attention should be paid to this aspect.

Birds

Predation is noticeably absent as a cause of regulation in bird populations although it is suggested as a limiting factor in some gallinaceous and grouse species (Newton 1980). Many studies have recorded the whole life cycle and it is unlikely that predation has been overlooked. These studies have, however, been in temperate regions and it is quite possible that predation could be a regulating process in the tropics.

Lack (1966) proposed that bird populations were limited by food shortage. Despite much uncritical and circumstantial evidence concerning the role of food supply, there seems to be support for Lack's view that food is an ultimate limiting factor (Newton 1980). Whether food shortage acts as a direct regulating factor is not so clear. Table 7.2 shows that foods shortage was the cause in 53% of the cases where density-dependence was recorded, while competition for territorial space or nesting sites accounted for 33% of the cases.

The cases where space is regulating are included as 'social' causes of regulation together with feeding interference (Goss-Custard 1980). Territoriality has been cited as a cause of regulation, particularly in grouse populations (Watson 1967; Zwickel 1980; Hannon, Sopuck & Zwickel 1982). The implication is that territoriality prevents some birds breeding (demonstrated by removal of breeding birds and observing others taking over) and thereby depresses recruitment in a density-dependent way. This predicts an increasing proportion of non-breeders as density increases, and has been demonstrated in song sparrows (Fig. 7.10). If all those without territories die, as in red grouse (Watson & Moss 1980; Watson 1985), then one can conclude that competition for territorial space is regulating. If food addition experiments show that territory decreases in size, (Watson, Moss & Parr 1984) then, as suggested by Lack (1966), the regulating factor is competition for food, and territoriality (or other forms of social behaviour) is the secondary, proximate mechanism which determines who obtains the food. In this case, territoriality (i.e. competition for space) is not a regulating factor. Therefore, both food and territoriality are limiting factors (Watson & Moss 1970; Goss-Custard 1980) but not necessarily regulating factors. Food and spacing behaviour may alternate in their effects depending on density (Boag & Schroeder 1987). Parasites (helminths) and spacing behaviour may act synergistically to regulate red grouse populations, and parasites may be producing cycles in grouse numbers (Hudson, Dobson Newborn 1985) in a way reminiscent of diseases acting on insect populations (Anderson & May 1980).

In many studies, it is concluded that territoriality regulates breeding numbers (Brown 1969; Patterson 1980) but often there is little knowledge about the non-breeding population. Without information on the non-territory holders, one cannot conclude that territorial behaviour regulates the population, for there could be a compensating density-

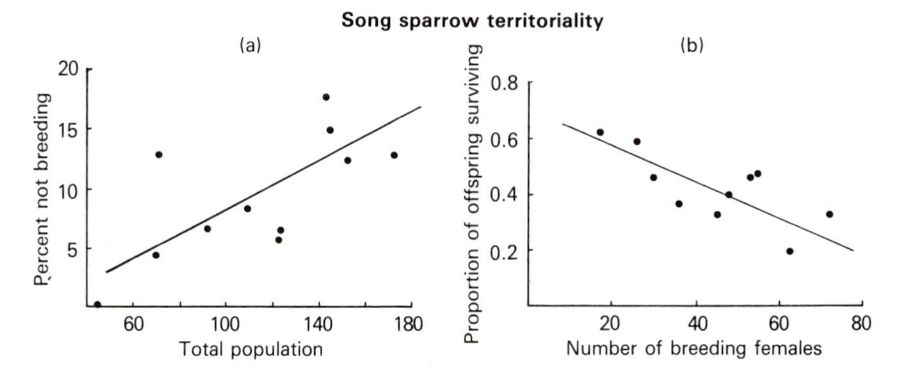

FIG. 7.10 Two periods of density dependence during the year for song sparrows. (Data from P. Arcese & J. N. M. Smith, personal communication.)
(a) Proportion of total population (males plus females) on Mandarte Island, British Columbia, excluded from breeding. During the breeding season territoriality excludes an increasing proportion as density increases.
(b) Proportion of offspring surviving in the non-breeding season declines with density of females.

dependent mortality either amongst the non-territory holders or in the non-breeding season (Hannon & Zwickel 1987). For example, in the song sparrow territoriality produces a density-dependent proportion of non-breeders (Fig. 7.10) (P. Arcese & J. N. M. Smith, personal communication), but there is also a density-dependent mortality in the non-breeding season which is equally effective in regulating the population. A food addition experiment allowed many non-breeders to become territory holders. Thus, much of the regulation took place before breeding and territoriality merely determined who could breed (Arcese & Smith 1988). A similar situation has been reported for mallards (Hill 1984). In summary, before regulation can be attributed to spacing behaviour, it is necessary to measure the mortality of birds in the non-breeding season; many of the studies claiming that territoriality of breeding birds regulates a population are, therefore, premature.

To conclude this section on causes of regulation, are there any clear generalities? At present the answer is no, partly because the picture is incomplete. Both predation and resource limitation are operating in some groups. Nevertheless predation has not been studied adequately in any animal group and the effects of disease and parasitism have been almost entirely neglected. The importance of epidemiological studies is now being recognized, but far more basic information is required to develop these aspects.

NON-EQUILIBRIUM CONDITIONS, MULTIPLE STABLE STATES AND THE WAY AHEAD

The high variance in numbers in many natural populations has challenged ecologists to explain the persistence of such populations (Connell & Sousa 1983; Schoener 1985; Ostfeld 1988). It is recognized that unregulated hypothetical populations describe random walks over time and eventually go extinct (Reddingius 1971). Since all species do go extinct, it is possible that present day populations are merely those following a random walk in transition to extinction. Strong (1984) argues that extinctions in random walk models are far too frequent to mimic the behaviour of real populations. However, there is no way that this statement can be tested. The strength of the regulation argument is that it predicts the existence of a process (density dependence) which can be measured in nature. The philosophical weakness of the argument is that it is very hard to reject; absence of density dependence can too easily be explained away on *ad hoc* methodological grounds.

The original alternative hypothesis to regulation was that of Andrewartha & Birch (1954): namely that density-independent factors changed sufficiently frequently that at least some sub-populations remained extant, and were able to recolonize other areas where sub-populations had become extinct. They considered that this process of chance events is particularly important at the edge of a species' range. The idea of many sub-populations recolonizing local extinctions and resulting in persistence without regulation was later developed more fully (e.g. Reddingius & Den Boer 1970). Although the idea continues to attract some workers (Ehrlich *et al.* 1972), the weight of evidence is in favour of the regulation hypothesis.

Nicholson (1958) recognized that where the environment fluctuates, populations may become extinct and are later reestablished; in these situations the regulatory factors would be relatively weak and masked by density-independent factors. Such a scenario can be illustrated by Fig. 7.11 which shows how stochasticity in the density-independent mortality results in greater changes in population size when the density-dependent mortality is weak than when it is strong. These ideas have been rephrased in recent years under the heading of 'density -vague regulation' by which Strong (1984, 1986) suggested there is stochasticity in both the density-dependent as well as the density-independent variables. In addition Strong (1984) stated that '... regulation at extremes of density can involve very different mechanisms from mechanisms around central or average densities'. Since average densities tend to occur in the centre of a species' range, this is not saying much more than that originally

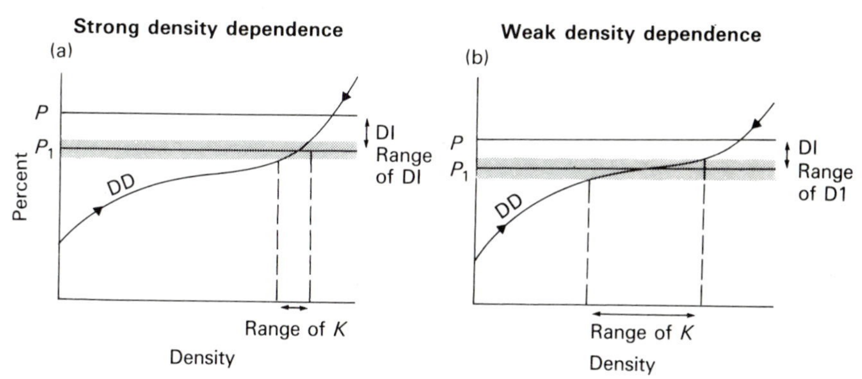

FIG. 7.11 Effect of stochasticity in the density-independent mortality (DI), the range of which is indicated by the shaded area, results in less fluctuation in the equilibrium K, when the density-dependent mortality (DD) is strong (A) than when it is weak (B).

conceived by Nicholson, and which is illustrated by the curvilinear density dependence in Fig. 7.1(b). Although different causes of regulation may operate at the centre and edge of a species' distribution (e.g. Randall (1982) described how the moth, *Coleophora alticolella*, is regulated at low altitude by predators, mid-altitude by competition for food, and at high altitude is subjected to fluctuation and local extinction by climate), such differences in mechanisms are not a necessary condition for curvilinear density dependence. In short, density vague phenomena may simply be caused by curvilinear density dependence so that at some densities regulation is strong, at others weak (e.g. Whittaker 1971; Elliott 1987).

Nevertheless, it is quite conceivable that multiple equilibria exist for a population [Fig 7.1(c)]. The ideas concerning multiple stable states (Holling 1973; May 1977) can be applied to both ecosystems and populations. Indirect evidence for multiple equilibria comes from studies on the spruce budworm (*Choristoneura fumiferana*) (Peterman, Clark & Holling 1979) [Fig. 7.12(a)] and other forest insects (McNamee, McLeod & Holling 1981; Berryman, Stenseth & Isaev 1987), salmonids (Peterman, Clark & Holling 1979), marine fish stocks and plankton (Steele & Henderson 1981, 1984), and wildebeest (Walker & Noy-Meir 1982). Many of these examples suggest regulation by predators at lower densities and by competition for food at higher densities. Escape from lower stable points could occur through a variety of mechanisms; for example, through the growth of forest food supply for the spruce budworm (Peterman, Clark & Holling 1979), rapid immigration in other forest insects (Berryman, Stenseth & Isaev 1987), and by climatic effects on the grass food of wildebeest. In addition the existence of multiple equilibria can depend on the behaviour of a population. For example,

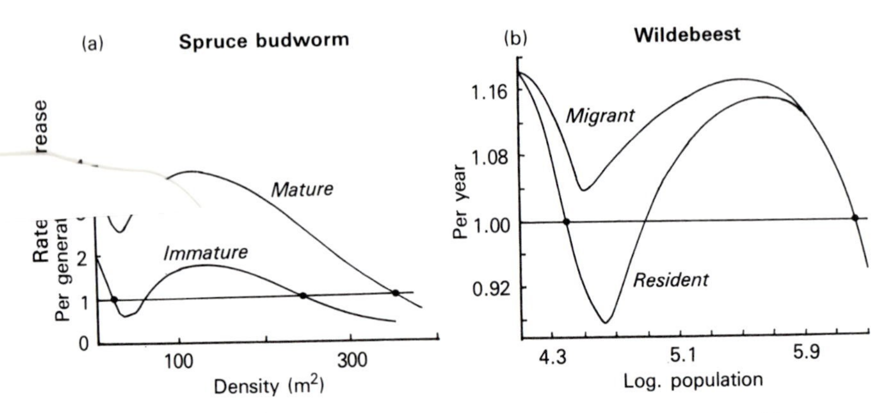

FIG. 7.12 Examples of possible multiple stable states, where the ratio N_{t+1}/N_t is plotted against density.
(a) Spruce budworm in immature conifer forests has two stable points (•). As the forest matures the lower point disappears, (redrawn from Peterman, Clark & Holling 1979). (b) Wildebeest resident populations could have two stable points, whereas migrant populations have only one. Modelled for the 25 000 km² Serengeti ecosystem, Tanzania. [After Fryxell, Greever & Sinclair (1988).]

migratory wildebeest in Serengeti [Fig. 7.12(B)] have only one upper equilibrium point, set by food supply as they possibly cannot be regulated by predators. In contrast, resident populations have two stable points and may be regulated by predators (Fryxell, Greever & Sinclair 1988). Hestbeck (1982, 1987) has suggested that small mammal populations can be regulated at different levels by spacing behaviour, predation and food supply.

These examples remain largely hypothetical interpretations of indirect evidence. Although there is some direct evidence of multiple stable states in jack pine sawfly (*Neodiprion swainei*) (McLeod 1979; Holling 1986) evidence on the causes of these equilibria is almost entirely lacking. There are also alternative interpretations of the spruce budworm data suggesting that there is no predator stability point (Royama 1984). The concept of multiple equilibria requires field testing both by field perturbation experiments (see above) or by measures of density dependence. The existence of multiple equilibria predicts: (i) that there are two (or more) densities where there is strong density dependence; (ii) that some of the percentage mortality at lower density should be higher than some of that at higher density, i.e. the two *k*-factors should overlap; and consequently (iii) there should be an inverse density-dependent effect at an intermediate density preceding a direct density-dependent effect at higher density.

These predictions have not yet been explicitly tested, but there is one example which has superficially similar phenomena. Stenning, Harvey &

Campbell (1988) showed for pied flycatchers (Fig. 7.13) two mortalities k_1, k_4 which conform to predictions (i) and (ii), and a third mortality k_3, which conforms to prediction (iii). True agreement with these predictions requires that there should be different underlying causes for the different slopes.

So far I have discussed multiple causes of regulation in terms of sequential effects at different densities. It is possible, however, that two

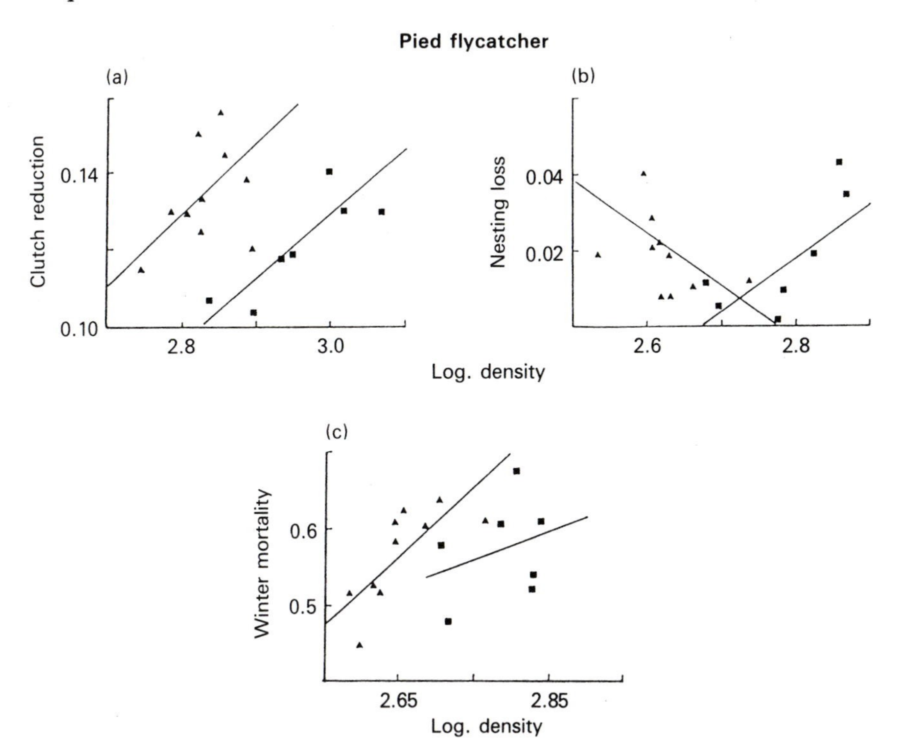

FIG. 7.13 Possible evidence for multiple stable states in the pied flycatcher (*Ficedula hypoleuca*) comes from changes in mortality in the time periods 1948–54. (triangles) and 1955–64 (squares). Density for 25 ha in the Forest of Dean, Gloucestershire. [After Stenning, Harvey & Campbell (1988).]
(a) k-value for clutch size reduction shows two densities where there is strong density dependence.
(b) k-value for mortality of nestlings shows inverse density dependence at lower densities and positive density dependence at higher densities.
(c) k-value for winter mortality also shows two densities with positive density dependence.

or more causes may be operating simultaneously and synergistically. This is an area of work that deserves far more attention, but some recent examples are pointing the way ahead. The introduced winter moth (*Operophtera brumata*) in Nova Scotia was originally thought to be controlled by introduced parasitoids. Recent re-analysis of the data has shown that pupal mortality is the regulating stage and this mortality is caused by beetle larvae but only after parasitoid introduction; in effect parasitized pupae grow slower and are thus more vulnerable to predation (Roland 1988). A similar interactive effect of predators and parasites is reported by Jones (1987), where larvae of the butterfly *Pieris rapae*, which were parasitized by braconid parasites, then became vulnerable to ant predation. Essentially parasitism slowed growth (analogous to the density-dependent growth caused by lack of food in many fish populations) and resulted in density-dependent predation. In other invertebrate species, including cockroaches, amphipods, isopods and ostracods, Moore (1984) demonstrated how acanthocephalan parasites alter the behaviour of hosts so that they become more vulnerable to predators and Curtis (1987) described how trematodes alter snail behaviour. Hudson, Dobson & Newborn (1985) have proposed an interaction of parasitic worms and behaviour as the regulating process in red grouse , and an interaction of disease and food could regulate some ungulate populations (Sinclair 1977). Predation not only mitigates competition but it can also promote it. Mittelbach & Chesson (1987) showed how predators concentrate small size classes of several species into a common protective habitat, thereby intensifying competition.

Apart from studies of fish populations, a few other examples suggest an interaction between food and predators; in snowshoe hares, one hypothesis proposes a simultaneous effect of food shortage and predation causing the decline phase of the 10-year cycle (Trostel *et al.* 1987). This predicts that at high density, animals that run out of food are not only slower to avoid predators but also expose themselves longer to predation; essentially this is density-dependent antipredator behaviour. In a remarkable series of studies, Goss-Custard and co-workers (e.g. Goss-Custard & Durell 1987; Selman & Goss-Custard 1988) have shown how shore bird populations may be regulated in winter through the interaction of feeding behaviour, age and social status, and food supply. McNamara & Houston (1987) have modelled the interaction of feeding and antipredator behaviour in terms of the individual; we now need to incorporate the effects of population density in the models.

CONCLUSIONS

Perhaps the first thing that is needed in future work is a more rigorous attention to terminology to avoid the continuing confusion of meaning in the literature. A survey of the literature has shown that we still have a poor understanding of where density dependence occurs in the life cycle of almost every group of animals. There are very few studies in the tropics so far, a major gap since it is likely that dynamic processes there could be very different from those in temperate regions; our knowledge of regulation in marine and aquatic life cycles is also minimal.

Further work is needed on the causes of density dependence in the rarer insect species, nearly all fish populations and in the effects of predation on birds and mammals. We need to pay attention to single or multiple causes of regulation, and if multiple, how the various factors interact, for example whether sequentially, synergistically or antagonistically.

Studies are beginning to show that regulation is not always constant. Long-term studies indicate temporal changes in the strength of regulation and other dynamic behaviour in the same population, while other studies have recorded spatial differences in regulation. For the most part, we know nothing about the underlying causes of these changes; are they local or global, short or long term? Are they due to qualitative differences in habitat and species composition or to quantitative shifts in variables? McCauley & Murdoch (1987) provide one approach for *Daphnia* but we have hardly started to explore these aspects. They are important from the point of view of biological control, for we need to know the conditions when control could operate and when it might break down.

There is still a paucity of good perturbation experiments, both to demonstrate regulation and to understand its mechanism. Aquatic species are difficult subjects for the detection of density dependence, but equally they are very suitable for perturbation experiments; progress could be made in this area.

Despite these gaps in our knowledge we have advanced since the 1957 Coldspring Harbor meeting. Unlike that symposium we now have a general model that accommodates all the current ideas on regulation — curvilinear, delayed and inverse density dependence, stochastic effects, and multiple stable states. Understanding of regulation may have been slow, but we should recognize that it takes a long time. We now have a considerable body of evidence demonstrating regulation in many groups; in 1957 there was almost none. Furthermore our ideas have become more sophisticated leading to a recent resurgence of interest in the dynamic behaviour of populations.

In reviewing the history of ideas on regulation, one sees how theories evolve. Theories have not been discarded because of a critical test or series of tests, but rather they wither away either because they are untestable [e.g. Thompson's (1939) multifactor hypothesis] or not heuristic, or there is lack of evidence in their support (e.g. group selection), while others expand because of more evidence in their favour or their power to generate new ideas. Nicholson's theories of equilibrium led in different ways to the works of both Wynne-Edwards (1962) and Lack (1966) and to the outburst of field studies. It is perhaps not appreciated that by 1957 Nicholson had anticipated most of the modern concepts on regulation; local extinctions, curvilinear density dependence and density-vague phenomena.

Perhaps the most interesting and challenging area for future work lies in a better understanding of both the mechanisms and causes producing multiple stable states. At present, we have only circumstantial evidence for these phenomena. This aspect could well be the most important advance we make in population dynamics in the next few decades. Such an understanding is vital not only for the management of ecosystems on a regional scale but also for the understanding of global changes that are now confronting us in stark reality, such as the change in atmospheric carbon dioxide and the ozone layer, the decline of tropical forests, the spread of aridity, the human population explosion in Africa. These are all examples which may well lead to complex, non-linear reponses in populations and ecosystems and which we must be able to predict.

ACKNOWLEDGMENTS

I thank numerous colleagues and students for their helpful comments, in particular Buzz Holling, Judy Myers, Bill Neill, Don Ludwig, Charles Krebs, Rudy Boonstra, Brad Anholt, John Richardson and Locke Rowe. Dennis Chitty was invaluable with the earlier literature and perceptive comments, and Ted Gullison helped with everything.

REFERENCES

Addison, J. T. (1986). Density-dependent mortality and the relationship between size composition and fishing effort in lobster populations. *Canadian Journal of Fisheries and Aquatic Sciences*, **43**, 2360–2367.

Alatalo, R. V. & Lundberg, A. (1984). Density dependence in breeding success of the pied flycatcher (*Ficedula hypoleuca*). *Journal of Animal Ecology*, **53**, 969–977.

Alexander, R. D. & Tinkle, D. W: (eds) (1981). *Natural Selection and Social Behaviour.* Chiron Press, New York.

Anderson, R. M. & May R. M. (1980). Infectious diseases and population cycles of forest insects. *Science*, **210**, 658–661.

Andrewartha, H. G. (1958). The use of conceptual models in ecology. *Cold Spring Harbor Symposia on Quantitative Biology*, **22**, 219–236. Waverly Press, Baltimore.

Andrewartha, H. G. (1961). *Introduction to the Study of Animal Populations.* Methuen, London.

Andrewartha, H. G. & Birch, L. C. (1954). *The Distribution and Abundance of Animals.* University of Chicago Press, Chicago.

Arcese, P. & Smith, J. N. M. (1988). Effects of population density and supplemental food on reproduction in Song Sparrows. *Journal of Animal Ecology*, **57**, 119–136.

Bell, G. (1987). (Review of) Evolution through group selection. V. C. Wynne-Edwards. *Heredity*, **59**, 145–147.

Bergerud, A. T. (1988). Caribou, wolves and man. *Trends in Ecology and Evolution*, **3**, 68–72.

Bergerud, A. T., Wyett, W. & Snider, B. (1983). The role of wolf predation in limiting a moose population. *Journal of Wildlife Management*, **47**, 977–988.

Berryman, A. A., Stenseth, N. C. & Isaev, A. S. (1987). Natural regulation of herbivorous forest insect populations. *Oecologia*, **71**, 174–184.

Birch, L. C. (1958). The role of weather in determining the distribution and abundance of animals. *Cold Spring Harbor Symposia on Quantitative Biology*, **22**, 203–218. Waverly Press, Baltimore.

Blank, T. H., Southwood, T. R. E. & Cross, D. J. (1967). The ecology of the partridge: 1. Outline of the population processes with particular reference to chick mortality and nest density. *Journal of Animal Ecology*, **36**, 649–656.

Boag, D. A. & Schroeder, M. A (1987). Population fluctuations in spruce grouse: what determines their numbers in spring? *Canadian Journal of Zoology*, **65**, 2430–2435.

Boonstra, R. & Boag, P. T. (1987). A test of the Chitty Hypothesis: inheritance of life-history traits in Meadow Voles *Microtus pennsylvanicus. Evolution*, **41**, 929–947.

Boonstra, R. & Rodd, F. H. (1983). Regulation of breeding density in *Microtus pennsylvanicus. Journal of Animal Ecology*, **52**, 757–780.

Brown, J. L. (1969). Territorial behavior and population regulation in birds, a review and re-evaluation. *Wilson Bulletin*, **81**, 293–329.

Bulmer, M. G. (1975). The statistical analysis of density dependence. *Biometrics*, **31**, 901–911.

Caputi, N. & Brown, R. S. (1986). Relationship between indices of juvenile abundance and recruitment in the western rock lobster (*Panulirus cygnus*) fishery. *Canadian Journal of Fisheries and Aquatic Sciences*, **43**, 2131–2139.

Chambers, R. J. Wellings, P. W. & Dixon, A. F. G. (1985). Sycamore aphid numbers and population density II. Some processes. *Journal of Animal Ecology*, **54**, 425–442.

Charnov, E. L. (1986). Life history evolution in a 'recruitment population': why are adult mortality rates constant? *Oikos*, **47**, 129–134.

Chittleborough, R. G. & Phillips, B. F. (1975). Fluctuations of year-class strength and recruitment in the western rock lobster *Panulirus longipes* (Milne-Edwards). *Australian Journal of Marine and Freshwater Research*, **26**, 317–328.

Chitty, D. (1960). Population processes in the vole and their relevance to general theory. *Canadian Journal of Zoology*, **38**, 99–113.

Chitty, D. (1967). The natural selection of self-regulation behaviour in animal populations. *Proceedings of the Ecological Society of Australia*, **2**, 51–78.

Chitty, D. (1987). Social and local environments of the vole *Microtus townsendii. Canadian Journal of Zoology*, **65**, 2555–2566.

Christian, J. J. (1961). Phenomena associated with population density. *Proceedings of the National Academy of Science, USA*, **47**, 428–429.

Christian, J. J. (1980). Endocrine factors in population regulation. *Biosocial Mechanisms of Population Regulation* (Ed. by M. N. Cohen, R. S. Malpass & H. G. Klein), pp. 55–115. Yale University Press, New Haven.

Christian, J. J. & Davis D. E.(1964). Endocrines, behaviour and population. *Science*, **146**, 1550–1560.

Clark, L. R., Geier, P. W., Hughes, R. D. & Morris, R. F. (1967). *The Ecology of Insect Popu-*

lation in Theory and Practice. Methuen, London.

Clutton-Brock, T. H., Major, M. & Guinness, F. E. (1985). Population regulation in male and female red deer. *Journal of Animal Ecology*, **54**, 831–846.

Cockburn, A. (1981). Population processes of the Silky Desert Mouse *Pseudomys apodemoides* (Rodentia), in mature heathlands. *Australian Journal of Zoology*, **8**, 499–514.

Connell, J. H. & Sousa, W. P. (1983). On the evidence needed to judge ecological stability or persistence. *American Naturalist*, **121**, 789–824.

Curtis, L. A. (1987). Vertical distribution of an estuarine snail altered by a parasite. *Science*, **235**, 1509–1511.

Cushing, D. H. (1981). *Fisheries Biology, 2nd edition.* University of Wisconsin Press, Madison.

Cushing, D. H. & Harris, J. G. K. (1973). Stock and recruitment and the problem of density dependence. *Fish Stocks and Recruitments* (Ed. by B. B. Parrish), pp.142–155. Conseil International pour I'Exploration de la mer, Charlottenlund Slot, Denmark.

Darwin, C. (1859). *On the Origin of Species by Means of Natural Selection, or the Preservation of Favoured Races in the Struggle for Life.* John Murray, London.

Dempster, J. P. (1983). The natural control of populations of butterflies and moths. *Biological Reviews*, **58**, 461–481.

Diamond, J. M. (1969). Avifaunal equilibria and species turnover rates on the Channel Islands of California. *Proceedings of the National Academy of Science, USA*, **64**, 57–63.

Doherty, P. J. (1983). Tropical territorial damselfishes: is density limited by aggression or recruitment? *Ecology*, **64**, 176–190.

Ebling, F. J. & Stoddart, D. M. (eds) (1978). *Population Control by Social Behavior.* Praeger Publishers, New York.

Ehrlich, P. R., Breedlove, D. E., Brussard, P.F. & Sharp M. A. (1972). Weather and 'regulation' of subalpine populations. *Ecology*, **53**, 243–247.

Ekman, J. (1984). Density-dependent seasonal mortality and population fluctuations of the temperate zone willow tit *(Parus montanus). Journal of Animal Ecology*, **53**, 119–134.

Elliott, J. M. (1987). Population regulation in contrasting populations of trout *Salmo trutta* in two Lake District streams. *Journal of Animal Ecology*, **56**, 83–98.

Elton, C.S. (1949). Population interspersion: an essay on animal community patterns. *Journal of Ecology*, **37**, 1–23.

Elton, C. S. (1955). Natural control of insect populations. *Nature*, **176**, 419.

Enright, J. T. (1976). Climate and population regulation: The biogeographer's dilemma. *Oecologia*, **24**, 295–310.

Erlinge, S., Goransson, G., Hansson, L., Hogstedt, G., Liberg, O., Nilsson, I. N., Nilsson, T., von Schantz, T. & Sylven, M. (1983). Predation as a regulating factor in small rodent populations in Southern Sweden. *Oikos*, **40**, 36–52.

Ewald, P. W. (1987). Pathogen-induced cycling of outbreak insect populations. *Insect Outbreaks* (Ed. by P. Barbosa & J. C. Schultz), pp. 269–286. Academic Press, San Diego.

Faeth, S. H. & Simberloff, D. (1981). Population regulation of a leaf-mining insect, *Cameraria* sp. Nov., at increased field densities. *Ecology*, **62**, 620–624.

Fedigan, L. M. (1983). Dominance and reproductive success in primates. *Yearbook of Physical Anthropology*, **26**, 91–129.

Fogarty, M. J. & Idoline, J. S. (1986). Recruitment dynamics in an American lobster *(Homarus americanus)* population. *Canadian Journal of Fisheries and Aquatic Sciences*, **43, 2368–2376.**

Fowler, C. W. (1981). Density dependence as related to life history strategy. *Ecology*, **62**, 602–610.

Fowler, C. W. (1987). A review of density dependence in populations of large mammals. *Current Mammalogy*, Vol 1 (Ed. by H. H. Genoways), pp. 401–441. Plenum Press, New York.

Fowler, C. W. & Smith, T. D. (eds) (1981). *Dynamics of Large Mammal Populations*, John Wiley & Sons, New York.

Fryxell, J. M., Greever, J. & Sinclair, A. R. E. (1988). Why are migratory ungulates so abundant? *American Naturalist*, **131**, 781–798.

Gage, S. H., Miller, C. A. & Mook, L. J. (1970). The feeding response of some forest birds to the black-headed budworm. *Canadian Journal of Zoology*, **48**, 359–366.

Galindo, C. & Krebs, C. J,. (1987). Population regulation in deermice: the role of females. *Journal of Animal Ecology*, **56**, 11–23.

Gaston, K. J. & Lawton, J. H. (1987). A test of statistical techniques for detecting density-dependence in sequential censuses of animal populations. *Oecologia*, **74**, 404–410.

Goss-Custard, J. D. (1980). Competition for food and interference among waders. *Ardea*, **68**, 31–52.

Goss-Custard, J. D. & Durell S. E. A. le V. dit. (1987). Age-related effects in oystercatchers, *Haematopus ostralegus*, feeding on mussels, *Mytilus edulis*. II. The effect of interference on overall intake rate. *Journal of Animal Ecology*, **56**, 549–558.

Greig-Smith, P. W. (1985). Winter survival, home ranges and feeding of first-year and adult bullfinches. *Behavioural Ecology* (Ed. by R. M. Sibly & R. H. Smith), pp. 387–392. Blackwell Scientific Publications, Oxford.

Hancock, D. A. (1973). The relationship between stock and recruitment in exploited invertebrates. *Fish Stocks and Recruitment* (Ed. by B. B. Parrish), pp. 113–131. Conseil International pour l'Exploration de la mer, Charlottelund Slot, Denmark.

Hannon, S. J., Sopuck, L. G. & Zwickel F. C. (1982). Spring movements of female blue grouse: evidence for socially induced delayed breeding in yearlings. *Auk*, **99**, 687–694.

Hannon S. J. & Zwickel, F. C. (1987). Spacing behavior and population regulation in female blue grouse. *Auk*, **104**, 344–345.

Harvey, P. (1985). Intrademic group selection and the sex ratio. *Behavioural Ecology* (Ed. by R. M. Sibly & R. H. Smith), pp. 59–71. Blackwell Scientific Publications, Oxford.

Hassell, M. P. (1975). Density dependence in single-species populations. *Journal of Animal Ecology*, **44**, 283–295.

Hassell, M. P. (1985). Insect natural enemies as regulating factors. *Journal of Animal Ecology*, **54**, 323–334.

Hassell, M. P. (1987). Detecting regulation in patchily distributed animal populations *Journal of Animal Ecology*, **56**, 705–713.

Hassell, M. P., Southwood, T. R. E. & Reader, P. M. (1987). The dynamics of the viburnum whitefly (*Aleurotrachelus jelinekii*): a case study of population regulation. *Journal of Animal Ecology*, **56**, 283–300.

Healey, M. C. (1967). Aggression and self-regulation of population size in deermice. *Ecology*, **48**, 378–392.

Hestbeck, J. B. (1982). Population regulation of cyclic mammals: the social fence hypothesis. *Oikos*, **39**, 157–163.

Hestbeck, J. B. (1987). Multiple regulation states in populations of small mammals: a state-transition model. *American Naturalist*, **129**, 520–532.

Hill, D. A. (1984). Population regulation in the mallard (*Anas platyrhynchos*). *Journal of Animal Ecology*, **53**, 191–202.

Hill, M. G. (1988). Analysis of the biological control of *Mythimna separata* (Lepidoptera: Noctuidae) by *Apantiles ruficrus*, (Braconidae: Hymenoptera) in New Zealand. *Journal of Applied Ecology*, **25**, 197–208.

Holden, M. J. (1973). Are long-term sustainable fisheries for elasmobranchs possible? *Fish Stocks and Recruitment* (Ed. by B. B. Parrish), pp. 360–367. Conseil International pour l'Exploration de la mer, Charlottelund Slot, Denmark.

Holling, C. S. (1973). Resilience and stability of ecological systems. *Annual Review of Ecology and Systematics*, **4**, 1–23.

Holling, C. S. (1986). The resilience of terrestrial ecosystems: local surprise and global change. *Sustainable Development of the Biosphere* (Ed. by W. C. Clark & R. E. Munn), pp. 292–317. Cambridge University Press, Cambridge.

Houston D. B. (1982). *The Northern Yellowstone Elk.* Macmillan Publishing Company, New York.

Howard, L. O. & Fiske, W. F. (1911). The importation into the United States of the parasites of the gipsy moth and the brown-tail moth. *Bulletin of the US Bureau of Entomology*, No. **91**.

Hudson, P. J., Dobson, A. P. & Newborn, D. (1985). Cyclic and non-cyclic populations of red grouse: a role for parasitism? *Ecology and Genetics of Host-Parasite Interactions* (Ed. by D. Rollinson & R. M. Anderson) p. 77–89. Academic Press, London.

Iles, T. D. (1973). Interaction of environment and parent stock size in determining recruitment in the Pacific sardine as revealed by density dependent O-group growth. *Fish Stocks and Recruitment* (Ed. by B. B. Parrish), pp. 229–240. Conseil International pour l'Exploration de la mer, Charlottenlund Slot, Denmark.

Ims, R. A. (1987). Responses in spatial organization and behaviour to manipulations of the food resource in the vole *Clethrionomys rufocanus. Journal of Animal Ecology*, **56**, 585–596.

Jamieson, G. S. (1986). Implications of fluctuations in recruitment in selected crab populations. *Canadian Journal of Fisheries and Aquatic Sciences* **43**, 2085–2098.

Jannett, F. J. (1978). The density dependent formation of extended maternal families of the montane vole, *Microtus montanus nanus. Behavioural Ecology and Sociobiology*, **3**, 245–263.

Jones, R. (1973). Density dependent regulation of the numbers of cod and haddock. *Fish Stocks and Recruitment* (Ed. by B. B. Parrish), pp. 156–173. Conseil International pour l'Exploration de la mer, Charlottenlund Slot, Denmark.

Jones, R. E. (1987). Ants, parasitoids, and the cabbage butterfly *Pieris rapae. Journal of Animal Ecology*, **56**, 739–749.

Jones, R. E., Nealis, V. G., Ives, P. M. & Scheermeyer, R. (1987). Seasonal and spatial variation in juvenile survival of the cabbage butterfly *Pieris rapae*: evidence for patchy density dependence. *Journal of Animal Ecology*, **56**, 723–737.

Keith, L. B. (1974). Some features of population dynamics in mammals. *11th International Congress of Game Biologists, Stockholm*, pp. 17–57.

Keith, L. B., Todd, A. W., Brand, C. J., Adamcik, R. S. & Rusch, D. H. (1977). An analysis of predation during a cyclic fluctuation of snowshoe hares. *13th International Congress of Game Biologists*, 151–174.

Kenmore, P. E., Carino, F. O., Perez, C. A., Dyck, V. A. & Gutierrez, A. P (1984). Population regulation of the rice brown planthopper (*Nilaparvata lugens* Stal) within rice fields in the Philippines. *Journal of Plant Protection in the Tropics*, **1**, 19–37.

Kerfoot, W. C., DeMott, W. R. & DeAngelis, W. R. (1985). Interactions among cladocerans: food limitation and exploitative competition. *Food Limitation and the Structure of Zooplankton Communities* (Ed. by W. Lampert). *Archives fur Hydrobiologie*, **21**, 431–451.

Kidd, N. A. C. & Lewis, G. B. (1987). Can vertebrate predators regulate their prey? A reply. *American Naturalist*, **130**, 448–453.

Klomp, H. (1962). The influence of climate and weather on the mean density level, the fluctuations and the regulation of animal populations. *Archives Neerlandaises Zoologie*, **15**, 68–109.

Kluyver, H. N. (1971). Regulation of numbers in populations of Great Tits (*Parus m. major* L.). *Dynamics of Populations* (Ed. by P. J. den Boer & G. R. Gradwell), pp. 507–523 Pudoc, Wageningen.

Krebs, C. J. (1978). A review of the Chitty Hypothesis of population regulation. *Canadian Journal of Zoology*, **56**, 2463–2480.

Krebs, C. J. (1985). *Ecology: the Experimental Analysis of Distribution and Abundance*, 3rd edition. Harper & Row, New York.

Krebs, C. J., Gaines, M. S., Keller, B. L., Myers, J. H. & Tamarin, R. H. (1973). Population cycles in small rodents. *Science*, **179**, 35–41.

Krebs, C. J. & Myers, J. (1974). Population cycles in small mammals. *Advances in Ecological Research*, **8**, 267–399.

Krebs, J. R. (1970). Regulation of numbers in the great tit (Aves: Passeriformes). Journal of Zoology (London), **162**, 317–333.

Kruuk, H. H. & Parish, T. (1987). Changes in the size of groups and ranges of the European badger (*Meles meles L.)* in an area in Scotland. *Journal of Animal Ecology*, **56**, 351–364.

Lack, D. (1954). *The Natural Regulation of Animal Numbers*. Clarendon Press, Oxford.

Lack, D. (1955). Mortality factors affecting adult numbers. *The Numbers of Man and Animals*. (Ed. by J. B. Cragg & N. W. Pirie, pp. 47–56. The Institute of Biology, Oliver & Boyd, London.

Lack, D. (1966). *Population Studies of Birds*. Clarendon Press, Oxford.

Larkin, P. A. (1973). Some observations on models of stock and relationships for fishes. *Fish Stocks and Recruitment* (Ed. by B. B. Parrish), pp. 316–324. Conseil International pour l'Exploration de la mer, Charlottelund Slot, Denmark.

Lasker. R. (1985). What limits clupeoid production? *Canadian Journal of Fisheries and Aquatic Sciences*, **44**, (Suppl. 1) 31–38.

Le Cren, E. D. (1987). Perch (*Perca fluviatilis*) and pike (*Esox lucius*) in Windermere from 1940 to 1985: studies in population dynamics. *Canadian Journal of Fisheries and Aquatic Sciences*, **44**, (Suppl. 2), 216–228.

Lidicker, W. (1975). The role of dispersal in the demography of small mammals. *Small Mammals: Productivity and Population Dynamics* (Ed. by K. Petrusewicz, F. B. Golley & L. Ryszkowski), pp. 103–128. Cambridge University Press, Cambridge.

Maelzer, D. A. (1970). The regression of log N_{n+1} on log N_n as a test of density dependence: an exercise with computer-constructed, density independent populations. *Ecology*, **51**, 810–822.

Malthus, T. R. (1798). *An Essay on the Principle of Population*. Johnson, London.

May, R. M. (1977). Thresholds and breakpoints in ecosystems with a multiplicity of stable states. *Nature*, **269**, 471–477.

May, R. M., Conway, G. R., Hassell, M. P. & Southwood, T. R. E. (1974). Time delays, density dependence and single species oscillations. *Journal of Animal Ecology*, **43**, 747–770.

Maynard Smith, J. (1964). Group selection and kin selection. *Nature*, **201**, 1145–1147.

Maynard Smith, J. (1976). Group selection. *Quarterly Review of Biology*, **51**, 277–283.

McCauley, E. & Murdoch, W. W. (1987). Cyclic and stable populations: plankton as a paradigm. *American Naturalist*, **129**, 97–121.

McCleery, R. H. & Perrins, C. M. (1985). Territory size, reproductive success and population dynamics in the great tit, *Parus major. Behavioural Ecology* (Ed. by R. M. Sibly & R. H. Smith,) pp. 353–373. Blackwell Scientific Publications, Oxford.

McGovern, M. & Tracy, C. R. (1981). Phenotypic variation in electromorphs previously considered to be genetic markers in *Microtus ochrogaster. Oecologia*, **51**, 276–280.

McLaren, I. A. & Smith, T. G. (1985). Population ecology of seals: retrospective and prospective views. *Marine Mammal Science*, **1**, 54–83.

McLeod, J. M. (1979). Discontinuous stability in a sawfly life system and its relevance to pest management strategies. *Selected Papers in Forest Entomology from the 15th International Congress of Entomology* (Ed. by W. E. Waters), pp. 68–81. US Forest Service General Technical Report WO-8.

McNamara, J. M. & Houston, A. I. (1987). Starvation and predation as factors limiting population size. *Ecology*, **68**, 1515–1519.

McNamee, P. J., McLeod, J. M & Holling, C. S. (1981). The structure and behaviour of defoliating insect/forest systems. *Researches in Population Ecology*, **23**, 280–298.

Milne, A. (1957). The natural control of insect populations. *Canadian Entomologist*, **89**, 193–213.

Milne, A. (1958). Theories of natural control of insect populations. *Cold Spring Harbor Symposia on Quantitative Biology*, **22**, 253–271. Waverly Press, Baltimore.

Mittelbach, C. G. & Chesson, P. L. (1987). Predation risk: indirect effects on fish populations. *Predation* (Ed. by W. C. Kerfoot & A. Sih), pp. 315–332. University Press of New England, Hanover.

Moore, J. (1984). Parasites that change the behaviour of their host. *Scientific American*, **250**, 108–115.

Morris, R. F. (ed.) (1963). The dynamics of epidemic spruce budworm populations. *Memoirs of the Entomological Society of Canada*. No. **21**.

Murdoch, W. W. (1970). Population regulation and population inertia. *Ecology*, **51**, 497–502.

Murray, B. G. (1982). On the meaning of density dependence. *Oecologia*, **53**, 370–373.

Myers, J. H. (1987). Population outbreaks of introduced insects: Lessons from the biological control of weeds. *Insect Outbreaks* (Ed. by P. Barbosa, & J. C. Schultz), pp. 173–194. Academic Press, San Diego.

Myers, J. H. (1988). Can a general hypothesis explain population cycles of forest lepidoptera? *Advances in Ecological Research* (Ed. by A. Macfadyen, E. D. Ford, A. H. Fitter & M. Begon) pp.179–234.Academic Press, London .

Myers, J. & Krebs, C. J. (1971). Genetic, behavioral, and reproductive attributes of dispersing field voles *Microtus pennsylvanicus* and *Microtus ochrogaster. Ecological Monographs*, **41**, 53–78.

Newton, I. (1980). The role of food in limiting bird numbers. *Ardea*, **68**, 11–30.

Newton, I. & Marquiss, M. (1986). Population regulation in Sparrowhawks. *Journal of Animal Ecology*, **55**, 463–580.

Nicholson, A. J. (1933). The balance of animal populations. *Journal of Animal Ecology*, **2**, 132–178.

Nicholson, A. J. (1954a). Compensatory reactions of populations to stresses, and their evolutionary significance. *Australian Journal of Zoology*, **2**, 1–8.

Nicholson, A. J. (1954b). An outline of the dynamics of animal populations. *Australian Journal of Zoology*, **2**, 9–65.

Nicholson, A. J. (1958). The self-adjustment of populations to change. *Cold Spring Harbor Symposia on Quantitative Biology*, **22**, 153–173. Waverly Press, Baltimore.

Nicholson, A. J. & Bailey, V. A. (1935). The balance of animal populations. *Proceedings of the Zoological Society of London*, **3**, 551–598.

Nilsson, S. G. (1987). Limitation and regulation of population density in the nuthatch *Sitta europaea* (Aves) breeding in natural cavities. *Journal of Animal Ecology* , **56**, 921–937.

Ostfeld, R. S. (1985). Limiting resources and territoriality in Microtine rodents. *American Naturalist*, **126**, 1–15.

Ostfeld, R. S. (1988). Fluctuations and constancy in populations of small rodents. *American Naturalist* **131**, 445–452.

Parrish, B. B. (ed.) (1973). *Fish Stocks and Recruitment.* Conseil International pour l'Exploration de la mer, Charlottenlund Slot, Denmark.

Patterson, I. J. (1980). Territorial behaviour and the limitation of population density. *Ardea*, **58**, 53–62.

Penn, J. W. & Caputi, N. (1985). Stock recruitment relationships for the tiger prawn, *Penaeus esculentus*, fishery in Exmouth gulf, Western Australia, and their implications for management. *Second Australian National Prawn Seminar* (Ed. by P. C. Rothlisberg, B. J. Hill & D. J. Staples), pp. 165–173. NPS2, Cleveland, Queensland, Australia.

Perrins, C. M. (1965). Population fluctuations and clutch-size in the Great Tit, *Parus major. Journal of Animal Ecology*, **34**, 601–647.

Peterman, R. M., Clark, W. C. & Holling, C. S. (1979). The dynamics of resilience: shifting stability domains in fish and insect systems. *Population Dynamics* (Ed. by R. M. Anderson, B. D. Turner & L. R. Taylor). pp. 321–341. Symposium of the British Ecological Society 20, Blackwell Scientific Publications, Oxford.

Pimentel, D. (1968). Population regulation and genetic feedback. *Science*, **159**, 1423–1437.

Podoler, H. & Rogers, D. (1975). A new method for the identification of key factors from life-table data. *Journal of Animal Ecology*, **44**, 85–114.

Porter, J. M. & Coulson, J. C. (1987). Long-term changes in recruitment to the breeding group, and the quality of recruits at a Kittiwake *Rissa tridactyla* colony. *Journal of Animal Ecology*,**56**, 675–689.

Price, P. W. (1987). The role of natural enemies in insect populations. *Insect Outbreaks* (Ed. by P. Barbosa & J. C. Schultz), pp. 287–312. Academic Press, San Diego.

Randall, M. G. M. (1982). The dynamics of an insect population throughout its altitudinal distribution: *Coleophora alticolella* (Lepidoptera) in northern England. *Journal of Animal Ecology*, **51**, 993–1016.

Reddingius, J. (1971). Gambling for existence. *Acta Biotheoretica, suppl. 1*, **20**, 1–208.

Reddingius, J. & den Boer, P. J. (1970). Simulation experiments illustrating stabilization of animal numbers by spreading of risk. *Oecologia*, **5**, 240–284.

Reynoldson, T. B. (1966). The distribution and abundance of lake dwelling triclads—towards a hypothesis. *Advances in Ecological Research*, **3**, 1–71.

Ricker, W. E. (1954). Stock and recruitment. *Journal of the Fisheries Research Board, Canada*, **11**, 559–623.

Ricklefs, R. E. (1979). *Ecology* (2nd edition). Chiron Press, New York.

Roland, J. (1988). Decline of winter moth populations in North America: direct *v* indirect effect of introduced parasites. *Journal of Animal Ecology*, **57**, 523–531.

Royama, T. (1977). Population persistence and density dependence. *Ecological Monographs*, **47**, 1–35.

Royama, T. (1984). Population dynamics of the spruce budworm *Choristoneura fumiferana. Ecological Monographs*, **54** 429–462.

Ryder, J. P. (1975). Egg-laying, egg size, and success in relation to immature–mature plumage of ring-billed gulls. *Wilson Bulletin*, **87**, 534–542.

Schluter, D. (1988). The evolution of finch communities on Islands and Continents. *Ecological Monographs*: Kenya vs. Galapagos **58**, 229–249.

Schoener, T. W. (1985). Are lizard population sizes unusually constant through time? *American Naturalist*, **126**, 633–641.

Selman, J. & Goss-Custard, J. D. (1988). Interference between foraging redshank, *Tringa totanus. Animal Behaviour*, **36**, 1542–1544.

Sibly, R. M. & Smith, R. H. (eds) (1985). *Behavioural Ecology.* Blackwell Scientific Publications, Oxford.

Sinclair, A. R. E. (1977). *The African Buffalo.* University of Chicago Press, Chicago.

Sinclair, A. R. E. (1985). Does interspecific competition or predation shape the African ungulate community? *Journal of Animal Ecology*, **54**, 899–918.

Sinclair, A. R. E., Dublin, H. & Borner, M. (1985). Population regulation of Serengeti

Wildebeest: a test of the food hypothesis. *Oecologia*, **65**, 266–268.

Skogland, T. (1985). The effects of density-dependent resource limitation on the demography of wild reindeer. *Journal of Animal Ecology*, **54**, 359–374.

Slade, N. A. (1977). Statistical detection of density dependence from a series of sequential censuses. *Ecology*, **58**, 1094–1102.

Smith, H. S. (1935). The role of biotic factors in the determination of population densities. *Journal of Economic Entomology*, **28**, 873–898.

Smuts, G. L. (1978). Interrelations between predators, prey and their environment. *BioScience*, **28**, 316–320.

Solomon, M. E. (1949). The natural control of animal populations. *Journal of Animal Ecology*, **18**, 1–35.

Solomon, M. E. (1959). A symposium on population ecology. *Ecology*, **40**, 325–326.

Solomon, M. E. (1964). Analysis of processes involved in the Natural Control of insects. *Advances in Ecological Research*, **2**, 1–54.

Solomon, M. E., Glen, D. M., Kendall, D. A. & Milsom, N. F. (1976). Predation of overwintering larvae of codling moth (*Cydia pomonella* (L.)) by birds. *Journal of Applied Ecology*, **13**, 341–352.

Southern, H. N. (1970). The natural control of a population of tawny owls (*Strix aluco). Journal of Zoology (London)*, **162**, 197–285.

Southwood, T. R. E. & Reader, P. M. (1976). Population census data and key factor analysis for the viburnum whitefly, *Aleurotrachelus jelinekii* (Fraunenf.), on three bushes. *Journal of Animal Ecology*, **45**, 313–325.

Southwood, T. R. E. & Reader, P. M. (1988). The impact of predation on the viburnum whitefly, (*Aleurotrachelus jelinekii). Oecologia*, **74**, 566–570.

St Amant, J. L. S. (1970). The detection of regulation in animal populations. *Ecology*, **51**, 823–828.

Steele, J. H. & Henderson, E. W. (1981). A simple plankton model. *American Naturalist*, **117**, 676–691.

Steele, J. H. & Henderson, E. W. (1984). Modelling long-term fluctuations in fish stocks. *Science*, **224**, 985–987.

Stenning, M. J., Harvey, P. H. & Campbell, B. (1988). Searching for density-dependent regulation in a population of pied flycatchers *Ficedula hypoleuca* Pallas. *Journal of Animal Ecology*, **57**, 307–317.

Strong, D. R. (1984). Density-vague ecology and liberal population regulation in insects. *A New Ecology* (Ed. by P. W. Price, C. N. Slobodchikoff & W. S. Gaud), pp. 313–327. John Wiley & Sons, New York.

Strong, D. R. (1986). Density-vague population change. *Trends in Ecology and Evolution*, **1**, 39–42.

Strong, D. R., Lawton, J. H. & Southwood, T. R. E. (1984). *Insects on Plants: Community Patterns and Mechanisms.* Harvard University Press, Cambridge, Massachusetts.

Stubbs, M. (1977). Density dependence in the life-cycles of animals and its importance in *K*- and *r*-strategies. *Journal of Animal Ecology*, **46**, 677–688.

Taitt, M. J., Gipps, J. H.W., Krebs, C. J. & Dundjerski, Z. (1981). The effect of extra food and cover on declining populations of *Microtus townsendii. Canadian Journal of Zoology*, **59**, 1593–1599.

Taitt, M. J. & Krebs, C. J. (1981). The effect of extra food on small rodent populations: 11. Voles (*Microtus townsendii*). *Journal of Animal Ecology*, **50**, 125–137.

Tamarin, R. H. (1978). Dispersal, population regulation, and *k*-selection in field mice. *American Naturalist*, **112**, 545–555.

Tamarin, R. H. (1980). Dispersal and population regulation in rodents. *Biosocial Mechanisms of Population Regulation* (Ed. by M. N. Cohen, R. S. Malpass & H. G. Klein), pp. 117–133. Yale University Press, New Haven.

Thompson, W. R. (1929). On natural control. *Parasitology*, **21**, 269–281.

Thompson, W. R. (1939). Biological control and the theories of the interactions of populations. *Parasitology*, **31**, 299–388.

Tinbergen, J. M., Van Balen, J. H. & Van Eck, H. M. (1985). Density dependent survival in an isolated great tit population: Kluyver's data reanalysed. *Ardea*, **73**, 38–48.

Torok, J. & Toth L. (1988). Density dependence in reproduction of the collared flycatcher (*Ficedula albicollis*) at high population levels. *Journal of Animal Ecology*, **57**, 251–258.

Trostel, K., Sinclair, A. R. E., Walters, C. J. & Krebs, C. J. (1987). Can predation cause the 10-year hare cycle? *Oecologia*, **74**, 185–192.

Uvarov, B. P. (1931). Insects and climate. *Transactions of the Entomological Society of London*, **79**, 1–249.

Varley, G. C. (1947). The natural control of population balance in the knapweed gall-fly (*Urophora jaceana*). *Journal of Animal Ecology*, **16**, 139–187.

Varley, G. C. & Gradwell, G. R. (1960). Key factors in population studies. *Journal of Animal Ecology*, **29**, 399–401.

Varley, G. C. & Gradwell, G. R. (1968). Population models for the winter moth. *Insect Abundance* (Ed. by T. R. E. Southwood), pp. 132–142. Oxford University Press, London.

Vickery, W. L. & Nudds, T. D. (1984). Detection of density-dependent effects in annual duck censuses. *Ecology*, **65**, 96–104.

Walker, B. H. & Noy-Meir, I. (1982). Aspects of the stability and resilience of savanna ecosystems. *Ecology of Tropical Savannas* (Ed. by B. J. Huntley & B. H. Walker), pp. 555–590. Springer-Verlag, Berlin.

Watson, A. (1967). Social status and population regulation in the red grouse (*Lagopus lagopus scoticus*). *Proceedings of the Royal Society Study Group*, **2**, 22–30.

Watson, A. (ed.) (1970). *Animal Populations in Relation to their Food Resources*. Symposium of the British Ecological Society No. 10. Blackwell Scientific Publications, Oxford.

Watson, A. (1985). Social class, socially-induced loss, recruitment and breeding of red grouse. *Oecologia*, **67**, 493–498.

Watson, A. & Moss, R. (1970). Dominance, spacing behaviour and aggression in relation to population limitation in vertebrates. *Animal Populations in Relation to Their Food Resources* (Ed. by A. Watson), pp. 167–220. Blackwell Scientific Publications, Oxford.

Watson, A. & Moss, R. (1980). Advances in our understanding of the population dynamics of red grouse from a recent fluctuation in numbers. *Ardea*, **68**, 103–111.

Watson, A., Moss, R. & Parr, R. (1984). Effects of food enrichment on numbers and spacing behaviour of red grouse. *Journal of Animal Ecology*, **53**, 663–678.

Whittaker, J. B. (1971). Population changes in *Neophilaenus lineatus* (L.) (Homoptera: Cercopidae) in different parts of its range. *Journal of Animal Ecology*, **40**, 425–443.

Wiens, J. A. (1966). On group selection and Wynne-Edwards' hypothesis. *American Scientist*, **54**, 273–287.

Wilson, D. S. (1983). The group selection controversy: history and current status. *Annual Review of Ecology and Systematics*, **14**, 159–187.

Wynne-Edwards, V. C. (1962). *Animal Dispersion in Relation to Social Behaviour*. Oliver & Boyd, Edinburgh.

Wynne-Edwards, V. C. (1965). Social organization as a population regulator. *Social Organization of Animal Communities* (Ed. by P. E. Ellis), pp. 173–178. Symposium of the Zoological Society of London, No. 14.

Wynne-Edwards, V. C. (1986). *Evolution through Group Selection*. Blackwell Scientific Publications, Oxford.

Zwickel, F. C. (1980). Surplus yearlings and the regulation of breeding density in blue grouse. *Canadian Journal of Zoology*, **58**, 896–905.

8 COMPETITION

R. LAW[1] AND A.R. WATKINSON[2]
[1]*Department of Biology, University of York, York Y01 5DD, UK*
[2]*School of Biological Sciences, University of East Anglia, Norwich NR4 7TJ, UK*

INTRODUCTION

Competition between organisms is readily demonstrated. We have only to remove the neighbours of a sedentary organism such as a terrestrial plant to observe often an increase in its growth and reproductive capacity (e.g. Clements, Weaver & Hanson 1929; Goldberg 1987; Miller & Werner 1987). Between mobile organisms with more transitory interactions it is a little less easy to demonstrate competition; nonetheless, many field experiments have shown that, by changing the average population density of members of the same or of other species, the average performance of an individual is altered (Connell 1983; Schoener 1983, 1985). Yet, of all the ecological concepts under examination in this symposium, competition has had a particularly chequered history. Despite the recognition of competition as a potential force in nature, going back to Charles Darwin and beyond, the contemporary debate over the part played by competition in organizing the natural world attests to our continuing uncertainty over its role (Schoener 1982; Strong *et al.* 1984; Diamond & Case 1986a). This history is one in which the British Ecological Society has played a prominent part, from the first presidential address by Tansley in 1914 onwards.

To assess the contribution which the concept of competition has made to our understanding of the natural world is quite an undertaking. At times we wondered if the task was possible, and we have therefore deemed it necessary to restrict the scope of this essay in certain ways. Following an outline of what ecologists mean by competition, and a brief historical sketch of the subject, we have chosen three areas for analysis: the role of competition in evolution, the influence of competition on population dynamics, and the competitive structure of communities. The focus is on the empirical literature; we are concerned with what is known about competition in the real world, rather than with theoretically expected outcomes given the assumption of competition (Roughgarden 1986). We have limited the frame of reference further by concentrating on studies in which competition is measured rather than assumed. This is intended in part to prevent overlap with Chapter 4 on the niche, but it also reflects the view that we need a firm foundation of measured interaction strengths, on which to build an analysis of their effects on the structure and dynamics of communities. The scope of the essay is further restricted to competition between species — intraspecific competition being considered only insofar as it is relevant to interspecific competition.

DEFINITIONS — FROM CAUSE TO EFFECT

There are, broadly speaking, two ways in which competition has been defined by ecologists. The first focuses on causes — resources required by different individuals which are shared among them in some way — and leads to definitions in the spirit of Milne (1961) 'Competition is the endeavour of two (or more) animals to gain the same particular thing, or to gain the measure each wants from the supply of a thing when that supply is not sufficient for both (or all)'.

Definitions of this kind have the virtue of drawing attention explicitly to the resource(s) for which individuals compete. They have the drawback that the precise resource or combination of resources may be difficult to determine, and not wholly relevant to certain kinds of study. Some authors have recognized that competition could be one step removed from the resources, coming about from direct conflict between individuals (interference), rather than through their consumption of the same resources (exploitative competition). Birch (1957), for example, added the qualifier '. . . if the resources are not in short supply competition occurs when the animals seeking that resource nevertheless harm one or another in the process'.

Although these definitions are couched in terms of animals, plant ecologists of an ealier age were well aware of these notions of exploitative and interference competition (e.g. Warming 1909, p.366; Tansley 1914; Clements, Weaver & Hanson 1929, p.316). It is worth adding that the exploitative competitive ability of an organism depends both on its ability to reduce the performance of others (competitive effect) and its ability to continue to perform relatively well in the presence of competitors (competitive response) (Goldberg & Fleetwood 1987). Grime (1979) and Tilman (1982) respectively concentrate on only one of these aspects of exploitative competitive ability and as a consequence reach rather different conclusions about the role of competition in determining plant community structure (Thompson 1987; Tilman 1987; Thompson & Grime 1988).

The second approach to defining competition focuses on effects on population dynamics — the reduction in the contribution made by individuals to future generations when they are brought together. This leads to definitions such as that of Odum (1959, pp.226–227) in which two species are said to be in competition if populations of each adversely affect the growth of the other in the struggle for common resources, and that of Williamson (1957) in which two species are in competition when they have a controlling factor in common (a controlling factor being one which restricts growth more severely as population density increases). The merit of this kind of definition is that the ecologist is not reduced to paralysis when confronted with the range of potential interactions in species-rich communities, in which many different combinations of resources could be implicated. Moreover, it opens up the possibility of measuring the strength of competition through manipulation of the densities of the species involved (Bender, Case & Gilpin 1984).

These two ways of defining competition reflect the different kinds of interests of ecologists, from detailed ecophysiology to broad properties of community structure. Being by no means equivalent to one another, it is perhaps inevitable that they have had a somewhat uneasy coexistence in the minds of ecologists and in the literature (e.g. Milne 1961; Williamson 1972, pp.96–97; Tilman 1982, p.6). Here we make use of a fusion of the definitions. The species among which competition could occur are constrained to those which exploit the same class of resources in a similar way, i.e. the guild as defined by Root (1967). Within the guild, competition is said to occur when species reciprocally inhibit each other's population growth. This fusion is not altogether satisfactory, because inhibition may come about for reasons other than the exploitation of common resources, for example, through direct interference or through indirect effects of interactions with species not in the guild (Holt 1977; Jeffries & Lawton 1984). However, we think that this amalgam accurately reflects a widespread usage of the term 'competition' in contemporary literature. Another drawback of the fusion is that resource dynamics are not made explicit (Tilman 1982). We are mindful that, as a result, the state space may not be dynamically sufficient to describe the dynamics of the guild (Lewontin 1974, p.8), but at least this leaves us with a state space which is reasonably clearly defined by the set of species, rather than one which also contains a potentially unbounded collection of resources.

BRIEF HISTORY

The operative word here is 'brief'; we can scarcely do justice to all the ramifications of the subject, and have chosen simply to highlight a small number of books and papers which we believe to have been important in guiding the thoughts of ecologists on competition. For a detailed history prior to 1928, see Clements, Weaver & Hanson (1929), and for recent reviews see Hutchinson (1978), Jackson (1981), Kingsland (1985) and McIntosh (1985). The history of competition is closely linked with that of succession and the niche, and the reader should refer to Chapter 4 for more information.

The roots of the concept of competition lie deep in history. Clements, Weaver and Hanson (1929) suggested that the earliest recognition of its importance, at least for good thinning practice of forest trees, was due to Petrus de Crescentiis in 1305. Malthus (1798, p.76) identified competition for limited resources as the force that checks the growth of animal and plant populations, a concept later put to good use by Darwin (1859). The importance of competition in plant communities was evidently well understood by DeCandolle (1820), who envisaged a battle for space among plants which live together. A more sophisticated view of this battle emerged with the publication of Nägeli's (1865) work on the distribution of the calcicole *Achillea atrata* and its calcifuge congener *A. moschata*. He observed that each species could grow in the soil preferred by its congener as long as the congener was absent; this work thus bound plant competition to soil type, plant distribution and ultimately to what later became known as the niche.

In the history of competition, Darwin (1859) played a central part in two respects. First, he identified competition as a crucial element in the struggle for existence (Darwin 1859, p. 116), and thereby set the scene for competition as a major driving force of evolution by natural selection. Secondly, he commented that the struggle for existence is generally more severe between closely related organisms than between more distantly related ones because of their greater similarity in habits (Darwin 1859, pp. 126–127). Much of the recent interest in competition represents an attempt to work out the implications of this latter comment, given the evolutionary role of competition (Diamond 1978).

However, it was not until the emergence of ecology as a scientific discipline in its own right at the turn of the century that the systematic investigation of competition was started. To Warming (1909, p. 94), one of the founding fathers of ecology, competition was a major feature of plant communities, being most severe between species with similar requirements; species with disparate needs being complementary to one another in their utilization of the soil. The early British ecologists were acutely aware of the potential importance of competition. Indeed, the first presidential address to the British Ecological Society considered competition in some detail and argued that it is one of the most important processes in plant communities (Tansley 1914). Tansley, stimulated by Nägeli's (1865) paper, was also responsible for an early experimental demonstration of competitive exclusion involving two closely related species, *Galium saxatile* (a calcifuge) and *G. sylvestre* (a calcicole), when grown together in contrasting soils (Tansley 1917). He and some other ecologists took an explicitly dynamic view of plant ecology, in which competition featured prominently. For example, the competitive interactions between plants were of enduring interest to Salisbury (1929), as discussed in his presidential address 'the biological equipment of species in relation to competition' to the British Ecological Society. Among the many interesting features of this prescient paper is a nice distinction between what we would now call the fundamental and realized niche '... plants grow not where they would but rather where they must'. At much the same time Watt (1925) was also involved in analysis of the community dynamics of woodlands, in which competition was thought to be heavily implicated. It was not until his own presidential address, when he described the replacement of aggregates of plant species (patches) by others in an orderly sequence, depending on their competitive relations, that this work made its full impact (Watt 1947).

Particularly influential in the first half century was the work of Clements (1904, 1905). Clements saw competition as one of six basic processes in communities, controlling the direction of succession (Clements 1905, p. 199; Clements, Weaver & Hanson 1929, p. 327). He gave two laws of competition : (i) that competition is most keen between most similar individuals; and (ii) that the closeness of competition depends directly on similarity of form (Clements 1904, p. 167). The latter represented a shift from Darwin's (1859) emphasis on taxonomic relatedness towards functional similarity (Gause 1934, p. 15), which is echoed in some contemporary literature (e.g. Goldberg & Werner 1983). Strongly competing species can clearly be unrelated, as for example Brown, Davidson & Reichman (1979) demonstrated for competition between seed-eating rodents and ants. It is not at all clear what, if any, relationship exists between taxonomic relatedness and competition.

Clements also made important advances in the experimental study of competition; being one of the first people to investigate competition by manipulating the density of plants in the field (Clements, Weaver & Hanson 1929, pp. 130 *et seq.*). Moreover, he anticipated modern niche concepts by envisaging the 'division of labor' among coexisting species [see Clements, Weaver & Hanson (1929, pp. 314–315) which includes a breathtaking misquotation of Warming (1909)]. That this has not been recognized, may be attributable to the fact that it was closely tied to the much criticized notion of the community as a complex organism

(Simberloff 1982; see also McIntosh 1985, pp. 80–81). Although the tradition of dynamic plant community ecology of which Clements formed a part still has its intellectual descendants, the recent resurgence of interest in the role of competition in plant community dynamics owes more to the inspiration of agronomy and animal ecology than to earlier work in plant ecology (e.g. de Wit 1960; Grubb 1977; Harper 1977; Crawley 1986).

Animal ecology, which developed rather later than plant ecology, brought its own distinctive contribution to the study of competition, still very much in evidence in contemporary thinking. One of the sources of this was Gause's (1934) book *The Struggle for Existence*, well known for its statement that '. . . as a result of competition two similar [co-existing] species scarcely ever occupy similar niches, but displace each other in such a manner that each takes possession of certain peculiar kinds of food and modes of life in which it has an advantage over its competitor' (p. 19).

It should be clear from earlier comments that this view was held by a number of plant ecologists before Gause wrote his book, and Hutchinson (1975) suggests that it may have been widely known in animal ecology, as well. Nonetheless, through the influence of Darwin's (1859) conjectures, the literature on plant ecology (e.g. Clements, Weaver & Hanson 1929), the mathematical work of Volterra (1926) and Lotka (1932), and his own experimental and theoretical studies, Gause achieved a major synthesis and conceived the modern treatment of competition. The subject was born in 1944 at a Symposium of the British Ecological Society on the ecology of closely allied species (Anon 1944; Hardin 1960). Two papers read at the meeting, although controversial at the time, later proved to be particularly influential. Lack (1944) observed that closely-related bird species which live together have different feeding habits, or different sizes or both. Elton (1946) documented a conspicuous shortage of genera with more than one species living in the same community, for both animals and plants. The meeting in effect suggested a way in which competition could be instrumental in controlling the combinations of species which could live together in communities; it required some dissimilarity in their ways of life.

The effect of competition on the assembly of species into communities was developed, for the most part, in North America, through the stimulus of Hutchinson and his students. A pre-requisite for this development was to find a way of measuring the niche of a species; a formal definition of the niche was given by Hutchinson (1957), as the n-dimensional hypervolume of environmental axes within which a species occurs, and was later modified to become a resource utilization function (May & MacArthur 1972). This led to many field studies of the niches of co-existing species, starting with MacArthur's (1958) study of warblers in coniferous forests (see review by Hutchinson 1978). MacArthur & Levins (1967) suggested a method by which the overlap in niches of different species could be used to estimate the strength of competition between them, the key issue on which their co-existence would ultimately depend. This linked field studies to the mathematical analysis of interacting species, and made it possible to predict the limits of similarity of competing species compatible with their co-existence. (May & MacArthur 1972). From this blend of field study and theory there emerged a picture of communities neatly ordered by competition into groups of compatible species (Diamond 1978), not altogether dissimilar to that suggested by plant ecologists of earlier generations (Woodhead 1906; Clements, Weaver & Hanson 1929, p. 314). This view is however currently the subject of much debate (Schoener 1982; Strong *et al.* 1984; Diamond & Case 1986a).

COMPETITION AND EVOLUTION

It is a peculiar feature of history that competition, envisaged by Darwin (1859) as a major force of natural selection, should have become at most a minority interest of evolutionary biologists. A probable reason for this is that the biological equipment of an individual in relation to competition is not easily identified; indeed it is not immediately clear which of the many phenotypic traits that confer competitive ability would be likely to undergo evolutionary change through competition. Arguably, the best trait to consider is competitive ability itself, because this integrates all aspects of the phenotype relevant to competition. Yet there are at least two problems in doing this likely to deter all but the most tenacious students of evolution. First, competitive ability, being defined only in the context of other species, is difficult to measure. Secondly, in common with other coevolutionary processes, one must bear in mind that the competitive abilities of the other species may be evolving as well. Notwithstanding these problems, there has been sufficient interest in the evolution of competitive ability to generate some insights.

It would be as well to establish at the outset that there exists genetic variation in competitive ability on which natural selection may act, but rather little is known about the genetic architecture of the trait. Lerner & Ho (1961) observed differences in the success of inbred lines of *Tribolium castaneum* and *T. confusum* when placed in interspecific competition, but were unable to obtain repeatable evolution of competitive ability in mass cultures with substantial genetic variation. This suggests that genetic variance was mostly non-additive, a conclusion also reached by Futuyma (1970) following an investigation of evolution in *Drosophila melanogaster* in competition with a standard inbred strain of *D. simulans*. On the other hand, Hedrick (1972) found a major difference in the competitive ability of two stocks of *D. melanogaster* of common ancestry, when placed in competition with *D. simulans*. Moreover, some attempts to select for competitive ability have been successful (see below), so an additive component to the genetic variation does occur in some circumstances.

A little information on genetic variation in competitive ability is also available from studies on plants. Genetic differences in yield between progenies of *Festuca pratensis* were found when grown in competition with a cultivar of *Lolium perenne* (Van Bogaert 1974), and McNeilly (1981) observed genetic differences in yield between populations of *Poa annua*, also grown with a cultivar of *L. perenne*. In a genetic analysis of four varieties of *Trifolium repens* in competition with two cultivars of *L. perenne*, Hill & Michaelson-Yeates (1987) showed that the variance in dry weight per unit area had a large genetic component, most of which was non-additive. Rarely is it possible to relate competitive ability to individual genes but in the case of the garden pea, *Pisum sativum*, it has been shown how isogenic lines, differing at only the *af* and *st* loci, vary in their competitive ability towards the wild oat, *Avena fatua* (Butcher 1983). Unfortunately, the more detailed information available on the genetic basis of competitive ability in most crop plants is confined to intraspecific interactions. (e.g. Sakai 1961), and so is not strictly relevant in the present context.

The most direct way to investigate the effect of competition between a pair of species, on their evolution, is to place them together in an arena and to leave them to interact. Table 8.1 summarizes the published literature on such experiments. We have included studies in which, following a period of interaction, controlled tests for evolutionary changes in competitive ability were carried out; such tests require the competitive ability of individuals following interspecific competition to be compared with that of individuals maintained without interspecific competition. The experiments fall into two classes. First, there are trials in which the populations were left to follow their own unimpeded course from generation to generation, without control of density. Secondly, there are trials in which the densities of interactants were reset to constant values at the start of each generation. These latter allow continued competition which might otherwise be rapidly terminated by exclusion of one or other party, but they are less realistic simulations of natural systems.

Species[1]	Type of trial[2]	Number of trials[3]	Evolution of competitive ability[4]	Comments	Reference
Drosophila melanogaster *D. simulans*	Variable	1	1(2)	20 trials run, but evolution of competitive ability was investigated in one in which competitive status of species changed	Moore (1952)
Tribolium castaneum *T. confusum*	Variable	4	0(1)*	28 trials run, but investigation confined to 4 trials in which *T.confusum* eliminated *T. castaneum*	Park & Lloyd (1955)
Tribolium castaneum *T. confusum*	Variable	4	0(1)*	*T. confusum* was always eliminated in competition. Evolutionary change investigated in 4 trials in which *T. confusum* survived longest	Lerner & Ho (1961)
Musca domestica *Phaenicia sericata*	Variable	1	1(2)	Argued that the rarer species underwent evolutionary increase in competitive ability	Pimentel *et al.* (1965)
Drosophila melanogaster	Constant	4	2(2)*	Argued that competition caused evolution of niche differences	Seaton & Antonovics (1967)

TABLE 8.1. Con't:

Species[1]	Type of trial[2]	Number of trials[3]	Evolution of competitive ability[4]	Comments	Reference
Drosophilia nebulosa *D. serrata*	Variable	2	3(4)		Ayala (1969)
Drosophilia melanogaster *D. simulans*	Constant	28	8(28)	*D. melanogaster* kept in competition with standard in-bred strain of *D. simulans*	Futuyma (1970)
Tribolium castaneum	Constant	2	0(4)		Sokal *et al.* (1970)
Musca domestica	Constant	2	0(4)		Sokal *et al.* (1970)
Drosophila melanogaster *D. simulans*	Constant	2	2(2)	Results given for generation 77, *D. melanogaster* only	Van Delden (1970)
Musca domestica	Constant	20→15	1(2)*	Incorporated head start for one strain at start of each generation	Bryant & Turner (1972)
Tribolium castaneum	Constant	4	1(2)*	Results given in detail for first experiment only	Dawson (1972)
Tribolium confusum	Constant	4	0(2)*	Results given in detail for first experiment only	Dawson (1972)
Drosophila melanogaster *D. simulans*	Variable	24?	—	Species did not usually coexist in long term. One strain of *D. melanogaster* had high competitive	Hedrick (1972)

Drosophila melanogaster *D. simulans*	Constant	4	0(2)*	Results here refer to first 10 generations; serious experimental difficulties later	Barker (1973)
Tribolium castaneum *T. confusum*	Constant	18	—	*T. castaneum* showed preference for eating eggs of species to which it had been previously exposed	Dawson (1979)
Drosophila melanogaster	Constant	2	1(4)*	Varied headstart for one strain at start of some generations	Sulzbach & Emlen (1979)
Drosophila melanogaster	Constant	13	1(2)*	Experiment run for 8 years. Evidence for increased interference	Pruzan-Hotchkiss *et al.* (1980)
Drosophila melanogaster	Constant	4	0(8)		Sulzbach (1980)

[1] Where only one species is given, reproductively isolated strains of the species were used.

[2] Trials are of two kinds; variable, in which population densities were allowed to fluctuate naturally, and constant, in which the densities were set to fixed values at the start of each generation.

[3] Each trial consists of a cage in which the species or strains were allowed to compete for a number of generations.

[4] The number of cases in which evolution of competitive ability was observed is given; this could be 0, 1 or 2 for each trial depending on the experiment. The total number of cases in which competitive ability could have evolved is given in brackets. In certain cases the average competitive ability over several trials is given; these are marked *.

Table 8.1 shows that evolution of competitive ability occurred in a minority of cases (21 out of 72). Our tally of the cases in which such evolution took place gives sixteen involving an increase in competitive ability and five involving a decrease (see Futuyma 1970; Dawson 1972; Pruzan-Hotchkiss *et al.* 1980). This tally is not altogether straightforward because the experiments are heterogeneous in design, and some interpretation of results has been necessary. Of the twenty trials which allowed both partners to evolve, only one gives putative evidence for niche divergence (Seaton & Antonovics 1967); this trial is itself open to reinterpretation because the increased yield of one strain following competition was evident in the absence of inter-strain competition, and also because the selected strains when placed in competition did not both give increased yields relative to their unselected stocks, as one would expect had niche differentiation been taking place.

Following Seaton & Antonovics (1967), there have been strenuous efforts to demonstrate experimentally the evolution of niche differences but with the exception of an experiment by Chen mentioned in Antonovics (1978), these attempts have failed. It may be objected that experiments in small arenas leave little opportunity for niche divergence, and there is certainly a need for a new generation of evolutionary competition experiments which incorporate some ecologically relevant heterogeneity in the environment. On the other hand, arenas which appear homogenous to a human observer are not necessarily so to the interacting species. Arthur (1987, p. 75), for example, found *Drosophila* medium 3 cm deep partitioned between species (see also Arthur 1986). It is also reasonable to object that the studies, being confined to a few species of flies and beetles, are liable to give a biased impression of the evolution of competitive ability; further experiments to widen the taxonomic basis of the data would clearly be helpful. Nevertheless, our inability to obtain repeatable experimental demonstrations of niche differentiation under conditions which in some respects strongly favour such evolution, casts serious doubt over the efficacy of this pattern of evolution.

The next step in analysis of the evolution of competitive ability would be to take the natural environment as the arena. Since there are then likely to be many more than two potentially competing species, it is not altogether surprising that this has not been attempted. The only study we know of which comes anywhere near to doing this was carried out by Aarssen & Turkington (1985) on three adjacent pastures of low species diversity. The grasslands had been sown with a mixture of *Dactylis glomerata*, *Lolium perenne* and *Trifolium repens* 2, 21 and 40 years earlier, and were taken to represent different (unreplicated) stages in the evolution of the plant community. Aarssen and Turkington took shoots of these three species from each pasture, together with *Holcus lanatus* and *Poa compressa* which occurred naturally, and carried out ten competition experiments involving all pairwise combinations of species from each pasture. In three pairs of species there was a tendency for yields of each partner to become more similar with increasing age of pasture, suggesting that the competitive abilities had become more evenly matched over the course of time. In one pair of species the reverse trend was found. In no pair of species did they find an increase in the summed yield of both species with time, as would be expected had niche differentiation been taking place. Comparisons involving commercial seed stocks were also made with somewhat different results, but the relationship of these stocks to those in the grassland is uncertain. These results may not, however, be robust because Aarssen (1988) was unable to repeat them in an experiment on four of the species using seed-derived samples.

In the absence of data on temporal changes in competitive ability in nature, what other less critical sources of information can be brought to bear on the evolution of competitive ability? One possibility is to investigate competition between species which, under natural conditions, have a history of sympatry in some locations but not in others. This method has been widely applied in niche studies (see below), but is rarely adopted as a basis for measuring change in competitive ability. An exception to this is some work on *Trifolium repens* taken from locations dominated by different species of grass within a single community (Turkington & Harper 1979). Shoots of *T. repens* characteristically gave a greater yield when grown in competition with the grass species with which they were naturally associated than when grown with other grasses, suggesting fine scale differentiation in competitive ability in at least one component species of the community. Gliddon & Trathan (1985) detected a similar response when *T. repens* was grown in a pot experiment with single genotypes of *Lolium perenne* with which it was naturally associated; the reciprocal response in *L. perenne* was not detected. However, it remains to be seen whether these differences are heritable (see Evans & Turkington 1988) and, if they are heritable, whether they represent differentiation of *T. repens* with respect to grass species or to the soil microflora associated with the grass species (Turkington *et al.* 1988).

Another possible way forward is to investigate the competitive interactions of a group of species at different locations. Hairston (1980, 1983) for instance, found that two species of salamander, *Plethodon glutinosus* and *Plethodon jordani*, competed strongly in one area of the Appalachian Mountains and weakly in another. Replacement of the native *P. jordani* by the non-native populations changed the interaction, indicating that the difference was a property of the populations rather than the environment. Although Hairston (1983) argued that evolution had been in the direction of increased competition, there remains some doubt about the evolutionary pathway in the absence of information on competition between the ancestral populations (but see Hairston, Nishikawa & Stenhouse 1987). This will always be a drawback of studies in which the evolutionary sequence is not measured directly with time.

Because information on the evolution of competitive ability in natural communities is so hard to come by, we offer some comments based on general ecological intuition (see also Haldane 1932, p.119; Huxley 1963, p.34; Connell 1980). In a community of sedentary organisms, close spatial proximity between individuals of the species experiencing selection and the species causing selection is necessary if the latter is to constitute a major selection pressure on the former. Consider, for example, the data on a diverse limestone grassland in Fig. 8.1, in which the density of various plant species is measured in the immediate neighbourhood of seven 'target' species – in effect the community as it appears to these species (Mahdi & Law 1987). Certain species feature more prominently in the neighbourhoods of all the target species than others and *a priori* may be expected to exert stronger selection pressures on the target species. A corollary is that target species which themselves differ substantially in their abundance as neighbours of other species (such as *Plantago lanceolata* and *Potentilla erecta*) may differ in the strength of selection they exert on one another. It is only the more abundant species (such as *P. lanceolata* and *Carex caryophyllea*) which are likely to have a substantial reciprocal influence on each other. Spatial proximity is only one of a number of conditions that needs to be satisfied for competition-driven evolution. Another requirement is that the species which causes the selection pressure should interact with the species experiencing selection (i.e. influence the rate of increase of the latter), but attempts to demonstrate such interactions failed in the case of the community in Fig. 8.1, as dicussed below. If competition-driven selection pressures are in operation in diverse communities like that in

FIG. 8.1 Density of plant species per m^2 in a limestone grassland as they appear to seven 'target' species; measurements are the mean ($\pm$ 1 S E) densities of neighbour species in neighbourhoods of radius 3 cm around the target species. Each row gives the apparent community of a target species. Within each row, neighbour species are divided into two groups; those which are also target species, and those which are not. Shaded bars are the densities of conspecifics (from Mahdi & Law 1987).

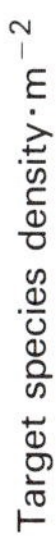
Target species density · m⁻²

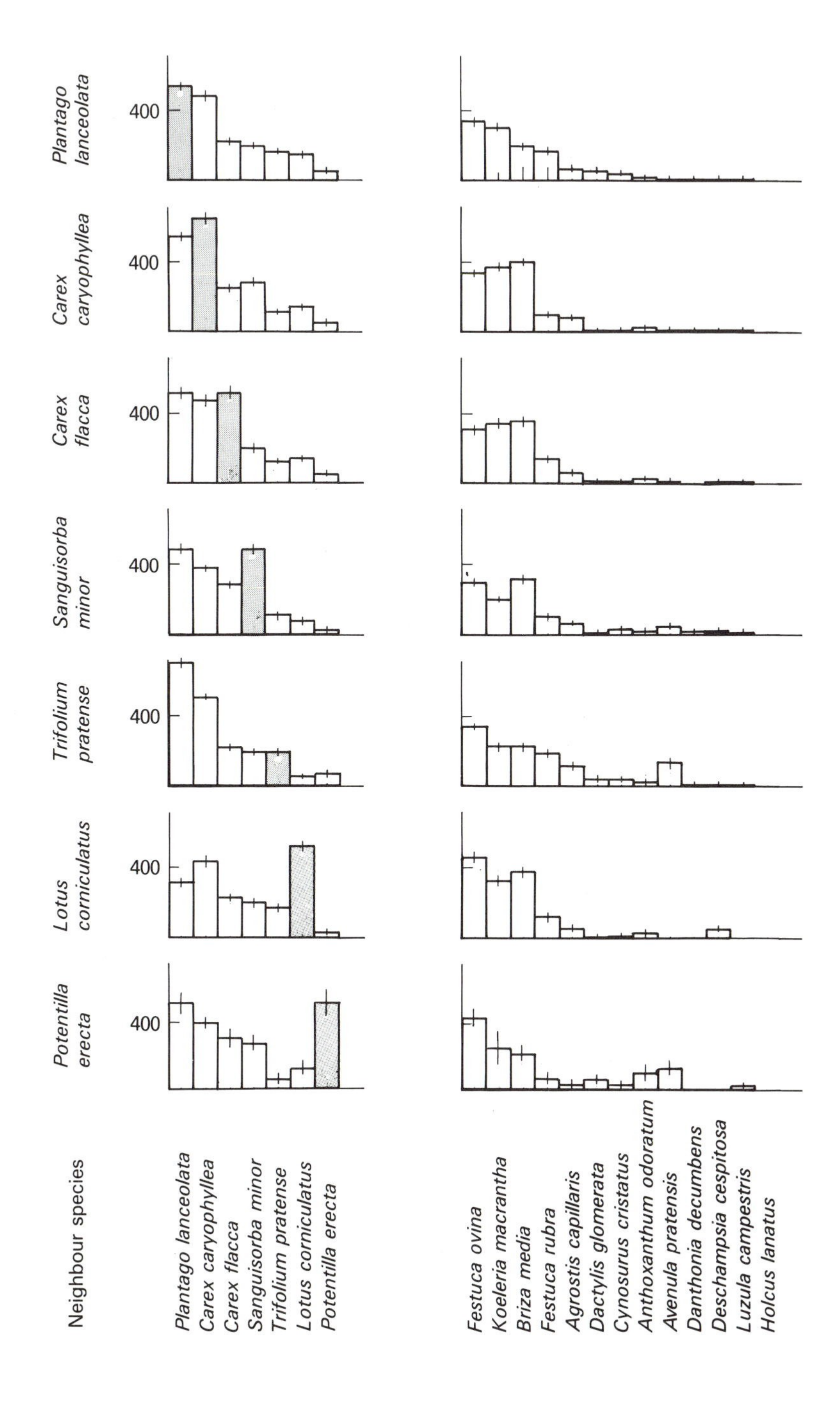
Plantago lanceolata
Carex caryophyllea
Carex flacca
Sanguisorba minor
Trifolium pratense
Lotus corniculatus
Potentilla erecta
400
Neighbour species
Plantago lanceolata
Carex caryophyllea
Carex flacca
Sanguisorba minor
Trifolium pratense
Lotus corniculatus
Potentilla erecta
Festuca ovina
Koeleria macrantha
Briza media
Festuca rubra
Agrostis capillaris
Dactylis glomerata
Cynosurus cristatus
Anthoxanthum odoratum
Avenula pratensis
Danthonia decumbens
Deschampsia cespitosa
Luzula campestris
Holcus lanatus

Fig. 8.1, perhaps they would be best envisaged in terms of the general competitive milieu of the community, rather than in terms of pairwise interspecific interactions.

In communities of mobile organisms on the other hand, one would expect a somewhat different distribution of competitive interactions. Competition now depends less on the stochastic forces that lead to spatial proximity, because the organisms are able to forage actively for their resources. Thus pairwise interactions between species with similar resource requirements are *a priori* expected to be stronger. It is still the case though, that selective pressures generated by such competition depend on the relative abundance of the species, so reciprocity of selection is likely to be the exception rather than the rule. The contemporary view is that competition in communities of this kind causes evolution of niche differences; since this literature is well known and discussed in Chapter 4 it is not considered here. We note, however, that this view is not supported by laboratory experiments and that there are as yet no data which unequivocally demonstrate such evolution in the field. Although the field data showing character displacement in sympatry (e.g. Fenchel 1975; Roughgarden, Heckel & Fuentes 1983, Malmquist 1985; Schluter, Price & Grant 1985) strongly suggest that this has happened, the critical step of demonstrating niche differentiation driven by competition, in the process of evolving, remains to be achieved.

In summary, there is no single result in the survey above that gives strong evidence for interspecific competition as an evolutionary force in the way that, for instance, the rabbit and myxoma virus give evidence for host–pathogen co-evolution (Fenner & Ratcliffe 1965). Some laboratory studies have shown that competitive ability can evolve, but the results are erratic and often point towards a lack of appropriate genetic variation. Data on evolution of competitive ability under more natural conditions are almost completely lacking, but the selective pressures generated by competition between species are likely to be weak and non-reciprocal in communities of sedentary organisms. In communities of mobile organisms such selection could be stronger, and there is some circumstantial evidence that competition has led to the evolution of niche differences in these systems. Overall, however, the results do not lend strong support to the view the interspecific competition is a major force in evolution, or that niche differentiation is the main outcome. Whether this is an indication of our present fragmentary knowledge or whether it represents the real state of nature, only time and patient study will tell.

POPULATION DYNAMICS OF COMPETING SPECIES

In his analysis of the struggle for existence, Gause (1934) was much concerned with the process of interspecific competition. He recognized that the struggle for existence and natural selection were matters concerning the dynamics of populations, and that the relationships between species in mixed populations would be of prime importance in determining those dynamics. However, his interests were not primarily of a evolutionary nature. Rather he was interested in how interspecific competition (and predation) might affect the abundance and dynamics of individual species' populations in mixed communities. 'It became evident that the processes of competition between different species of protozoa and yeast cells are sometimes subject to perfectly definite quantitative laws. But it has also been found that these processes are extremely complicated and that their trend often do not harmonize with the predictions of the relatively simple mathematical theory.' (Gause 1934, p.vii.)

Numerous competition experiments have since been carried out in which two species are grown both separately and together under carefully controlled conditions. These experiments fall into two groups. First, there are those where the competing populations are counted at regular intervals over a number of generations. It is only these experiments which allow the effects of competition on the dynamics of the two species to be observed directly. Secondly, there are those experiments carried out on longer-lived organisms where the species are grown together for only a generation or part of a generation, but with various combinations of densities, so that the dynamics of the competing species can then be inferred indirectly. Amongst this second group, three types of experiment can be identified. First, there are the numerous replacement series experiments based on the experimental design of de Wit (1960) in which the densities of the species are initially constrained so that they sum to the same total density, only their relative proportions being varied. Ratio diagrams can then be constructed to predict qualitatively the long-term outcome of competition, although strictly speaking these predictions apply only if the total density is constrained as it was in the experiment (Law & Watkinson 1987). Secondly, there are additive experiments, where the density of one species is kept constant and the other is allowed to vary. These can only be used to predict the outcome of competition under a very restrictive set of circumstances such as when a weed invades a crop growing at constant density (Watkinson 1981; Firbank &

Watkinson 1986). It is only the third type of experiment, the additive series, where the density of both species is allowed to vary, which may be used to predict the population trajectories of both species over time (Ayala, Gilpin & Ehrenfeld 1973; Antonovics & Fowler 1985; Law & Watkinson 1987).

Here we shall restrict ourselves primarily to a consideration of those experiments where the dynamics of the competing species can be observed directly. These have been carried out on a wide range of organisms including Protozoa (e.g. Gause 1934, 1935), diatoms (e.g. Tilman, Mattson & Langer 1981), yeasts (Gause 1934), freshwater cladocera (e.g. Frank 1957), hydrozoans (e.g. Slobodkin 1964) and a number of beetles and flies, particularly *Tribolium* spp. (e.g. Park 1954; Leslie, Park & Mertz 1968) and *Drosophila* spp. (e.g. Miller 1964; Ayala 1969). Higher plants have rarely been used in such experiments but the study of Clatworthy & Harper (1962) on *Lemna* and *Salvinia* provides a notable exception.

Much of the later work has served to confirm or elaborate the various observations made by Gause. On the basis of experiments involving mixtures of either Protozoa or yeasts, he demonstrated that the rate of increase of one species' population was adversely affected by the presence of a second species, when resources were limiting; that complete competition resulted in the survival of only one species, and that coexistence might occur only when there was a partitioning of resources. Moreover, it was shown that the outcome of competition was strongly dependent on the environmental conditions and in some cases on the initial starting densities of the two species. These observations led to the formulation of the competitive exclusion principle. Subsequent experiments demonstrated that the outcome of competition between two species might also depend on the genetic structure of the starting populations (Lerner & Ho 1961); chance ecological accidents (Mertz, Cawthorn & Park 1976); particular aspects of the biology of the individual species (Clatworthy & Harper 1962); and the presence of other organisms such as predators (Slobodkin 1964) and parasites (Park 1948). Predicting the outcome of competition in mixtures from single species population studies has proved problematical. In some cases it is possible (Gilpin, Carpenter & Pomerantz 1986), but in others it is not (Clatworthy & Harper 1962; Gill 1972).

In quantifying the effects of interspecific competition in his experiments, Gause drew strongly on the theoretical analysis of interspecific competition by Volterra (1926) and Lotka (1932). The predictions from their theoretical work correspond to all the major observations made by

Gause in his competition experiments (i.e. the predictable exclusion of one species by another; exclusion dependent upon initial densities; and stable coexistence). The Lotka–Volterra model of competition relates the rates of increase of two species to their densities (N_1, N_2) by the equations

$$\frac{dN_1}{dt} = N_1\, g_1\,(N_1, N_2) = r_1\, N_1 \left(\frac{K_1 - N_1 - \alpha_{12}\, N_2}{K_1}\right)$$

$$\frac{dN_2}{dt} = N_2\, g_2\,(N_1, N_2) = r_2\, N_2 \left(\frac{K_2 - N_2 - \alpha_{21}\, N_1}{K_2}\right)$$

where r_i is the intrinsic rate of natural increase, K_i is the carrying capacity, g_i is the *per capita* rate of increase of species i ($i = 1, 2$) and α_{12}, α_{21} are competition coefficients which describe the competitive interactions between the species in terms of species' equivalents. The competition coefficients are perhaps best understood in terms of the partial derivatives of the *per capita* rates of increase of each species with respect to changes in the density of the two species (Fig. 8.2). Thus α_{12} is obtained from the *per capita* rate of increase of species 1, and is the ratio of the gradient arising from changes in the density of species 2 to the gradient arising from changes in the density of species 1

$$\frac{\partial g_1}{\partial N_2} / \frac{\partial g_1}{\partial N_1}.$$

The coefficient α_{21} is the corresponding ratio for species 2. Gause calculated the competitive relationships between pairs of species in his experiments from estimates of the values of r and K in single species' populations and the rates of population change at particular densities in the species' mixtures. For two yeast species he obtained competition coefficients ranging from 0.439 to 3.15 (Gause 1934, p.81), whilst for two *Paramecium* species he obtained values ranging from -1 to 1.64 (Gause 1934, pp.108–109). It should be noted that there are problems in calculating the competition coefficients from the Lotka–Volterra model in the case of the protozoan data (Williamson 1972, p.103).

Competition coefficients of a wide range of interacting species have now been estimated under controlled conditions, motivated by the Lotka–Volterra equations in their original or modified form (e.g. Ayala, Gilpin & Ehrenfeld 1973; Hassell & Comins 1976; Pacala & Silander 1987). The most detailed set of estimates for the value of the competition coefficient between pairs of species comes from studies on *Drosophila*. Shorrocks & Rosewell (1987) list fifty-two values of α for larval

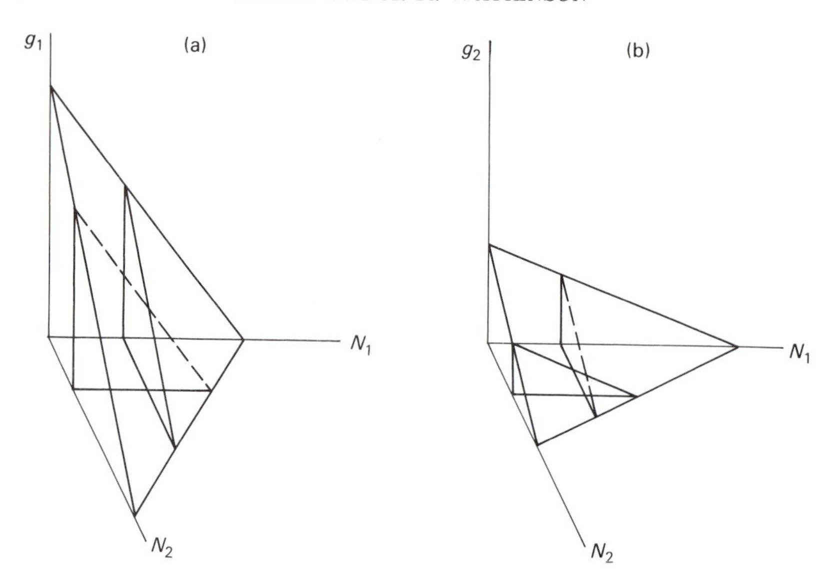

FIG. 8.2 Graphical interpretation of the competition coefficients of the Lotka–Volterra model of competition. N_1 and N_2 are the densities of species 1 and 2, respectively; g_1 and g_2 are their *per capita* rates of increase. The competition coefficient α_{12} is the gradient of the continuous heavy line divided by the discontinuous heavy line in (a); α_{21} is the corresponding ratio in (b).

competition in seven drosophilid species maintained under laboratory conditions. They range from less than one to just over twelve with the majority of values lying between 0.5 and 2.5. Frequently the competition coefficients are such that under experimental conditions one species can be expected to oust another (Gilpin, Carpenter & Pomerantz 1986) with the outcome of competition depending strongly on the conditions. But the probability of coexistence increases with the amount of food resource as this then allows a reduction in competition through resource partitioning both in the depth of larval distribution in the resource medium (Arthur 1986) and in its temporal usage. Nevertheless Gilpin, Carpenter & Pomerantz (1986) show that no more that three species are likely to persist when ten species are put together. Coexistence in guilds of between five and fifteen species is, however, possible in the field, even in the absence of resource partitioning (see later), if spatial aggregation is taken into consideration (Shorrocks & Rosewell 1987). Unfortunately there are no data available on the competitive interactions of *Drosophila* in the field.

For plants, the most comprehensive set of data on competitive ability is perhaps for wheat in competition with a variety of weeds, including *Agrostemma githago* (Watkinson 1981; Firbank & Watkinson 1986), *Bromus sterilis* (Firbank, Mortimer & Putwain 1985) and *Elymus repens* (Firbank & Mortimer 1985). These data are from both greenhouse and field experiments. The estimates of the competition coefficients range from 0.41 to 1.5 for the effect of wheat on the weed and from 0.06 to 1.63 for the effect of the weed on the wheat. Again the value of α varies with the conditions and in particular on the relative emergence time of crop and weed. This latter observation highlights the difficulty in forecasting the outcome of competition. On the basis of data on the competitive relationships between wheat and *Bromus sterilis* collected between October 1982 and December 1983, Firbank, Mortimer & Putwain (1985) attempted to predict seedling densities of the weed in December 1984 for a range of cultivation and herbicide treatments. On average the predictions underestimated the actual level of weed infestation by approximately 23% (range −37 to 68%) due to a complex interaction between the time of emergence of the weed and the time of cultivation. This is the only example that we are aware of where experimental field data on competition have been used to predict how the numbers of competing species will change in the future.

Competitive interactions formulated in terms of Lotka–Volterra competition coefficients have the property of being independent of the densities of the interactants. Although this is a convenient theoretical property, several studies have suggested that competition coefficients need to be envisaged as functions of density (Ayala, Gilpin & Ehrenfeld 1973; Smith-Gill & Gill 1978; Law & Watkinson 1987); indeed, Gause's estimates of the coefficients in yeasts themselves varied with density, but he assumed that they should be constant and ascribed the variations 'to an imperfect method of their calculation' (Gause 1934, p.81). The reasons for density-dependence are usually unclear, but a glimpse of the potential complexities is obtained from the interaction between *Drosophila hydei* and *Drosophila melanogaster* at low resource levels (Arthur 1986). Arthur found that *D. hydei* needed *D. melanogaster* to keep the resource in a utilizable form, and concluded that there was a 'competitive' + ↑ − interaction, which he called contramensalism. However, it seems likely that at high densities of *D. melanogaster*, the interaction would revert to − ↑ −, in which case the competition coefficient for *D. hydei*, at least, would be density dependent.

Estimates of competition coefficients and the other parameters needed to describe the dynamics of competing species have for the most part been obtained in controlled environments far removed from the conditions that the species would experience in nature (but see Firbank, Mortimer & Putwain 1985). It would be most surprising, therefore, if they were to provide reliable indicators of the dynamics of species under natural conditions; for this it is necessary to turn to the field. Broadly speaking, two kinds of information are available from natural populations, both of which are based on perturbations in population density of potentially competing species. The first involves 'natural' invasions and extinctions, of which there are numerous examples in the literature, showing that an introduced species has been associated with the decline or extinction of one or more native species (see reviews in Diamond & Case 1986b; Moulton & Pimm 1986; Werner 1986). The second involves the deliberate perturbation of field populations, an increasingly popular experimental method in community ecology; 72 studies reviewed by Connell (1983) and 164 studies by Schoener (1983).

As a source of information on the dynamics of competing species in nature, the data on natural and experimental perturbations of populations have certain drawbacks. Certainly some provide evidence for effects of competition on the dynamics of species within a community (e.g. Brown *et al.* 1986), but the results are primarily qualitative showing the presence or absence of effects following changes in population density. Also, there can be serious difficulties in demonstrating both a causal link from the perturbed population to the responding one and vice versa. It is, for example, somewhat chastening to find that this problem arises in the interaction in Great Britian between the native red (*Sciurus vulgaris*) and introduced grey (*Sciurus caroliniensis*) squirrels, one of the best documented species displacements (Reynolds 1981, 1985). Evidence of direct competition between the squirrels for resources is not strong, and it is not known what effect the absence of the red squirrel would have on invasion of the grey squirrel; indeed it is likely that interaction between the two species can at most provide only a partial explanation of red squirrel decline.

Certain studies involving experimental manipulations have attempted to estimate the competition coefficients (e.g. Seifert & Seifert 1976, 1979; Hairston 1980) and others have shown temporal variability in competition, reversals in the rank order of competitive superiority and frequent asymmetry in competition (Connell 1983), but we do not know of any experiment in which all the information required to describe the dynamics of the interacting species in nature has been obtained. Few experimental studies have even attempted to investigate intraspecific interactions within each of the competing species (Connell 1983;

Underwood 1986) notwithstanding the crucial role that these may play in population dynamics.

Thus, although we have a detailed understanding of the dynamics of a range of competing species under controlled conditions stimulated by the work of Gause, the role of competition in the dynamics of single species in nature remains largely unexplored. The task is daunting. At its simplest, in the unlikely event that Lotka–Volterra dynamics operate in a species, we are faced with the need to estimate the single-species' parameters (r and K), together with its competition coefficients with all potentially competing species, and also the population densities of these species. This, of course, ignores the possibility discussed earlier of density-dependent competition coefficients, and of modification by other competing species of the competitive interactions between pairs of species (e.g. Haizel & Harper 1973). Neither does it take into account the effects of interactions with other trophic levels (e.g. Lubchenco 1986), or the effects of the age and size structure of the populations (e.g. Watt 1981). In view of this it is surprising how successful field studies of single-species population dynamics can be in the absence of information on interspecific competition. It has often been possible to describe accurately the population dynamics of a species simply in terms of the intraspecific modification of its *per capita* rate of increase, notwithstanding the potentially competitive community in which the species is embedded (e.g. Watkinson & Davy 1985). This means either that competitive interactions between species are rarer than might at first appear, or that there is a large degree of equivalence in the competitive effect of species (Goldberg & Werner 1983; Watkinson 1985), or both. These issues are discussed in the next section.

COMPETITIVE STRUCTURE OF COMMUNITIES

> In order to determine the [competitive] powers of the different species [in a plant community], we must resort to experiment, i.e. we must give the individuals of the various species a freer field than they have in the restricted conditions of a closed association. The simplest way in which this can be done is by clearing a patch of ground of some or all of the species present and seeing what happens. [Tansley 1914]

Depending on your disposition, this quotation may indicate the foresight of a great ecologist or the subsequent failure of community ecologists to come to terms with one of the central questions of the subject. Tansley clearly understood the need to manipulate the densities of potentially competing species to determine the strength of their interactions, yet 75 years later we are hard put to find a natural or semi-natural community in which the competitive interactions of a substantial number of the constituent species have been measured with precision. The importance of manipulation experiments is now widely appreciated, but such experiments generally involve the perturbation of no more than one or two species in a community which, *a priori* are thought to be important in competition (Connell 1983; but see also Schoener 1985). Our purpose here is not to review again the evidence from these experiments that competition between some species does occur in nature (Connell 1983; Schoener 1983, 1985); it is rather to consider the distribution of interactions within guilds (*sensu* Root 1967). What is known about the competitive structure of guilds of species, and what bearing if any does this information have on the large body of theory concerning the dynamics of competing species?

First we comment on some concepts underpinning manipulation experiments. We envisage a vector of population densities $\boldsymbol{N}(t) = [N_1(t), \ldots, N_k(t)]$ describing a k-species community at time t; the density of the i th species at time $t+1$ is given by $N_i(t+1) = N_i(t)\, g_i(\boldsymbol{N}(t))$, for $i = 1, \ldots, k$. The dynamics of the community are governed by the species-dependent *per capita* rates of increase g_i, each one representing a surface in $k+1$ dimensional space. A manipulation experiment endeavours to obtain information on the shape of these surfaces. The experiment does not have to make assumptions about the shape of the surfaces or that the community is at equilibrium before perturbation, although interpretation of the results may be a good deal harder without such assumptions (cf. Bender, Case & Gilpin 1984). Fig. 8.3(a,b) illustrates the principle involved for two species. Three values of g_i are obtained for each species based on separate perturbations of the density of each species together with an unperturbed control. A significant difference between a control and perturbed value of g_i indicates that the perturbed species has an effect on the dynamics of species i. Whether this is a direct effect, or an indirect one mediated through interactions with other species, depends on the dynamics of the species concerned and details of experimental design (Bender, Case & Gilpin 1984).

The limited information available on the shapes of the g_i terms suggests that they are highly non-linear, being much more sensitive to changes in density when density is low than when it is high (as sketched in Fig. 8.3). For example, individual plant weight and seed production have inverse relationships with density in competition experiments involving two species (Watkinson 1985; Goldberg & Fleetwood 1987; Law & Watkinson 1987). A similar shape has been observed in natural plant

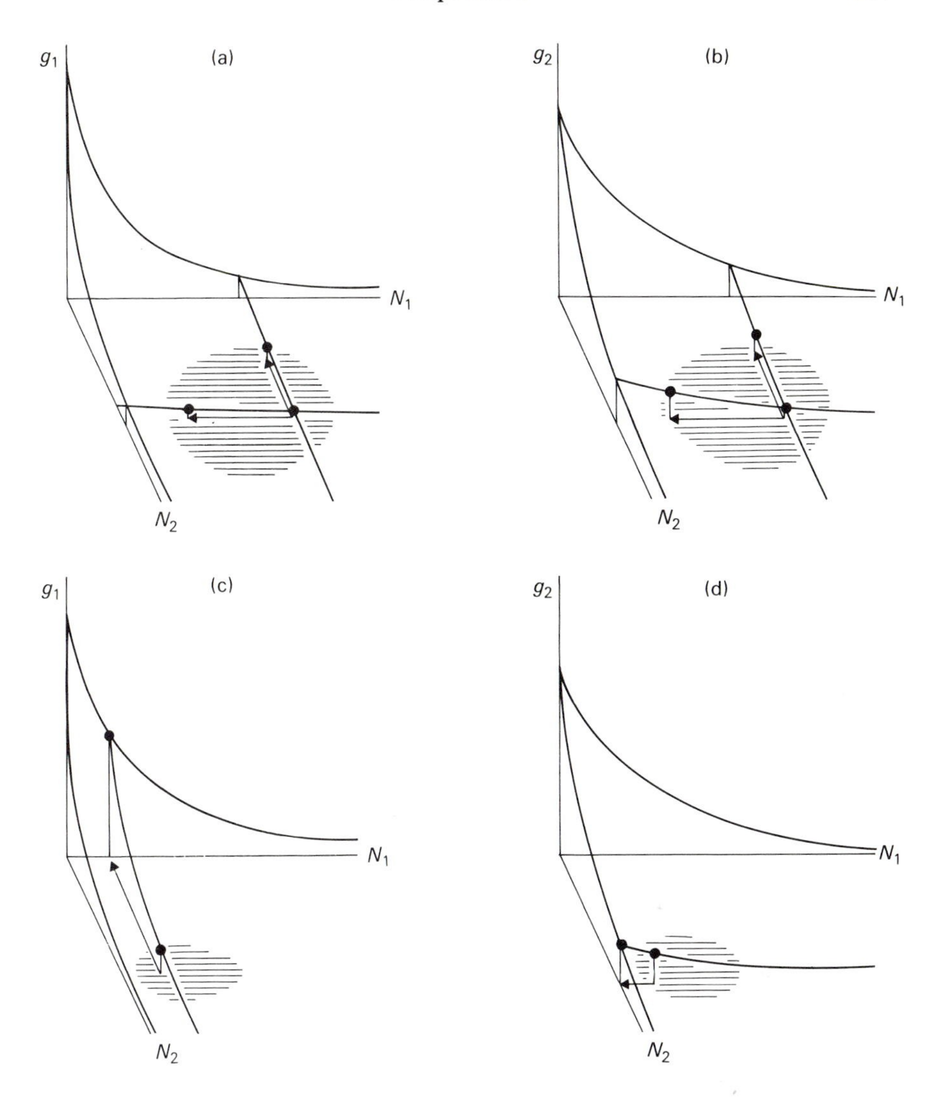

FIG. 8.3. Manipulation experiments on a pair of species, of densities N_1 and N_2, with *per capita* rates of increase g_1 and g_2, respectively. Filled circles are the values of g_i before and after manipulation, arrows indicate the manipulations applied. The densities observed in nature are shown by hatching. (a) and (b) represent an experiment involving partial removal of species 1 and 2. (c) and (d) represent an experiment involving complete removal of species 1 and 2, from a community in which species 1 is relatively rare.

communities when removals have been made so that no more than two species remain (Goldberg 1987; Miller & Werner 1987). At present there is no single function that adequately describes these non-linearities (Antonovics & Kareiva 1988). Bearing in mind that in all but the simplest communities, it will only be feasible to obtain information on a small fragment of the phase space, perturbations which lie within the range of densities observed in nature are therefore likely to prove most informative.

Table 8.2 summarizes the results of manipulation experiments on natural and semi-natural communities. It includes all guilds known to us in which: (i) at least four single-species manipulations have been carried out; and (ii) the responses of at least four species have been monitored, generating a minimum of twelve interspecific responses. The precise conditions for inclusion are given in the footnotes to Table 8.2. The lower limit on the number of manipulated species is intended to reduce a probable bias towards removing the most abundant species when few species are used; appropriate data are so hard to come by that we have allowed manipulations of species-pairs in certain cases (see footnotes to Table 8.2). The choice of this lower limit was also guided by a need for information on pairwise responses (i.e. the response of species A following manipulation of species B and vice versa). When four species are manipulated, a maximum of six pairwise responses can be measured, as opposed to only three when three species are manipulated. From these data a very rough indication of the frequency distribution of single and pairwise responses can be obtained. An analysis of this kind runs an obvious risk of bias through inappropriate delimitation of guilds, for example, through inclusion of species between which competition is not expected *a priori*. The guilds considered here, however, are sufficiently clear-cut for this not to be a serious problem.

The most striking feature of Table 8.2 is how little information it contains; Tansley would surely have been disappointed to see how little progress has been made. Generalization from these studies, so few in number and so unrepresentative of the natural world, is therefore bound to be hazardous. While bearing in mind this caveat, the guilds do share some intriguing features. First, most responses to manipulation, if they occur at all, are below the limits of detection. Second, in cases where pair-wise responses which have been investigated, most commonly no interaction is detected. Those pairwise responses which are not 00 are usually asymmetric of the form -0 or $+0$; several authors have commented on the prevalence of the -0 asymmetry in nature (Lawton & Hassell 1981; Connell 1983; Schoener 1983). Significant responses by both species are rare (but see Seifert & Seifert 1976); in particular − − interactions, usually taken to indicate competition, are conspicuous by their absence. These data do not support the view that competition is a major force driving the present-day dynamics of species within communities.

It is much easier to criticize manipulation experiments than to carry them out, but the experiments in Table 8.2 do have certain biases which need to be kept in mind. Manipulations usually involve complete removal of the most abundant species, probably on the basis that the larger the perturbation the greater the chance of observing a response. Moreover, such manipulations are likely to shift communities outside the natural range of density of the manipulated species; this would not matter if the g_is were linear functions of density, but as noted above the functions are probably much more sensitive to changes in density when density is low than when it is high. Thus the response observed following complete removal of an abundant species is likely to overestimate the response in the natural range of density [Fig. 8.3(c)]. Another consequence of the complete removal of a species is that the size of perturbation depends on its abundance prior to removal. Species of different abundance may therefore appear to have asymmetric interactions because their densities are perturbed to different extents [Fig. 8.3(c,d)]. There is also a somewhat vexed question so as to what the response of the species to removal of an abundant species tells us about their interactions with this dominant. Allen & Forman (1976), for example, found that large responses were usually associated with: (i) removal of species which after elimination, left large areas of bare ground; and (ii) responding species with well-developed powers of clonal growth, enabling them to invade the bare areas. Such results may tell us more about the innate capacity for growth and reproduction of single species than about their interactions with other species. In addition to these problems, there are serious practical difficulties to be overcome in the design of manipulation experiments (Underwood 1986).

Mahdi's experiment (Table 8.2) was designed to reduce some of the potential biases of earlier plant perturbation studies, being based on the removal of individual ramets rather than on the complete elimination of species. The rationale behind this is that, in guilds of sedentary organisms in which interactions occur primarily with immediate neighbours, the overall response of one species to another is an average over all the local neighbourhood responses. It should be possible to remove a sample of ramets large enough to indicate the response, yet small enough to cause neligible disruption to the whole guild. It is interesting that the number of

TABLE 8.2. Results of manipulation experiments on communities of potentially competing species[1]

Community	Methods			Results[4]										
				Single-species responses				Pairwise responses						
	Design[2]	Manipulation[3]	Responses	0	−	+	(Σ)	00	−0	+0	+−	++	−−	(Σ)
Terrestrial plants in an old field (Allen & Forman 1976) 'Upper field'	Press (3 months)	5(0)	% cover of species at end, in manipulated treatment relative to controls	18	1	1	(20)	8	1	1	0	0	0	(10)
As above: 'Middle field'	Press (3 months)	5(0)	As above	17	3	0	(20)	7	3	0	0	0	0	(10)
As above: 'Lower field'	Press (3 months)	4(0)	As above	11	1	0	(12)	5	1	0	0	0	0	(6)
Herbivorous insects in water-filled bracts of *Heliconia wagneriana* (Seifert & Seifert 1976)	Pulse (≤8 days)	4 (various)	*Per capita* change in density between the beginning and end of experiment.	9	2	1	(12)	3	2	1	0	0	0	(6)
As above: *Heliconia imbricata*	Pulse (≤8 days)	(various)	As above	5	3	4	(12)	1	2	1	1	1	0	(6)
Herbivorous insects in water-filled bracts of *Heliconia bihai* (Seifert & Seifert 1979)	Pulse (≤8 days)	4 (various)	As above	10	1	1	(12)	4	1	1	0	0	0	(6)

Terrestial plants in mown grassland (Fowler 1981)	Press (*c.* 2 years)	7(0)	% cover of species at the end in manipulated treatment relative to controls	52	9	3	(64)	8 4	1	0	0	0	(13)
Terrestial plants in coastal community 'swale' (Silander & Antonovics 1982)	Press (1 year?)	4(0)	As above	15	4	1	(20)	1 0	0	0	0	0	(1)
Terrestial plants in limestone grassland (Mahdi 1988)	Pulse (*c.*2 years)	8[5] (see text	Change in density over *c.* 15 months in perturbed neighbourhoods relative to controls	98	4	2	(104)	26 1	1	0	0	0	(28)

FOOTNOTES ON TABLE 8.2

[1] With the exceptions below, experiments have to satisfy four conditions for inclusion: (a) manipulations are carried out within a single community; (b) manipulations are on single species as opposed to broader categories such as all dicotyledons; (c) manipulations of at least four species are carried out, one at a time; (d) the response of at least four species is monitored (papers which record significant responses without making explicit how many species were tested cannot be included). The exceptions are: Allen & Forman (1976) in which two of the manipulations were at the genus level, Silander & Antonovics (1982) in which two manipulations involved pairs of species from different genera, and Seifert & seifert (1976, 1979) in which replicated communities were set to a range of different starting densities.

[2] A 'press' experiment is one in which the density of a perturbed species is held at its initial perturbed value; a 'pulse' experiment is one in which the density of the perturbed species is free to change (Bender, Case & Gilpin 1984). The duration of the experiment is given in parentheses.

[3] The number of manipulation treatments applied to the community is given. The density of the manipulated species after the initial perturbation is given in parentheses.

[4] Interspecific responses significant at the 5% level according to the investigator's analysis are assigned non-zero values; following the usual convention, a negative response involves a reduction of the responding species as the density of the perturbed species increases. Pairwise responses can be given only when species are: (a) manipulated; and (b) measured in response to the manipulations of other species. Single-species responses include responses by non-manipulated species.

[5] The species manipulated are the seven species for which apparent communities are given in Fig. 8.1, plus *Ranunculus bulbosus*.

apparently significant responses in this experiment turned out to be approximately the same as the number expected by chance in carrying out *c.*† 100 statistical tests, so there are no grounds to conclude that the 'significant' responses are real. On the other hand, the experiment did run a risk of underestimating the number of significant responses, because the samples of ramets removed, which varied in size from twenty-four to seventy-six, may have been too small to allow detection of responses. Added to this, the community changed during the course of the experiment, with an increase in certain grasses and *Trifolium repens* both in the gaps and in the neighbourhood of control ramets, further decreasing the sensitivity of the experiment. Although we think it likely that earlier experiments have overestimated the responses, improvements in experimental design are needed to resolve this issue.

If the prevalence of weak interactions is real rather than a reflection of poor powers of resolution of manipulation experiments, several explanations are possible. Below we evaluate these explanations in the light of the results from the plant perturbation experiments described in Table 8.2. (Even we balk at the prospect of only three experiments by a single pair of investigators on which to base a discussion of animal communities.)

Ghosts of competition past

The most widely quoted theory is that strong competition at earlier times has acted as a sieve, leaving behind species with different niches. As a result, competition in contemporary time is weak. This theory, which may mean that we cannot distinguish the effects of competition from the effects of its absence, must rate as community ecology's most celebrated impasse. Insofar as it depends on evolution of niches driven by competition, it has been called the 'ghost of competition past' (Connell 1980), but qualitatively similar results would also be expected in the assembly of competing species into communities in the absence of evolution (Rummel & Roughgarden 1985). Guilds structured in this way should have the properties of: (i) niche differences among the constituent species; and (ii) interspecific competition which is weak relative to intraspecific competition.

Various authors have commented on the rather limited opportunities for niche differentiation between autotrophic plant species, all of which have the same basic requirements of light, water, carbon dioxide and some mineral nutrients (e.g. Harper 1968; Grubb 1977; Aarssen 1983). To rescue the theory, plant ecologists have looked for other ways in which niche space could be partitioned, including complementarity of life form, separation through time and space in resource use, differences in the conditions for regeneration, and differences in the ratios of limiting resources required (Grubb 1977, 1986; Tilman 1982; Cody 1986). Those who have searched for niche differences in plant communities have usually been able to find them, including, for example, differences in the phenology of growth (Al-Mufti *et al.* 1977; Rogers & Westman 1979; Sydes 1984), in soil depth (Sharitz & McCormick 1973), in soil moisture (Werner & Platt 1976; Barcśak, Fekete & Précsényi 1981), in the spatial and temporal patterns of root activity (Veresoglou & Fitter 1984), and in the conditions for regeneration (Grubb 1984). It is debatable, however, whether such differences provide evidence of a history of strong competition, because we do not know whether the absence of differences would influence the competitive interactions of the species.

Equally, the absence of niche differences does not prove that competition has not been important in the past. Mahdi, Law & Willis (1989), for example, were unable to find any niche differences in the grassland manipulated by Mahdi (1988), between eight species along four out of seven niche axes. The three remaining axes were only able to separate pairs of species at opposite ends of the axes. These were the eight species subsequently manipulated (Table 8.2), and between which competition could not be detected (Mahdi 1988). The lack of niche differentiation is not conclusive evidence against a history of strong competition, because there remain other unmeasured niche axes on which differences might be found. Indeed, since there are always unmeasured axes (the dimensionality of the niche is not bounded), the theory cannot be disproved by measuring niches alone.

More persuasive would be information on the relative strengths of intra- and interspecific competition, because niche differentiation should reduce competition between species relative to that within species. For obvious reasons manipulation experiments involving complete removal of species do not lend themselves to estimation of the strength of intraspecific competition. The only data (on plants) come from Mahdi (1988), who was unable to detect inter- and intraspecific competition in the eight species manipulated.

Spatial aggregations within species

Niche differentiation between species is not the only process which would lead to a reduction in interspecific competition relative to intraspecific competition. The same outcome would be expected if conspecifics are spatially aggregated and aggregations of different species occur at different places. A theory based on these premises has been developed to describe the dynamics of guilds of *Drosophila* species living on ephemeral resources (Atkinson & Shorrocks 1981; Shorrocks *et al.* 1984; Shorrocks & Rosewell 1987), but the theory may be more widely applicable. In particular, terrestrial plant species characteristically show spatial clustering of conspecifics (Pielou 1974). A tendency towards spatial clustering was observed in five out of the eight species manipulated in Mahdi's (1988) experiment (Fig. 8.1), but in spite of this, as noted above attempts to detect intraspecific competition in this experiment failed.

Non-interactive guilds

Another possible explanation is that population densities are so low that individuals within a guild rarely interact with one another. Populations, rather than being regulated by interactions within the guild, could be kept at much lower levels by other mechanisms such as climatic extremes or the action of natural enemies. This has been suggested as an explanation for the lack of competition in guilds of folivorous insects (Lawton & Strong 1981; Strong 1982, 1984). Since the integrity of at least two of the guilds in Table 8.2 requires regular cropping of the taller growing species, natural enemies are strongly implicated as mediators of the within-guild interactions (Fowler 1981; Mahdi 1988). On the other hand, in guilds of this kind, characterized by continuous plant cover, there is evidence that the reproductive performance of individual plants is much suppressed by the presence of neighbours (Clements, Weaver & Hanson 1929, p. 130 et seq.; Goldberg 1987; Miller & Werner 1987). It would be unrealistic to suppose that the guilds in Table 8.2 are non-interactive.

Non-linearities of the per capita rates of increase

We are left with the paradox that there are probably strong interactions within the guilds, but at the same time little response can be seen following manipulation of the density of the constituent species. The resolution of this paradox could lie in the non-linearities of the g_i functions, given that they take the general shape shown in Fig. 8.3. Although the *per capita* rate of increase rapidly declines as density starts to increase from zero, it may change much less at the higher densities found in natural communities (see also Miller & Werner 1987). A curious consequence of this is that although a guild may be highly interactive, species-specific interactions would then play no more than a minor role in the trajectory of the guild. Instead, other factors that cause one species' *per capita* rate of increase to be greater than another's would assume a greater importance in the dynamics of the guild. We think, for instance, of environmental conditions which change over the course of time, sometimes favouring one species and sometimes favouring another. If species–specific interactions play a minor role, the contemporary perception of community dynamics could be seriously flawed; perhaps autecology rather than interactions dominate the trajectory (Strong 1986), leaving a Gleasonian plot unfolding against a Clementsian background.

It is, however, premature to reach firm conclusions about the cause of apparently weak interactions. Manipulation experiments have yet to be accurately calibrated, and are liable to various biases which could either mask or exaggerate responses. Detailed reconstruction of the dynamics of guilds from perturbation of their component species has barely been started. Until there is a well-documented body of data representing the major kinds of communities in nature, the interactive structure of guilds, whether competitive or otherwise, will remain uncertain.

CONCLUSION — PAST, PRESENT AND FUTURE

It is difficult to perceive a real sense of progress in our understanding of the role of competition in nature. We cannot point to an accretion of knowledge, to empirical questions clearly framed and unequivocally answered; rather we are left with an impression of a subject moving from one world view to another depending on the influence of the prevailing protagonists. This has brought with it much valuable machinery for thinking about competition, including for example a formal mathematical framework for modelling the process and a well-developed theory for the evolution of the niche, but it has left ecologists largely in the dark about basic empirical matters such as those discussed in this essay.

We think that a reason for this state of affairs is that the empirical foundation of community ecology is not altogether secure. The basic building blocks of the subject are the strengths of interactions between species, just as the rates of birth, death and migration form the empirical basis of population ecology. It is a fundamental dilemma of community ecology that these interaction strengths are so difficult to measure in natural communities. Faced with this dilemma, ecologists interested in competition have for the most part diverted their attention towards other more easily measured traits thought to indicate the outcome of competition when species live together, such as the niche, relative growth rate, body size and tolerance to resource depletion. We are reminded of the search for the philosopher's stone. It is to be hoped that the search will prove more productive and that some character will be found which predicts accurately the competitive ability of organisms, but a set of measured interaction strengths will be needed against which to test the reliability of such a character, for without this information, the role of competition in nature is a matter of conjecture. A method does exist for measuring competition in nature, based on the response to manipulated population densities. However, there remain serious difficulties in turning it into a reliable quantitative tool, and it remains to be seen whether the results it has generated to date are accurate indicators of interaction strengths.

Gimingham's (1987) gentle admonishment that the 'current and future thrusts [of ecology] will be strengthened and enlightened by a sound understanding of the road it has travelled to reach its present position' is particularly timely for the study of competition. The 'road' of data on measured competitive interactions turns out to be remarkably short. There is a great deal of work ahead to place the subject on a firm empirical foundation.

ACKNOWLEDGMENTS

We are grateful to the British Ecological Society for providing us with the opportunity to research into the history of ecology; our perspective of the subject has been much enhanced by this study. A. Mahdi kindly allowed us to make use of his detailed unpublished analysis of interactions between plant species in a limestone grassland. We thank A. H. Fitter, J. H. Lawton, W. J. Sutherland, C. R. Townsend, P. H. Warren and M. H. Williamson for helpful discussions during the preparation of the paper.

REFERENCES

Aarssen, L. W. (1983). Ecological combining ability and competitive combining ability in plants: toward a general evolutionary theory of coexistence in systems of competition. *American Naturalist*, **122**, 707–731.

Aarssen, L. W. (1988). 'Pecking order' of four plant species from pastures of different ages. *Oikos*, **51**, 3–12

Aarssen, L. W. & Turkington, R. (1985). Competitive relations among species from pastures of different ages. *Canadian Journal of Botany*, **63**, 2319–2325.

Al-Mufti, M. M., Sydes, C. L., Furness, S. B., Grime, J. P. & Band, S. R. (1977). A quantitative analysis of shoot phenology and dominance in herbaceous vegetation. *Journal of Ecology*, **65**, 759–791.

Allen, E. B. & Forman, R. T. T. (1976). Plant species removals and old-field community structure and stability. *Ecology*, **57**, 1233–1243.

Anon. (1944). British Ecological Society Easter Meeting 1944: Symposium on 'the ecology of closely allied species'. *Journal of Animal Ecology*, **13**, 176–177.

Antonovics, J. (1978). The population genetics of mixtures. *Plant Relations in Pastures* (Ed. by J. R. Wilson), pp.233–252. CSIRO, East Melbourne, Australia.

Antonovics, J. & Fowler, N. L. (1985). Analysis of frequency and density effects on growth in mixtures of *Salvia splendens* and *Linum grandiflorum* using hexagonal fan designs. *Journal of Ecology*, **73**, 219–234.

Antonovics, J. & Kareiva, P. (1988). Frequency dependent selection and competition: empirical approaches. *Philosophical Transactions of the Royal Society London, B*, **319**, 601–613.

Arthur, W. (1986). On the complexity of a simple environment: competition, resource partitioning and facilitation in a two-species *Drosophila* system. *Philosophical Transactions of the Royal Society London, B*, **313**, 471–508.

Arthur, W. (1987). *The Niche in Competition and Evolution.* John Wiley & Sons, Chichester.

Atkinson, W. D. & Shorrocks, B. (1981). Competition on a divided and ephemeral resource: a simulation model. *Journal of Animal Ecology*, **50**, 461–471.

Ayala, F. J. (1969). Evolution of fitness. IV. Genetic evolution of interspecific competitive ability in *Drosophila. Genetics*, **61**, 737–747.

Ayala, F. J., Gilpin, M. E. & Ehrenfeld, J. G. (1973). Competition between species: theoretical models and experimental tests. *Theoretical Population Biology*, **4**, 331–356.

Barcsák, Z., Fekete, G. & Précsényi, I. (1981). Niche and compositional structure in natural and influenced grasslands. *Man and the Biosphere Programme: Survey of Ten Years Activity in Hungary* (Ed. by P. Stefanovitz, A. Berczik, G. Fekete & M. Seidl), pp.67–102. Budapest.

Barker, J. F. S. (1973). Natural selection for coexistence or competitive ability in laboratory populations of *Drosophila. Egyptian Journal of Genetics and Cytology*, **2**, 288–315.

Bender, E. A., Case, T. J. & Gilpin, M. E. (1984). Perturbation experiments in community ecology: theory and practice. *Ecology*, **65**, 1–13.

Birch, L. C. (1957). The meanings of competition, *American Naturalist*, **91**, 5–18.

Brown, J. H., Davidson, D. W. & Reichman, O. J. (1979). An experimental study of competition between seed-eating desert rodents and ants. *American Zoologist*, **19**, 1129–1143.

Brown, J. H., Davidson, D. W., Munger, J. C. & Inouye, R. S. (1986). Experimental community ecology: the desert granivore system. *Community Ecology* (Ed. by J. Diamond & T. J. Case), pp.41–61. Harper & Row, New York.

Bryant, E. H. & Turner, C. R. (1972). Rapid evolution of competitive ability in larval mixtures of the housefly. *Evolution*, **26**, 161–170.

Butcher, R. E. (1983). *Studies on Interference between Weeds and Peas.* PhD thesis, University of East Anglia.

Clatworthy, J. N. & Harper, J. L. (1962). The comparative biology of closely related species living in the same area: V. Inter- and intraspecific interference within cultures of *Lemna* spp. and *Salvinia natans. Journal of Experimental Botany*, **13**, 307–324.

Clements, F. E. (1904). The development and structure of vegetation. *Report of the Botanical Survey of Nebraska* **7**, 5–175.

Clements, F. E. (1905). *Research Methods in Ecology.* University Publishing Company, Lincoln, Nebraska.

Clements, F. E., Weaver, J. E. & Hanson, H. C. (1929). *Plant Competition: An Analysis of Community Functions.* Carnegie Institution, Washington.

Cody, M. L. (1986). Structural niches in plant communities. *Community Ecology* (Ed. by J. Diamond & T. J. Case), pp.381–405. Harper & Row, New York.

Connell, J. H. (1980). Diversity and the coevolution of competitors, or the ghost of competition past. *Oikos*, **35**, 131–138.

Connell, J. H. (1983). On the prevalence and relative importance of interspecific competition: evidence from field experiments. *American Naturalist*, **122**, 661–696.

Crawley, M. J. (ed.)(1986). *Plant Ecology*. Blackwell Scientific Publications, Oxford.

Darwin, C. (1859). *The Origin of Species by Means of Natural Selection*. Penguin Books, Harmondsworth (reprint 1968).

Dawson P. S. (1972). Evolution in mixed populations of *Tribolium*. *Evolution*, **26**, 357–365.

Dawson P. S. (1979). Evolutionary changes in egg-eating behavior of flour beetles in mixed-species populations. *Evolution*, **33**, 585–594.

DeCandolle, A. P. (1820). Essai elementaire de géographie botanique. *Dictionnaire des Sciences Naturelles*, **18**, 359–436.

Diamond, J. M. (1978). Niche shifts and the rediscovery of competition. *American Scientist*, **66**, 322–331.

Diamond, J. & Case, T. J. (eds) (1986a). *Community Ecology*. Harper & Row, New York.

Diamond, J. & Case, T. J. (1986b). Overview: introductions, extinctions, exterminations, and invasions. *Community Ecology* (Ed. by J. Diamond & T. J. Case), pp.65–79. Harper & Row, New York.

Elton, C. (1946). Competition and the structure of ecological communities. *Journal of Animal Ecology*, **15**, 54–68.

Evans, R. C. & Turkington, R. (1988). Maintenance of morphological variation in a biotically patchy environment. *New Phytologist*, **109**, 369–376.

Fenchel, T. (1975). Character displacement and coexistence in mud snails (Hydrobiidae). *Oecologia*, **20**, 19–32.

Fenner, F. & Ratcliffe, F. N. (1965). *Myxomatosis*. Cambridge University Press, Cambridge.

Firbank, L. G. & Mortimer, A. M. (1985). Weed-crop interactions and the optimal timing of weed control. *Proceedings of the British Crop Protection Conference — Weeds*,pp.879–887.

Firbank, L. G., Mortimer, A. M. & Putwain, P. D. (1985). *Bromus sterilis* in winter wheat: a test of a predictive population model. *Aspects of Applied Biology*, **9**, 59–66.

Firbank, L. G. & Watkinson, A. R. (1986). Modelling the population dynamics of an arable weed and its effect upon crop yield. *Journal of Applied Ecology*, **23**, 147–159.

Fowler, N. (1981). Competition and coexistence in a North Carolina grassland. II. The effects of the experimental removal of species. *Journal of Ecology*, **69**, 843–854.

Frank, P. W. (1957). Coactions in laboratory populations of two species of *Daphnia*. *Ecology*, **38**, 510–519.

Futuyma, D. J. (1970). Variation in genetic response to interspecific competition in laboratory populations of *Drosophila*. *American Naturalist*, **104**, 239–252.

Gause, G. F. (1934). *The Struggle for Existence*. Hafner, New York.

Gause, G. F. (1935). *Vérifications Expérimentales de la Théorie Mathematique de la Lutte pour la Vie*. Actualites scientifiques et industrielles, 277. Hermann, Paris.

Gill, D. E. (1972). Intrinsic rates of increase, saturation densities, and competitive ability. I. An experiment with *Paramecium*. *American Naturalist*, **106**, 461–471.

Gilpin, M. E., Carpenter, M. P. & Pomerantz, M. J. (1986). The assembly of a laboratory community: multispecies competition in *Drosophila*. *Community Ecology* (Ed. by J. Diamond & T. J. Case), pp.23–40. Harper & Row, New York.

Gimingham, C. H. (1987). Foreword. *Seventy-five years of the British Ecological Society* (J. Sheail), p.xi. Blackwell Scientific Publications, Oxford.

Gliddon, C. & Trathan, P. (1985). Interactions between white clover and perennial ryegrass in an old permanent pasture. *Structure and Functioning of Plant populations*. **2**. *Phenotypic and Genotypic Variation in Plant Populations* (Ed. by J. Haeck & J. W. Woldendorp), pp.161–169. North-Holland Publishing Company, Amsterdam.

Goldberg, D. E. (1987). Neighborhood competition in an old-field plant community. *Ecology*, **68**, 1211–1223.

Goldberg, D. E. & Fleetwood, L. (1987). Competitive effect and response in four annual plants. *Journal of Ecology*, **75**, 1131–1143.

Goldberg, D. E. & Werner, P. A. (1983). Equivalence of competitors in plant communities: a null hypothesis and a field experimental approach. *American Journal of Botany*, **70**, 1098–1104.

Grime, J. P. (1979). *Plant Strategies and Vegetation Processes.* John Wiley & Sons, Chichester.

Grubb, P. J. (1977). The maintenance of species-richness in plant communities: the importance of the regeneration niche. *Biological Reviews*, **52**, 107–145.

Grubb, P. J. (1984). Some growth points in investigative plant ecology. *Trends in Ecological Research for the 1980's* (Ed. by J. H. Cooley & F. B. Golley), pp.51–74. Plenum Press, New York.

Grubb, P. J. (1986). Problems posed by sparse and patchily distributed species in species-rich plant communities. *Community Ecology* (Ed. by J. Diamond & T. J. Case), pp.207–225. Harper & Row, New York.

Hairston, N. G. (1980). Evolution under interspecific competition: field experiments on terrestrial salamanders. *Evolution*, **34**, 409–420.

Hairston, N. G. (1983). Alpha selection in competing salamanders: experimental verification of an a priori hypothesis. *American Naturalist*, **122**, 105–113.

Hairston, N. G., Nishikawa, K. C. & Stenhouse, S. L. (1987). The evolution of competing species of terrestrial salamanders: niche partitioning or interference? *Evolutionary Ecology*, **1**, 247–262.

Haizel, K. A. & Harper, J. L. (1973). The effects of density and timing of removal on interference between barley, white mustard and wild oats. *Journal of Applied Ecology*, **10**, 23–31.

Haldane, J. B. S. (1932). *The Causes of Evolution.* Longmans, Green & Co., London.

Hardin, G. (1960). The competitive exclusion principle. *Science*, **131**, 1292–1297.

Harper, J. L. (1968). The regulation of numbers amd mass in plant populations. *Population Biology and Evolution* (Ed. by R. C. Lewontin), pp. 139–158. Syracuse University Press, New York.

Harper, J. L. (1977). *Population Biology of Plants.* Academic Press, London.

Hassell, M. P. & Comins, H. N. (1976). Discrete time models for two-species competition. *Theoretical Population Biology*, **9**, 202–221.

Hedrick, P. W. (1972). Factors responsible for a change in interspecific competitive ability in *Drosophila. Evolution*, **26**, 513–522.

Hill, J. & Michaelson-Yeates, T. P. T. (1987). Effects of competition upon the productivity of white clover – perennial ryegrass mixtures. Genetic analysis. *Plant Breeding*, **99**, 239–250.

Holt, R. D. (1977). Predation, apparent competition and the structure of prey communities. *Theoretical Population Biology*, **12**, 197–229.

Hutchinson, G. E. (1957). Concluding remarks. *Cold Spring Harbor Symposia on Quantitative Biology*, **22**, 415–427.

Hutchinson, G. E. (1975). Variations on a theme by Robert MacArthur. *Ecology and Evolution of Communities* (Ed. by M. L. Cody & J. M. Diamond), pp.492–521. Harvard University Press, Cambridge, Massachusetts.

Hutchinson, G. E. (1978). *An Introduction to Population Ecology.* Yale University Press, New Haven.

Huxley, J. (1963). *Evolution the Modern Synthesis*, 2nd edition. Geoge Allen & Unwin, London.

Jackson, J. B. C. (1981). Interspecific competition and species' distributions: the ghosts of theories and data past. *American Zoologist*, **21**, 889–901.

Jeffries, M. H. & Lawton, J. H. (1984). Enemy free space and the structure of ecological communities. *Biological Journal of the Linnean Society*, **23**, 269–286.

Kingsland, S. E. (1985). *Modeling Nature: Episodes in the History of Population Ecology*. University of Chicago Press, Chicago.

Lack, D. (1944). Ecological aspects of species-formation in passerine birds. *Ibis*, **86**, 260–286.

Law, R. & Watkinson, A. R. (1987). Response-surface analysis of two-species competition: an experiment on *Phleum arenarium* and *Vulpia fasciculata. Journal of Ecology*, **75**, 871–886.

Lawton, J. H. & Hassell, M. P. (1981). Asymmetrical competition in insects. *Nature*, **289**, 793–795.

Lawton, J. H. & Strong, D. R. (1981). Community patterns and competition in folivorous insects. *American Naturalist*, **118**, 317–338.

Lerner, I. M. & Ho, F. K. (1961). Genotype and competitive ability of *Tribolium* species. *American Naturalist*, **95**, 329–343.

Leslie, P. H., Park, T. & Mertz, D. B. (1968). The effect of varying the initial number on the outcome of competition between two *Tribolium* species. *Journal of Animal Ecology*, **37**, 9–23.

Lewontin, R. C. (1974). *The Genetic Basis of Evolutionary Change*. Columbia University Press, New York.

Lotka, A. J. (1932). The growth of mixed populations: two species competing for a common food supply. *Journal of the Washington Academy of Sciences*, 22, 461–469.

Lubchenco, J. (1986). Relative importance of competition and predation: early colonization by seaweeds in New England. *Community Ecology* (Ed. by J. Diamond & T. J. Case), pp. 537–555. Harper & Row, New York.

MacArthur, R. H. (1958). Population ecology of some warblers of northeastern coniferous forests. *Ecology*, **39**, 599–619.

MacArthur, R. H. & Levins, R. (1967). The limiting similarity, convergence, and divergence of coexisting species. *American Naturalist*, **101**, 377–385.

Mahdi, A. (1988). *The Plant Ecology of a Limestone Grassland Community: Spatial Organization and Coexistence*. PhD Thesis, University of Sheffield.

Mahdi, A. & Law, R. (1987). On the spatial organization of plant species in a limestone grassland community. *Journal of Ecology*, **75**, 459–476.

Mahdi, A., Law, R. & Willis, A. J. (1989). Large niche overlaps among coexisting plant species in a limestone grassland. *Journal of Ecology*, **77**.

Malmquist, M. G. (1985). Character displacement and biogeography of the pigmy shrew in northern Europe. *Ecology*, **66**, 372–377.

Malthus, T. R. (1798). *An Essay on the Principle of Population*. Penguin Books. Harmondsworth (reprint 1982).

May, R. M. & MacArthur, R. H. (1972). Niche overlap as a function of environmental variability. *Proceedings of the National Academy of Sciences of the USA*, **69**, 1109–1113.

McIntosh, R. P. (1985). *The Background of Ecology: Concept and Theory*. Cambridge University Press, Cambridge.

McNeilly, T. (1981). Ecotypic differentiation in *Poa annua*: interpopulation differences in response to competition and cutting. *New Phytologist*, **88**, 539–547.

Mertz, D. B., Cawthorn, D. A. & Park, T. (1976). An experimental analysis of competitive indeterminacy in *Tribolium. Proceedings of the National Academy of Sciences of the USA*, **73**, 1368–1372.

Miller, R. S. (1964). Larval competition in *Drosophila melanogaster* and *D. simulans*. *Ecology*, **45**, 132–148.

Miller, T. E. & Werner, P. A. (1987). Competitive effects and responses between plant species in a first-year old-field community. *Ecology*, **68**, 1201–1210.

Milne, A. (1961). Definition of competition among animals. *Symposia of the Society for Experimental Biology*, **15**, 40–61.

Moore, J. A. (1952). Competition between *Drosophila melanogaster* and *Drosophila simulans*. II. The improvement of competitive ability through selection. *Proceedings of the National Academy of Sciences of the USA*, **38**, 813–817.

Moulton, M. P. & Pimm, S. L. (1986). The extent of competition in shaping an introduced avifauna. *Community Ecology* (Ed. by J. Diamond & T. J. Case), pp.80–97. Harper & Row, New York.

Nägeli, C. (1865). *Bedingungen des Vorkommens von Arten und Varietaten innerhalb ihres Verbreitungsbezirkes*. Sitzungsber. Munchener Akad.

Odum, E. P. (1959). *Fundamentals of Ecology. 2nd edition*. Saunders, Philadelphia.

Pacala, S. W. & Silander, J. A. (1987). Neighbourhood interference among velvet leaf, *Abutilon theophrasti*, and pigweed, *Amaranthus reflexus*. *Oikos*, **48**, 217–224.

Park, T. (1948). Experimental studies of interspecific competition: I. Competition between populations of the flour beetles *Tribolium confusum* Duval and *Tribolium castaneum* Herbst., *Ecological Monographs*, **18**, 267–307.

Park, T. (1954). Experimental studies of interspecific competition: II. Temperature, humidity and competition in two species of *Tribolium*. *Physiological Zoology*, **27**, 177–238.

Park, T. & Lloyd, M. (1955). Natural selection and the outcome of competition. *American Naturalist*, **89**, 235–240.

Pielou, E. C. (1974). *Population and Community Ecology. Principles and Methods*. Gordon & Breach, New York.

Pimentel, D., Feinberg, E. H., Wood, P. W. & Hayes, J. T. (1965). Selection, spatial distribution and the coexistence of competing fly species. *American Naturalist*, **99**, 97–109.

Pruzan-Hotchkiss, A., Perelle, I. B., Hotchkiss, F. H. C., & Ehrman, L. (1980). Altered competition between two reproductively isolated strains of *Drosophila melanogaster*. *Evolution*, **34**, 445–452.

Reynolds, J. C. (1981). *The Interaction of Red and Grey Squirrels*. PhD thesis, University of East Anglia.

Reynolds, J. C. (1985). Details of the geographic replacement of the red squirrel (*Sciurus vulgaris*) by the grey squirrel (*Sciurus carolinensis*) in eastern England. *Journal of Animal Ecology*, **54**, 149–162.

Rogers, R. W. & Westman, W. E. (1979). Niche differentiation and maintenance of genetic identity in cohabiting *Eucalyptus* species. *Australian Journal of Ecology*, **4**, 429–439.

Root, R. B. (1967). The niche exploitation pattern of the blue-gray gnatcatcher. *Ecological Monographs*, **37**, 317–350.

Roughgarden, J. (1986). A comparison of food-limited and space-limited animal competition communities. *Community Ecology* (Ed. by J. Diamond & T. J. Case), pp. 492–516. Harper & Row, New York.

Roughgarden, J., Heckel, D. & Fuentes, E. R. (1983). Coevolutionary theory and the biogeography and community structure of *Anolis*. *Lizard Ecology, Studies of a Model Organism* (Ed. by R. B. Huey, E. R. Pianka & T. W. Schoener), pp.371–410. Harvard University Press, Harvard, Massachusetts.

Rummel, J. D. & Roughgarden, J. (1985). A theory of faunal buildup for competition communities. *Evolution*, **39**, 1009–1033.

Sakai, K-I. (1961). Competitive ability of plants: its inheritance and some related problems. *Symposia of the Society for Experimental Biology*, **15**, 245–263.

Salisbury, E. J. (1929). The biological equipment of species in relation to competition. *Journal of Ecology*, **17**, 197–222.

Schluter, D., Price, T. D. & Grant, P. R. (1985). Ecological character displacement in Darwin's finches. *Science*, **227**, 1056–1059.

Schoener, T. W. (1982). The controversy over interspecific competition. *American Scientist*, **70**, 586–595.

Schoener, T. W. (1983). Field experiments on interspecific competition. *American Naturalist*, **122**, 240–285.

Schoener, T. W. (1985). Some comments on Connell's and my reviews of field experiments on interspecific competition. *American Naturalist*, **125**, 730–740.

Seaton, A. P. C. & Antonovics, J. (1967). Population inter-relationships. I. Evolution in mixtures of *Drosophila* mutants. *Heredity*, **22**, 19–33.

Seifert, R. P. & Seifert, F. H. (1976). A community matrix analysis of *Heliconia* insect communities. *American Naturalist*, **110**, 461–483.

Seifert, R. P. & Seifert, F. H. (1979). A *Heliconia* insect community in a Venezuelan cloud forest. *Ecology*, **60**, 462–467.

Sharitz, R. R. & McCormick, J. F. (1973). Population dynamics of two competing annual plant species, *Ecology*, **54**, 723–740.

Shorrocks, B., Rosewell, J., Edwards, K. & Atkinson, W. D. (1984). Interspecific competition is not a major organizing force in many insect communities. *Nature*, **310**, 310–312.

Shorrocks, B. & Rosewell, J. (1987). Spatial patchiness and community structure: coexistence and guild size of drosophilids on ephemeral resources. *Organization of Communities Past and Present* (Ed. by J. H. R. Gee & P. S. Giller), pp.29–51. Blackwell Scientific Publications, Oxford.

Silander, J. A. & Antonovics, J. (1982). Analysis of interspecific interactions in a coastal plant community — a perturbation approach . *Nature*, **298**, 557–560.

Simberloff, D. (1982). A succession of paradigms in ecology; essentialism to materialism and probabilism. *Conceptual Issues in Ecology* (Ed. by E. Saarinen),pp.63–99. Pallas Paperbacks, Reidel, Dordrecht.

Slobodkin, L. B. (1964). Experimental populations of Hydrida. *Journal of Animal Ecology*, **33**, 131–148.

Smith-Gill, S. J. & Gill, D. E. (1978). Curvilinearities in the competition equations: an experiment with ranid tadpoles. *American Naturalist*, **112**, 557–570.

Sokal, R. R., Bryant, E. H. & Wool, D. (1970). Selection for changes in genetic facilitation: negative results in *Tribolium* and *Musca*. *Heredity*, **25**, 299–306.

Strong, D. R. (1982). Harmonious coexistence of hispine beetles on *Heliconia* in experimental and natural communities. *Ecology*, **63**, 1039–1049.

Strong, D. R. (1984). Exorcising the ghost of competition past: phytophagous insects. *Ecological Communities: Conceptual Issues and the Evidence* (Ed. by D. R. Strong, D. Simberloff, L. G. Abele & A. B. Thistle), pp.28–41. Princeton University Press, Princeton, New Jersey.

Strong, D. R. (1986). Density vagueness: abiding the variance in the demography of real populations. *Community Ecology* (Ed. by J. Diamond & T. J. Case), pp.257–268. Harper & Row, New York.

Strong, D. R. Simberloff, D., Abele, L. G. & Thistle, A. B. (eds) (1984). *Ecological Communities: Conceptual Issues and the Evidence*. Princeton University Press, Princeton, New Jersey.

Sulzbach, D. S. (1980). Selection for competitive ability: negative results in *Drosophila*. *Evolution*, **34**, 431–436.

Sulzbach, D. S. & Emlen, J. M. (1979). Evolution of competitive ability in mixtures of *Drosophila melanogaster*: populations with an initial asymmetry. *Evolution*, **33**, 1138–1149.

Sydes, C. L. (1984). A comparative study of leaf demography in limestone grassland. *Journal of Ecology*, **72**, 331–345.

Tansley, A. G. (1914). Presidential Address. *Journal of Ecology*, **2**, 194–202.

Tansley, A. G. (1917). On competition between *Galium saxatile L.* (*G. hercynicum* Weig.) and *Galium sylvestre* Poll. (*G. asperum* Schreb.) on different types of soil. *Journal of Ecology*, **5**, 173–179.

Thompson, K. (1987). The resource ratio hypothesis and the meaning of competition. *Functional Ecology*, **1**, 297–303.

Thompson, K. & Grime, J. P. (1988). Competition reconsidered – a reply to Tilman. *Functional Ecology*, **2**, 114–116.

Tilman, D. (1982). *Resource Competition and Community Structure.* Princeton University Press, Princeton, New Jersey.

Tilman, D. (1987). On the meaning of competition and the mechanisms of competitive superiority. *Functional Ecology*, **1**, 304–315.

Tilman, D., Mattson, M. & Langer, S. (1981). Competition and nutrient kinetics along a temperature gradient: an experimental test of a mechanistic approach to niche theory. *Limnology and Oceanography*, **26**, 1020–1033.

Turkington, R. & Harper, J. L. (1979). The growth, distribution and neighbour relationships of *Trifolium repens* in a permanent pasture: IV. Fine scale biotic differentiation. *Journal of Ecology*, **67**, 245–254.

Turkington, R., Holl, F. B., Chanway, C. P. & Thompson, J. D. (1988). The influence of microorganisms, particularly *Rhizobium*, on plant competition in grass-legume communities. *Plant Population Ecology* (Ed. by A. J. Davy, M. J. Hutchings & A. R. Watkinson), pp.343–366. Blackwell Scientific Publications, Oxford.

Underwood, T. (1986). The analysis of competition by field experiments. *Community Ecology: Pattern and Process* (Ed. by J. Kikkawa & D. J. Anderson), pp.240–268. Blackwell Scientific Publications, Melbourne.

Van Bogaert, G. (1974). The evaluation of progenies of meadow fescue (*Festuca pratensis* L.) in monoculture and in mixture with perennial ryegrass (*Lolium perenne* L.). *Euphytica*, **23**, 48–53.

Van Delden, W. (1970). Research notes. Drosophila. *Information Service*, **45**, 169.

Veresoglou, D. S. & Fitter, A. H. (1984). Spatial and temporal patterns of growth and nutrient uptake of five co-existing grasses. *Journal of Ecology*, **72**, 259–272.

Volterra, V. (1926). Variazioni e fluttuazioni del numero d'individui in specie animali conviventi. *Memoria della Regia Accademia Nazionale dei Lincei, ser. 6*, **2**, 31–113.

Warming, E. (1909). *Oecology of Plants.* Clarendon Press, Oxford.

Watkinson, A. R. (1981). Interference in pure and mixed populations of *Agrostemma githago*. *Journal of Applied Ecology*, **18**, 967–976.

Watkinson, A. R. (1985). Plant responses to crowding. *Studies on Plant Demography: A Festschrift for John L. Harper* (Ed. by J. White), pp. 275–289. Academic Press, London.

Watkinson, A. R. & Davy, A. J. (1985). Population biology of salt marsh and sand dune annuals. *Vegetatio*, **62**, 487–497.

Watt, A. S. (1925). On the ecology of British beechwoods with special reference to their regeneration. II. The development and structure of beech communities on the Sussex Downs. *Journal of Ecology*, **13**, 27–73.

Watt, A. S. (1947). Pattern and process in the plant community. *Journal of Ecology*, **35**, 1–22.

Watt A. S. (1981). Further observations on the effects of excluding rabbits from grassland A in East Anglian Breckland: the pattern of change and factors affecting it (1936–1973). *Journal of Ecology*, **69**, 509–536.

Werner, E. E. (1986). Species interactions in freshwater fish communities. *Community Ecology* (Ed. by J. Diamond & T. J. Case), pp.344–358. Harper & Row, New York.

Werner, P. A. & Platt, W. J. (1976). Ecological relationships of co-occurring goldenrods (*Solidago*: Compositae). *American Naturalist*, **110**, 959–971.

Williamson, M. H. (1957). An elementary theory of interspecific competition. *Nature*, **180**, 422–425.

Williamson, M. (1972). *The Analysis of Biological Populations*. Arnold, London.

Wit, C. T. de (1960). On competition. *Verslagen van Landbouwkundige Onderzoekingen*, **66**, 1–82.

Woodhead, T. W. (1906). Ecology of woodland plants in the neighbourhood of Huddersfield. *Journal of the Linnean Society*, **37**, 333–406.

9 LIFE-HISTORY STRATEGIES

H. CASWELL
Biology Department, Woods Hole Oceanographic Institution, Woods Hole, MA 02543, USA

THE CONCEPT AND ITS ORIGIN

The study of life-history strategies originated in the late 1940s and early 1950s from the combination of animal demography and evolutionary theory. Between 1920 and 1950, the application by Lotka, Pearl, Bodenheimer, Lack, Leslie, and Deevey, of life-table methods to animals had provided a powerful approach to analysing the quantitative demographic consequences of the life cycle. That analysis was based on age-specific rates of mortality and reproduction, and ecologists were well aware that those rates varied in interesting ways both within and among species. The development of population and quantitative genetics had simultaneously provided a rigorous basis for Darwinian arguments about the adaptive value of phenotypic traits.

It is well known that population ecology developed without paying serious attention to genetic variation, while population genetics largely ignored demography. The integration of the two has been one of the most long-standing problems in ecology (e.g. Roughgarden 1979), and has been fundamental to the study of life-history strategies. The basic approach, treating demographic traits as part of the phenotype and investigating their adaptive basis, was suggested by Fisher as early as 1930:

> There is something like a relic of creationist philosophy in arguing from the observation, let us say, that a cod spawns a million eggs, that *therefore* its offspring are subject to Natural Selection; and it has the disadvantage of excluding fecundity from the class of characteristics of which we may attempt to appreciate the aptitude. It would be instructive to know not only by what physiological mechanism a just apportionment is made between the nutriment devoted to the gonads and that devoted to the rest of the parental organism, but also what circumstances in the life history and environment would render profitable the diversion of a greater or lesser share of the available resources to reproduction. [Fisher 1930, p. 47]

This passage contains the fundamental premise of life history theory: that life-history traits are part of the phenotype, and hence as much (or as little) subject to adaptive explanation as are morphology, physiology , and behaviour. Putting this insight into practice, and thus inventing the modern study of life-history strategies, had to wait for a group of three seminal papers published in the late 1940s and early 1950s: Lack (1947) on clutch size, Medawar (1946, 1952) on senescence and Cole (1954) on parity. Each of these authors explicitly adopted the thesis that the pattern of the life cycle was subject to evolutionary explanation:

The thesis advanced here is that the reproductive rate of animals, like other characters, is a product of natural selection, hence that each species lays that number of eggs which results in the maximum number of surviving offspring. [Lack 1954, p. 154].

In this paper it will be regarded as axiomatic that the reproductive potentials of existing species are related to their requirements for survival; that any life history features affecting reproductive potential are subject to natural selection; and that such features observed in existing species should be considered adaptations, just as purely morphological or behavioral patterns are commonly so considered. [Cole 1954, p. 114]

I think many biologists would agree that Weismann was in principle correct, and that the process of senescence in the individual and the form of the age-frequency distribution of death that mirrors it statistically have been shaped by the forces of natural selection. [Medawar 1946, p. 34]

In this paper, I will survey some aspects of the connection between demography and evolutionary theory that makes the adaptive analysis of life histories possible. I will focus on the logical structure of life-history theory, some approaches to it, and what seems to be promising or problematical areas. The literature on life-history strategies is enormous, and I will not attempt to provide a review. Stearns (1976, 1977, 1980) has provided a series of such reviews. Comparing this paper with two recent essays on life histories (Partridge & Harvey 1988, Southwood 1988) will show clearly the diversity of approaches to this subject that are possible.

It is no accident that each of the original three papers on life-history strategies began with an apparent paradox. Lack (1947) was interested in clutch size patterns in birds. The data showed clearly that birds do not lay as many eggs as they are physiologically capable of doing. Yet, reproduction is clearly a component of fitness. Why should reproduction be curtailed? Lack rightly rejected the hypothesis that reproduction was adjusted to maintain population size against the prevailing force of mortality as incompatible with selection at the individual level:

Natural selection acts on the individual and its family rather than on the species as a whole. If some individuals of a species lay four eggs and others five, then, given that the difference is inherited, the five-egg birds are bound to predominate rapidly in the population, as they have more descendents, even if this leads to 'overproduction' – unless the laying of five eggs instead of four is a disadvantage in itself, either to the brood, or to the parents, or both. [Lack 1947, p. 318]

Lack proposed that the limited ability of parents to provide food for offspring meant that the survival of offspring in large clutches would be less than that in smaller clutches, and that the number of surviving offspring would be maximized at an intermediate clutch size.

Medawar (1946, 1952) addressed the issue of senescence. Survival is obviously a component of fitness, yet senescence is a conspicuous characteristic of the life cycle of most organisms. Why should survival be curtailed? Like Lack, Medawar rejected the explanation (proposed by Weismann) that selection would eliminate old and worn-out individuals because they are useless to the species.* Instead, he considered the total selective pressures operating on traits which have different effects at different ages. Using a simple model of a non-senescing population (of test tubes, although that makes no difference), he showed that the strength of selection on a trait decreased as the trait is expressed at older and older ages. Thus any trait which produces positive effects at young ages and negative effects at older ages will be favoured by selection, and detrimental effects will accumulate at older ages. The result is senescence.

Cole's (1954) paper is best known for its analysis of the problem of semelparity.† Reproduction is an obvious component of fitness, but many organisms forgo the opportunity for repeated reproduction, and die after a single reproduction episode:

> One feels intuitively that natural selection should favor the perennial reproductive habit because an individual producing seeds or young annually over a period of several years obviously has the potential ability to produce many more offspring than is the case when reproduction occurs but once. It is, therefore, a matter of some interest to examine the effect of iteroparity on the intrinsic rate of natural increase in order to see if we can find an explanation for the fact that repeated reproduction is not more general. [Cole 1954]

Using a simple model, Cole found that the gain in fitness resulting from iteroparity is equivalent to adding a single offspring to the semelparous clutch size. He concluded that the cost of the structures necessary to survive for more than one year probably outweighs the benefit of this small addition to clutch size, and found himself wondering why iteroparity should exist!

In each case, the resolution of the apparent paradox turned on the recognition of interactions between different parts of the life cycle (between egg production and juvenile survival, between early and late survival, and between early and later reproduction). This concern with the interaction of traits is fundamental to evolutionary demography. Indeed, it is well known that the problem of specifying an optimal life history in the absence of these interactions is trivial; the optimal organism is reproductively mature at birth, produces an infinite number of offspring continuously, and lives forever. The mere existence of organisms more

* That explanations of life-history traits couched in terms of the 'good of the species' can still be heard in discussions at symposia of learned ecological societies is a measure of how far the integration of ecology and evolution still has to go.

† Semelparous species are those which reproduce only once in their lifetime; iteroparous species reproduce repeatedly.

interesting than this is a measure of the importance of interaction between life-history traits.

REQUISITES FOR AN EVOLUTIONARY DEMOGRAPHY

If the life history is to be treated like any other part of the phenotype, the requirements for its evolutionary study are the same as they are for any other trait: phenotypic variation, fitness effects and heritability.

The existence of phenotypic variation in life-history traits is obvious. The existence of fitness effects follows immediately, since the vital rates are, as we shall see, components of fitness. The heritability of life-history variation is more problematic.

Genetic variation

Being intimately connected with fitness, the standard wisdom expects life-history traits to have been under such strong selection that all the additive genetic variance will have been consumed. This view is supported by noting that, in domesticated animals and plants, the narrow-sense heritabilities for reproductive traits are indeed often less than those for traits not closely related to fitness (Falconer 1981, table 10.1; Istock 1983). This is not universally true, however. Heritabilities of 0.5 or greater have, for example, been reported by Nguyen & Sleper (1983) for seed weight, number of panicles and seed yield in tall fescue grass; by Zimmermann, Rosielle & Waines (1984) for seed yield of beans in two different cropping systems, and by Wong & Baker (1986) for time to maturity in wheat. The crop science literature abounds with examples like these.

Studies of non-domesticated species have also revealed significant genetic variation in life-history traits, by means of correlations among relatives (e.g. Moss & Watson 1982; Berven 1987), common environment rearing experiments (e.g. Wyngaard 1986), or selection experiments (e.g. Rose & Charlesworth 1981; Rose 1984a; Luckinbill & Clare 1985). Substantial additive genetic variance for life-history traits appears to be the rule rather than the exception (Istock 1983).

Of the several possible explanations for this genetic variation (selective neutrality, mutation–selection balance, lack of equilibrium) the most reasonable and well supported seems to be antagonistic pleiotropy (Rose 1982, 1983), a pattern of gene action in which a gene has positive effects on one fitness component and negative effects on another. The result is a genetic 'trade-off' between life-history traits, precisely the pattern which Lack, Medawar and Cole proposed to escape the apparent paradoxes in observed life-history strategies.

The documentation of the genetic basis for variance in life-history traits is more common now than it was even a decade ago. When Stearns (1977) reviewed much of the empirical data on life-history traits, less than a third of the published studies had done so. Most life-history theorists made the tacit assumption that additive genetic variance would always be available, so that all that one had to do was model fitness effects, although some of them (at least the present author) felt uneasy about this. However, if additive genetic variance is as widespread as it now appears to be (Istock 1983), this assumption may not be so dangerous. Indeed, more sophisticated genetic analyses (below) suggest that it is not genetic variances but the genetic covariances that most urgently need to be examined.

Secondary theorem of natural selection

Selection arguments, in life-history theory as elsewhere, require a theoretical connection between phenotypic variation, fitness effects and genetic variance. This connection is provided by the so-called Secondary Theorem of Natural Selection*(Robertson 1968, elaborated by Emlen 1970 and Price 1970), which states that the rate of change in the mean phenotypic value z of some quantitative trait is proportional to the genetic covariance of the trait with fitness W:

$$\Delta \bar{z} = C_g(z, W) \tag{1}$$

where C_g denotes the genetic covariance. Since the slope of the regression of fitness on z is $C_g(z, W)/V_g(z)$, the rate of change in the trait can be approximated by

$$\Delta \bar{z} \approx \frac{\partial W}{\partial z} V_g(z), \tag{2}$$

where V_g is the genetic variance. The partial derivative of fitness with respect to z is the selection pressure (Emlen 1970) or selection gradient (Lande 1979) on z. This expression clearly reveals the two major requisites for an adaptive theory of life-history traits: fitness effects and genetic variance. Since $V_g \geq 0$, the direction of selection is determined by

* This result is much more useful than Fisher's Fundamental Theorem, which can be derived from it as a special case. Fisher, however, named his theorem first.

the selection gradient, and much of life-history theory consists of deriving expressions for the selection gradients of various life-history traits.

Equation (2) considers only a single trait z, but life history theory, as we have seen, must consider the action of selection on multiple, interacting traits. This important extension of equation (2) was provided by Lande (1979) and Lande & Arnold (1983). Consider a vector $\mathbf{z}$ of traits which are simultaneously under selection, with an additive genetic covariance matrix $\boldsymbol{G}$. Then

$$\Delta\bar{z} = \boldsymbol{G}\nabla\mathbf{W} \tag{3}$$

where the gradient vector ∇W is

$$\nabla W = \begin{pmatrix} \dfrac{\partial W}{\partial \bar{z}_1} \\ \dfrac{\partial W}{\partial \bar{z}_2} \\ \vdots \\ \dfrac{\partial W}{\partial \bar{z}_k} \end{pmatrix}. \tag{4}$$

This result encapsulates not only the interaction of genetic variance and the selection gradient in determining the change in a trait, but also the effect of the covariance between traits. The direction of change cannot be predicted from the selection gradient alone, but depends on the genetic interaction of the trait in question with other traits. Lack, Medawar and Cole could hardly have put it better.

ISSUES

What follows next is a personal selection of some important evolutionary issues which have been raised with particular clarity by the study of life-history strategies.

What is fitness?

The definition of fitness has always been problematic. In the simplest population genetic models, the relative fitnesses of alleles or genotypes are simply assigned as numbers. In consideration of life-history traits, however, it is crucial to have some way of combining age- or stage-specific demographic characters into some relevant measure of fitness.

Demographic measures of fitness

Fitness, whatever else one may say about it, has something to do with the rate at which genes are propagated into future generations. Fisher (1930) introduced an explicitly demographic definition of fitness, by discussing the Malthusian parameter m, now better known as the intrinsic rate of increase r, in the same chapter in which he derived the Fundamental Theorem of Natural Selection. In fact however, he did not consider selection in age-classified populations, and made little use of m in his derivations.

A definition of fitness appropriate for age-classified (or better still, stage-classified) populations is essential to the discussion of any trait the expression of which is related to age. Following a tradition begun by Norton (1928), Charlesworth (reviewed in his 1980 book) and Lande (1982) have provided a clear definition of fitness in age-classified situations, and Lande's argument has been extended to almost arbitrary stage-classifications (Caswell 1989b). Their answer agrees with Fisher's and justifies the practice of a generation of life-history theorists: given a constant environment, sufficient time for demographic ergodicity, and weak selection, fitness is measured by the intrinsic rate of increase r, or its discrete-time analogue $\lambda = e^r$. Thus in equation (3), the selection gradient vector is

$$\nabla \boldsymbol{W} = \nabla r \tag{5}$$

$$= \frac{1}{\lambda} \nabla \lambda . \tag{6}$$

This result* renders the selection gradient on life-history traits easily calculable from demographic models (e.g. as examples of several approaches, see Stearns & Crandall 1981; Caswell 1983, Sibley & Calow 1983, 1986). For example, if the demography is described by a population projection model

$$\boldsymbol{n}(t + 1) = \boldsymbol{A}\boldsymbol{n}(t), \tag{7}$$

with $\boldsymbol{n}$ a vector giving the abundance of individuals in each stage or age class, and $\boldsymbol{A}$ a population projection matrix, then λ is given by the dominant eigenvalue of $\boldsymbol{A}$ and the selection gradient on any element a_{ij} of $\boldsymbol{A}$ (i.e. on the rate of transition between any two stages in the life cycle) is

$$\frac{\partial \lambda}{\partial a_{ij}} = \frac{v_i w_j}{\langle \boldsymbol{w}, \boldsymbol{v} \rangle}, \tag{8}$$

* Note that this result is *not* an assumption about the use of λ, nor does it rely on group selection. It is a theorem of quantitative genetics.

(Caswell 1978, 1986), where v_i and w_j are the ith element of the left (reproductive value) eigenvector $\boldsymbol{v}$ and the jth element of the right (stable stage distribution) eigenvector $\boldsymbol{w}$, respectively, and $\langle \boldsymbol{w}, \boldsymbol{v} \rangle$ denotes the scalar product.

The measurement of λ or r as a demographic estimate of fitness is not trivial, since it requires the estimation of a complete set of vital rates. It is thus tempting to work instead with such components of fitness as reproductive output, survival, or expected lifetime reproductive output R_0 (e.g. the studies surveyed in Clutton-Brock 1988). Unfortunately, the results of such analyses can be misleading, because they are insensitive to the timing of events in the life cycle, which is potentially extremely important.

As an example, consider the data of Levin *et al.* (1987) on planktotrophic and lecithotrophic strains* of the polychaete *Streblospio benedicti*. Under the laboratory conditions of this study, the lecithotrophic strain was more fit ($\lambda = 1.319$, $R_0 = 89.9$) than the planktotrophic strain ($\lambda = 1.205$, $R_0 = 15.5$). However, the two strains differ in several ways; the assessment of the source of the lecithotrophic advantage differs markedly depending on whether λ or some less complete measure of fitness is used.

Figure 9.1 shows the differences (measured as lecithotrophic − planktotrophic) in age-specific survival (P_i) and fertility (F_i) between the two strains. The most conspicuous differences are a planktotrophic fertility advantage from age 15 to 25 weeks and a lecithotrophic survival advantage during the first 10 weeks of life. Clearly, these differences counteract each other. Any attempt to evaluate the selective value of larval development mode based on reproduction alone, or survival alone, must fail.

The simplest way to combine reproduction and survival is in terms of expected lifetime reproductive output, which is given by $R_0 = \Sigma_i L_i F_i$ where L_i is the probability of surviving to age class i and F_i is the fertility of age class i. The terms in this summation give the expected reproductive output at each age. From Fig. 9.2(a) which compares these terms for the two strains, it appears quite clear that the lecithotrophic advantage in R_0 arises from advantages in expected reproduction between 15 and 30 weeks of age.

Let us take this analysis one step further, by examining the contribution of individual age-specific survival and fertility differences to the

* Lecithotrophic larvae are provisioned with egg yolk by the parents and do not feed in the plankton. Planktotrophic larvae, generally smaller, rely on feeding in the plankton. Both strains occur in this species.

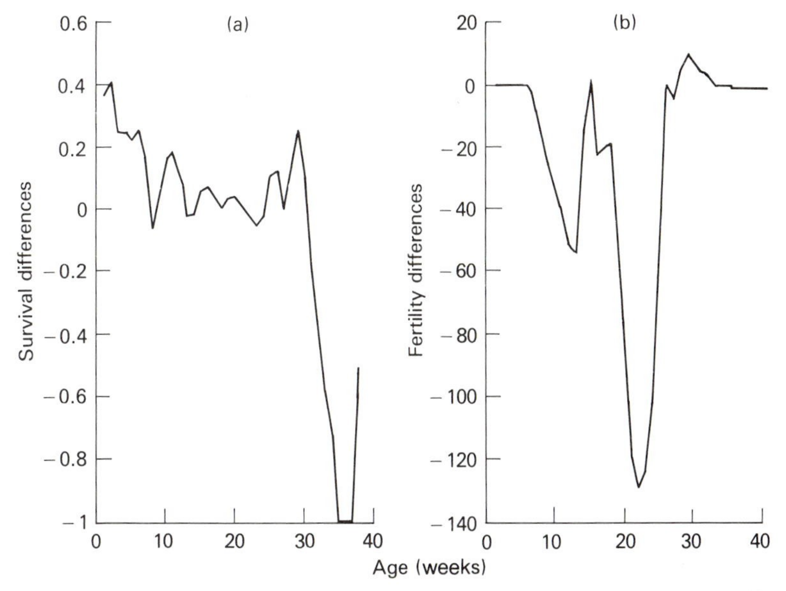

FIG. 9.1 Differences in: (a) age-specific survival probability (P_i) and (b) age-specific fertility (F_i) between lecithotrophic and planktotrophic strains of *Streblospio benedicti* (Levin *et al.* 1987). Positive values indicate a lecithotrophic advantage, and vice versa.

difference in R_0. The difference between strains in expected reproduction can be written

$$\Delta R_0 \approx \sum_i \Delta P_i \frac{\partial R_0}{\partial P_i} + \sum_i \Delta F_i \frac{\partial R_0}{\partial F_i}. \tag{9}$$

Each term in the summations is the contribution of an age- specific life history difference (ΔP_i or ΔF_i) to the difference in ΔR_0. These contributions are plotted in Figs 9.2(b,c). They show that ΔR_0 reflects a lecithotrophic survival advantage from ages 0–10 days and a planktotrophic fertility advantage, especially from ages 15 to 25 weeks.

Finally, let us compare these conclusions with the corresponding analysis of the difference in fitness as measured by λ. Decomposing the fitness differential $\Delta\lambda$ as

$$\Delta\lambda \approx \sum_i \Delta P_i \frac{\partial \lambda}{\partial P_i} + \sum_i \Delta F_i \frac{\partial \lambda}{\partial F_i}. \tag{10}$$

gives a set of contributions of age-specific survival and fertility differences to fitness (see Caswell 1989a for a detailed discussion of this an-

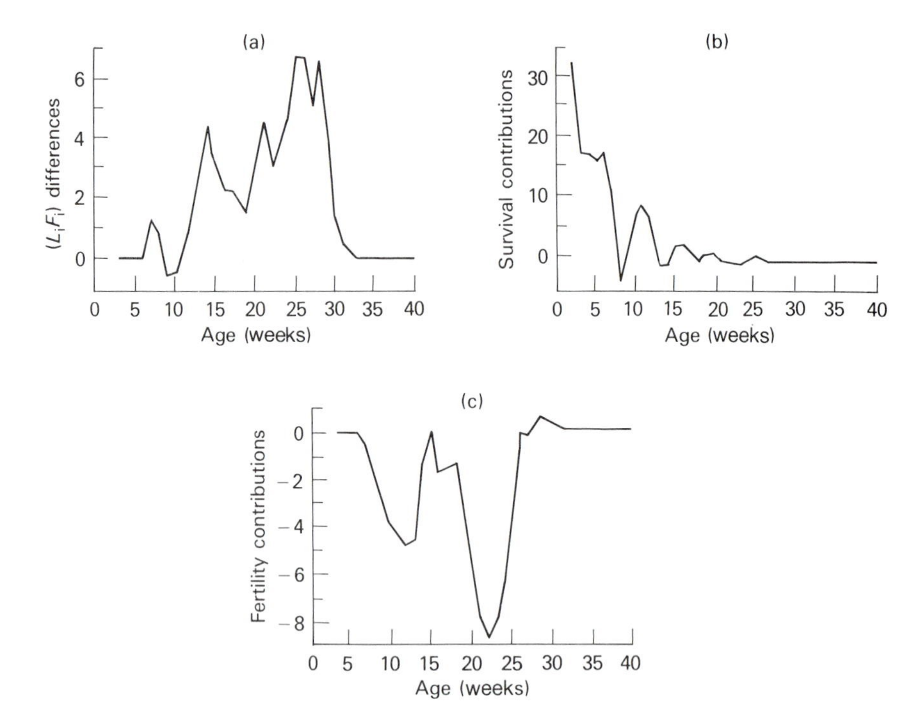

FIG. 9.2. Contributions of age-specific life-history differences to differences in expected life-time reproduction R_0 between lecithotrophic and planktotrophic strains of *Streblospio benedicti* (Levin *et al.* 1987). Positive values indicate a lecithotrophic advantage, and vice versa. (a) Differences in age-specific reproduction (L_iF_i). (b) Contributions of age-specific survival differences ΔP_i *to* ΔR_0 [see equation (9)]. (c) Contributions of age-specific fertility differences ΔF_i to ΔR_0 [see equation (9)].

alysis). These contributions are shown in Fig. 9.3. The fitness differences between the two strains are accounted for almost entirely by the lecithotrophic survival advantage during the first 8–10 weeks of life, and the planktotrophic fertility advantage during the first 15 weeks of life. The large planktotrophic fertility advantage after age 15, which has such a significant impact on R_0, makes a negligible contribution to the difference in fitness. The moral of the story is that calculations based on components of fitness or on expected offspring production miss the effect

---

FIG. 9.3. Contributions of age-specific life-history differences to differences in fitness λ between lecithotrophic and planktotrophic strains of *Streblospio benedicti* (Levin *et al.* 1987). Positive values indicate a lecithotrophic advantage, and vice versa. (a) Contributions of age-specific survival differences ΔP_i to $\Delta\lambda$ [see equation (10)]. (b) Contributions of age-specific fertility differences ΔF_i to $\Delta\lambda$ [see equation (10)].

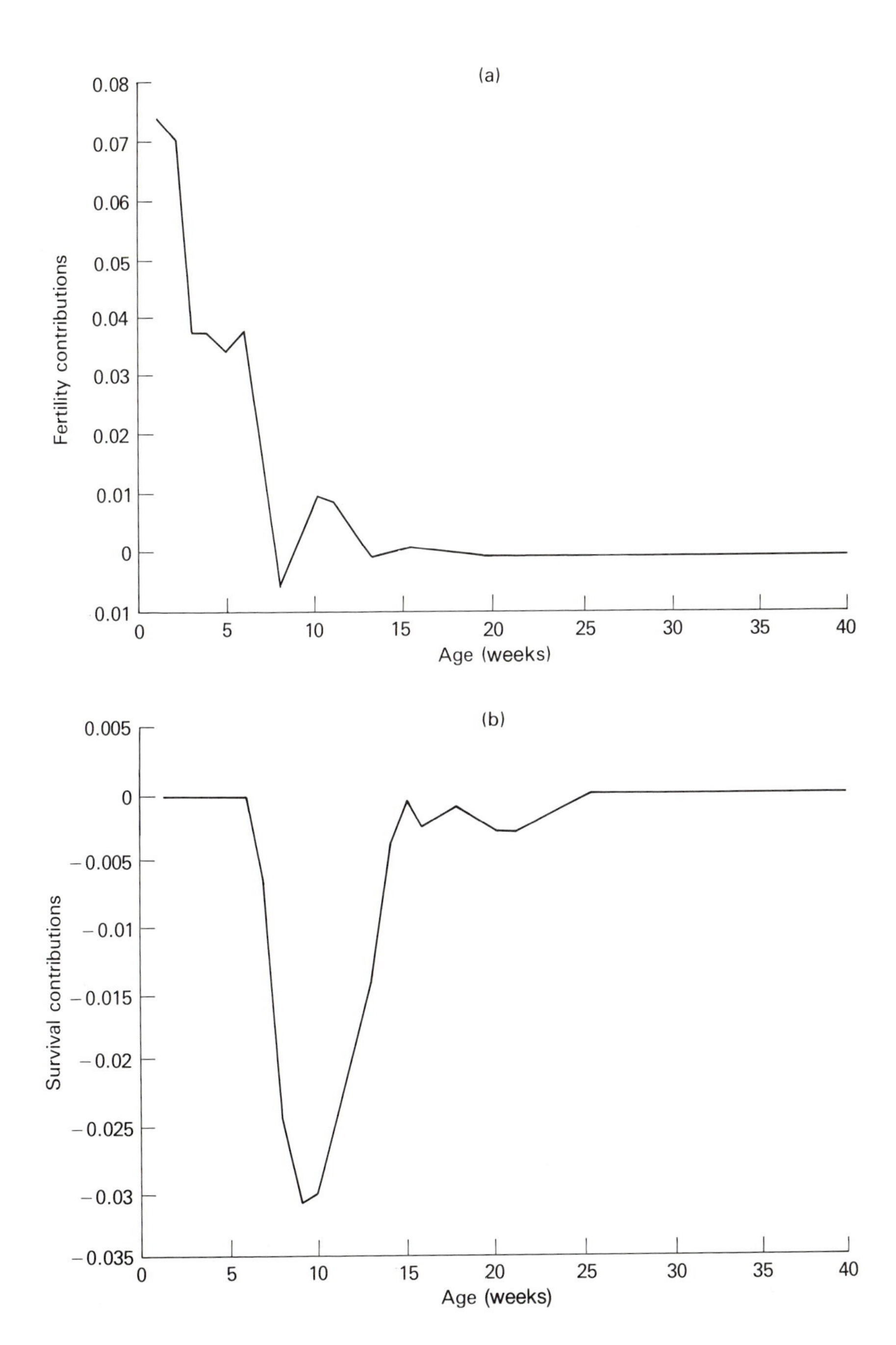
(a)
0.08
0.07
0.06
0.05
0.04
0.03
0.02
0.01
0
0.01
Fertility contributions
0
5
10
15
20
25
30
35
40
Age (weeks)
(b)
0.005
0
−0.005
−0.01
−0.015
−0.02
−0.025
−0.03
−0.035
Survival contributions
0
5
10
15
20
25
30
35
40
Age (weeks)

of the timing of events in the life cycle; this effect can be extremely important.

Density dependence, r and K

Dependence of vital rates on density is conspicuous by its absence in the theory just outlined. MacArthur (1962), MacArthur & Wilson (1967), Roughgarden (1971; see 1979 for review), Charlesworth (1971) and others have analysed the results of simple selection models in cases where the vital rates of a population without age structure (or any kind of structure) follow the logistic equation

$$\frac{\mathrm{d}N}{\mathrm{d}t} = r_0 N\left(\frac{K - N}{K}\right). \tag{11}$$

where N is population size, r_0 is the *per capita* rate of increase in the limit as $N \rightarrow 0$ and K is the equilibrium population size, or carrying capacity.

The well-known result of this analysis is that selection at low densities favours increases in the parameter r_0, while in the neighbourhood of the equilibrium, r_0 is neutral and selection favours increases in K. The two situations were called r-selection and K-selection, and it was asserted that the definition of fitness varied depending on density.

These results were widely discussed in the context of life-history theory (Pianka 1970), so that when Stearns (1976, 1977) reviewed the subject, one of his main concerns was whether studies had supported the predictions of r- and K-selection. Unfortunately for all this, the r/K theory *per se* has almost nothing to say about life-history traits, since it is based on the logistic equation, which includes no age- or stage-specific vital rates (Caswell 1982).

Even more unfortunately, the claim that r/K theory showed that the proper measure of fitness depends on density is false. In fact, neither r_0 nor K, the two 'traits' appearing in this simple model, can be equated with fitness. In the genetic models (Charlesworth 1971; Roughgarden 1971) it is quite clear that fitness is measured by

$$W = \frac{1}{N}\frac{\mathrm{d}N}{\mathrm{d}t},$$

regardless of density. It is the correlation of r_0 and K with fitness, and thus the selection gradients $\partial W / \partial r_0$, and $\partial W / \partial K$, which depend on density.

Its attempt to reduce the manifold variation in environmental conditions to a single axis, and to predict selective pressures as a function of this axis, made r/K theory a valuable contribution. Several authors have since added other axes (reviewed by Southwood 1988), particulary emphasizing adversity or stress selection.

Variable environments

Environmental variation adds additional complexity to the definition of fitness. In a stochastic environment, population growth is a random variable, and there is more than one way in which it can be measured. Demographic models in stochastic environments have been extensively studied by Cohen (e.g. Cohen 1979; Cohen, Christensen & Goodyear 1983) and Tuljapurkar (Tuljapurkar & Orzack 1980; Tuljapurkar 1982).

Two candidates for central tendency measures of population growth are the growth rate of expected population size and the expectation of the growth rate of population size. Due to the multiplicative character of population growth, these two measures are *not* identical. In fact, it is possible for expected population size to have a positive growth rate, and thus to increase to infinity, while the expected growth rate is negative, so that the probability of extinction increases asymptotically to unity (Lewontin & Cohen 1969).

Consider a population described by a discrete matrix projection model

$$\mathbf{n}(t+1) = A_t \, n(t), \tag{12}$$

where the matrices A_t are determined by some sort of stochastic process describing the variability of the environment. Let $N(t) = \Sigma n_i(t)$ denote population size and $E(\cdot)$ denote the expectation. Then using a mixture of the notation of Cohen (1979) and Tuljapurkar (1982), the two measures of long-term population performance are

$$\log \mu = \lim_{t \to \infty} \left(\frac{1}{t}\right) \log E[N(t)] \tag{13}$$

$$a = \lim_{t \to \infty} \left(\frac{1}{t}\right) E[\log N(t)] \tag{14}$$

The first of these ($\log \mu$) measures the rate of growth of average population size, the second (a) the average growth rate.

It is always the case that $a \leqslant \log \mu$. If the sequence of environments is uncorrelated, then μ is simply the dominant eigenvalue of $E(\mathbf{A})$, the average population projection matrix.* However, a cannot be calculated simply from the sequence of projection matrices (Cohen 1979). The question is, which (if either) of these measures is appropriate as a measure of fitness?

Tuljapurkar (1982) examined a simple one locus, two allele genetic model to answer this question. Using stochastic linearization, he determined that the necessary condition for invasion of a rare allele was determined by the value of a of that allele relative to the allele present in the population. In other words, fitness is determined by the average of the growth rates, not the growth rate of the average population size.

He also developed a useful approximation for a, valid for small fluctuations in the vital rates. Let $\lambda_{\bar{A}}$ denote the dominant eigenvalue of the mean matrix. Then, for uncorrelated[†] environments,

$$a \approx \log \lambda_{\bar{A}} - \frac{1}{2\lambda_{\bar{A}}^2} \sum_{i,j} \sum_{k,l} \frac{\partial \lambda}{\partial a_{ij}} \frac{\partial \lambda}{\partial a_{kl}} Cov(a_{ij}\, a_{kl}) \,. \qquad (15)$$

The summation term in this is for small fluctuations, the first order approximation to the variance $V(\lambda)$ of λ; thus

$$a \approx \ln \lambda_{\bar{A}} - \frac{V(\lambda)}{2\lambda_{\bar{A}}^2} \,. \qquad (16)$$

This is related to the use, by Real (1980) and Lacey *et al.* (1983), of the economic concept of *risk aversion*, which measures utility by expected fitness discounted by the variance in fitness. This is a more rigorous version of the ill-defined idea of 'bet-hedging', which was the major alternative to *r*/*K* theory at the time of Stearns' 1977 review. However, the first term in Tuljapurkar's approximation (16) for a is the fitness of the average life history, not the average fitness. Because λ is a non-linear function of the entries in the matrix **A**, these two averages are not equivalent. However, a clearly decreases with the variance in fitness; thus Tuljapurkar's genetic analysis shows that natural selection in uncorrelated variable environments is inherently risk averse.

In conclusion, it seems clear that quantitative genetic theory provides a clear demographic operational measure of fitness. It can be calculated from data, its derivatives can easily be calculated from the vital rates, and

* Demography in stochastic environments depends critically on patterns of autocorrelation in the environment; see Tuljapurkar (1982).

† Tuljapurkar (1982) also considered correlated environments; this adds another term to the expression and I will not consider it here, but it can significantly affect the outcome of selection; see Orzack (1985).

it applies equally to density-independent and density-dependent situations. However, there are enough assumptions and approximations made in deriving that genetic theory for the question of how fitness should be measured still to be worthy of attention.

Frequency dependence and evolutionary stability

The description of natural selection in equation (3) assumes that the selection gradients are independent of gene or trait frequencies. For some important classes of life-history traits, particularly those such as mating, which are concerned with interindividual interaction, this is not true. The benefit of allocating resources to male or female reproduction, for example, depends on the relative frequencies of males and females at the time. Such situations can be addressed most easily using the concept of the evolutionary stable strategy (ESS) (Maynard Smith 1982). Simply put, a strategy is evolutionary stable if it is proof against invasion by alternative strategies. Charnov (1982) provides an excellent example of the application of ESS methods to a life-history problem.

Trait covariance

As mentioned above, the interaction of traits has been fundamental to life-history theory from the beginning. It is only rather recently that a quantitative genetic theory for selection in such cases has been available (Lande 1979, 1982; Arnold & Wade 1984).

Lande's result in equation (3) shows quite clearly how the interaction of traits affects selection. Consider the response of an arbitrary trait, which I take as z_1 without loss of generality. Then

$$\Delta \bar{z}_1 = V_g(z_1)\frac{\partial \ln \lambda}{\partial z_1} + \sum_{j>1} C_g(z_1, z_j)\frac{\partial \ln \lambda}{\partial z_j}, \tag{17}$$

where V_g and C_g denote the additive genetic variance and covariance, respectively. However, the slope of the regression of z_j on z_1 is given by

$$\frac{\partial z_j}{\partial z_1} = \frac{C_g(z_1, z_j)}{V_g(z_1)}. \tag{18}$$

Thus

$$\Delta \bar{z}_1 = V_g(z_1)\left(\frac{\partial \ln \lambda}{\partial z_1} + \sum_{j>1} \frac{\partial \ln \lambda}{\partial z_j}\frac{\partial z_j}{\partial z_1}\right), \tag{19}$$

and we see that the change in z_1 due to selection has two components; a direct component due to the effect of z_1 on fitness and an indirect component due to the effects of changes in z_1 on the other traits and the resulting effects of those traits on fitness.

An example of the importance of trait covariances is provided by the data of Doyle & Hunte (1981) on the amphipod *Gammarus lawrencianus*. A population of this species was subjected to natural selection under laboratory conditions for twenty-six generations, producing marked changes in age-specific survival and reproduction, and increasing r by some 72% compared to the field population. Doyle and Hunte presented age-classified projection matrices for the selected and unselected populations, so it is possible to calculate the selection gradients for the P_i and F_i, as well as the realized selection responses ΔP_i and ΔF_i. If inter-trait correlations were not important, there should be a linear relationship between ΔP_i and $\partial\lambda/\partial P_i$ and between ΔF_i and $\partial\lambda/\partial F_i$. Figure 9.4 shows the relation between the selection gradients and the response to selection; clearly the response to selection is not well explained by the selection gradients alone.

Considerable recent effort has been devoted to quantifying the patterns of covariation among life-history traits. Dingle & Hegmann (1982), Rose (1984b), and Giesel (1986), for example, have used quantitative genetic methods in the laboratory. The agricultural literature contains a plethora of studies on the covariance structure of what ecologists would refer to as life-history traits.

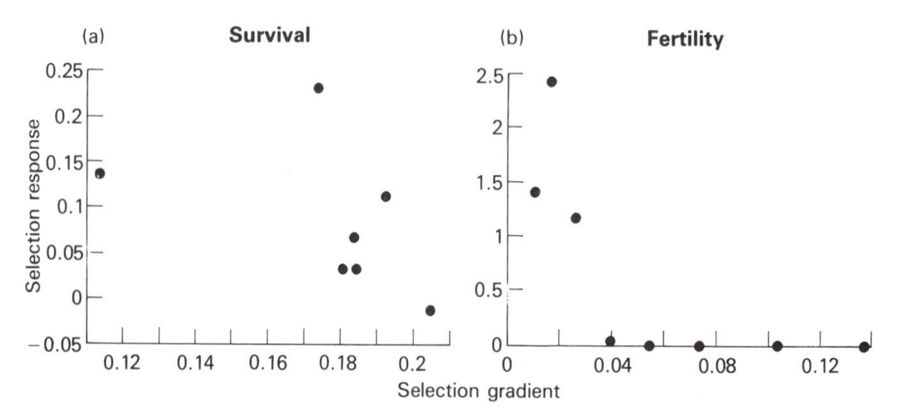

FIG. 9.4. Relation between the selection gradients for (a) survival probability ($\partial\lambda/\partial P_i$) and (b) fertility ($\partial\lambda/\partial F_i$) and the corresponding responses to selection, for *Gammarus lawrencianus*. [Data from Doyle & Hunte (1981)].

Two important points should be made about these attempts. First, care must be taken to distinguish between phenotypic and genetic correlations. It is well known that the magnitude and even the sign of the two correlations are often different (Falconer 1981). A positive phenotypic correlation between say, reproduction and survival, could result from positive environmental effects on both traits; the underlying genetic correlation can very well be negative.

Second, it is by now well known that genetic variances and heritabilities are specific to a particular population in a particular environment at a particular time; they are not invariant properties of the species or the trait (e.g. Feldman & Lewontin 1975). Genetic covariances are equally labile, and should not be thought of as fixed properties. They can vary with environmental conditions (Service & Rose 1985, Giesel 1986) as well as with such genetic phenomena as inbreeding (e.g. Giesel, Murphy & Manlove 1982, Rose 1984b).

Constraints on life-history evolution

Gould & Lewontin (1979) issued a devastating critique of the 'adaptationist program' in evolutionary biology, referring to the practice of assuming that observed phenotypes must be optimal solutions to problems of organismal design, and that our ability to explain them this way is limited only by our ingenuity in dreaming up adaptive stories. They emphasize the possible importance of genetic, morphological, developmental or phylogenetic constraints in preventing evolution from achieving optimal solutions.

Life-history theory has had to confront the issue of constraints from its very beginnings. Indeed, life-history theory is as much a theory of constraints as of anything. At one level, constraints arise because of genetic covariance between traits (Cheverud 1984); selection operates within these constraints according to the theory outlined above.

At a higher plane, there is the question of the taxomic level at which life-history traits should be interpreted. Stearns (1980) discussed this critically by questioning whether life-history tactics even exist. One approach to this problem has been to examine patterns of life-history traits in taxonomic lineages, particularly in relation to body size (e.g. Stearns 1983, 1984, May & Rubenstein 1984). The idea seems to be that if clear allometric relations appear between a life-history trait and body size (and Stearns' analysis of mammal data, for example, finds that about 50% of the variance in the traits he examined is explained by allometric variation in body size) the variation in the life-history trait is an epiphenomenon, merely reflecting phylogenetic evolution of body size.

This conclusion may be true, but it begs an important question: why is body size the primary trait? If say, generation time is correlated with body size within a lineage, why is not body size the epiphenomenon, constrained to follow the evolution of generation time? After all, demographic traits are closer to fitness than morphological traits.

Stearns (1983, 1984; but see Vitt & Siegel 1985 on the pitfalls of this kind of analysis) has taken a more thorough approach to this problem, by using factor analysis at several taxomic levels to see where clearly identifiable constellations of correlated life-history traits appear after body size is taken into account. Derrickson & Ricklefs (1988) have pointed out problems with this approach, and suggest a means of identifying phylogenetic constraints by looking for differences between inter-trait correlation patterns at different phylogenetic levels.

At an even higher plane, mechanical and morphological constraints may play an important role. Buss (1987), for example, has explored the profound consequences of the choice, at the cellular level, between ciliation and continued cell division, on constraining patterns of life-history evolution at the kingdom level; an example of life-history evolution on a grand scale.

The integration of life-history theory and developmental biology has long been encouraged (e.g. Stearns 1982). As it develops, the role of phylogenetic and developmental constraints will doubtless be clarified.

Effects of the environment

There is some irony in contrasting the preoccupation of life-history theory with demographic traits as results of selection, with the interest of ecology in those same traits as results of the environment. Both perspectives are valid. Food supply, temperature, etc. do change survival and reproduction, but only within limits; those limits are presumably the result of evolution.

This suggests a different approach to life-history strategies, which identifies the 'strategy' not as a particular demographic pattern, but as a pattern of response to the environment, the so-called *norm of reaction* for the trait (Schmalhausen 1949). Evolutionary biologists have recognized the importance of norms of reaction for some time (Gause 1942, Schmalhausen 1949, Bateson 1963) and Stearns & Koella (1986) have recently developed an optimization model for the norm of reaction of age and size at reproduction.

The alternative to identifying the norm of reaction as the trait under selection is to view the existence of the norm of reaction as 'phenotypic plasticity' (Bradshaw 1965, Caswell 1983, Scheiner & Goodnight 1984, Orzack 1985, Schlichting 1986). At one time, evidence that life-history differences were genetic, and not due to phenotypic plasticity, was considered necessary before the differences were worth discussing (Stearns 1977). The logic behind this position is clearly spurious; most of the models predicting life-history patterns, with which one would like to compare data, are based on hypothesized patterns of selection gradients and trait interactions. The models predict that, in a given environment, certain combinations of life-history traits produce higher fitness than other combinations. Whether the adaptive combinations are achieved by selection for the traits themselves or by achieving a norm of reaction which brings about the traits in response to the environment is irrelevant to the model (Bateson 1963).

Treating the norm of reaction as a potentially adaptive trait requires the existence of genetic variation. Such genetic variation appears as a genotype × environment interaction in quantitative genetic studies (Via & Lande 1985); such interactions are commonly documented in the agricultural literature. More work is needed in this area.

CONCLUSIONS

The analysis of life-history strategies follows naturally from the recognition that demographic patterns are subject to adaptive interpretation. Life-history theory has from its beginnings faced some of the most difficult problems in evolutionary biology: the definition of fitness, interaction of traits and constraints on evolution. Recent developments in quantitative genetics provide a more secure basis for theory, but raise new problems concerning the genetic structure of life-history traits. Demographic definitions of fitness are available, and are clearly superior to definitions based on only some components of fitness, but problems of stochasticity raise issues of risk aversion and extinction probability. Constraints can be incorporated in quantitative genetic terms, but analyses of phylogenetic constraint at higher levels are important. The effects of the environment, expressed as phenotypic plasticity or norms of reaction, are important and still need to be incorporated into the theory.

ACKNOWLEDGMENTS

The preparation of this paper was supported by NSF Grants OCE85-16177, BSR86-9395, and BSR87-4936. Woods Hole Oceanographic Institution Contribution 6775. I thank the participants in the symposium, two reviewers and S. D. Tuljapurkar for helpful comments.

REFERENCES

Arnold, S. J. & Wade, M. J. (1984). On the measurement of natural and sexual selection: theory. *Evolution*, **39**, 709–719.

Bateson, G. (1963). The role of somatic change in evolution. *Evolution*, **17**, 529–539.

Berven, K. A. (1987). The heritable basis of variation in larval developmental patterns within populations of the wood frog (*Rana sylvatica), Evolution*, **41**, 1088–1097.

Bradshaw, A. D. (1965). Evolutionary significance of phenotypic plasticity in plants. *Advances in Genetics*, **13**, 115–155.

Buss, L. W. (1987). *The Evolution of Individuality*. Princeton University Press, Princeton, New Jersey.

Caswell, H. (1978). A general formula for the sensitivity of population growth rate to changes in life history parameters. *Theoretical Population Biology*, **14**, 215–230.

Caswell, H. (1982). Life history theory and the equilibrium status of populations. *American Naturalist*, **120**, 317–339.

Caswell, H. (1983). Phenotypic plasticity in life history traits: demographic effects and their evolutionary consequences. *American Zoologist*, **23**, 35–46.

Caswell, H. (1986). Life cycle models for plants. *Lectures on Mathematics in the Life Sciences*, **18**, 171–233.

Caswell, H. (1989a). The analysis of life table response experiments. I. Decomposition of treatment effects on population growth rate. Submitted.

Caswell, H. (1989b). *Matrix Population Models: Construction. Analysis and Interpretation.* Sinauer Associates, Sunderland, Massachusetts.

Charlesworth, B. (1971). Selection in density-regulated populations. *Ecology*, **52**, 469–474.

Charlesworth, B. (1980). *Evolution in Age-structured Populations*. Cambridge University Press, Cambridge.

Charnov, E. L. (1982). *The Theory of Sex Allocation*. Princeton University Press, Princeton, New Jersey.

Cheverud, J. M. (1984). Quantitative genetics and developmental constraints on evolution by selection. *Journal of Theoretical Biology*, **110**, 155–171.

Clutton-Brock, T. H. (1988). *Reproductive Success*. University of Chicago Press, Chicago.

Cohen, J. E. (1979). Comparative statics and stochastic dynamics of age-structured populations. *Theoretical Population Biology*, **16**, 159–171.

Cohen, J. E., Christensen, S. W. & Goodyear, C. P (1983). A stochastic age-structured population model of striped bass (*Morone saxitalis*) in the Potomac River. *Canadian Journal of Fisheries and Aquatic Sciences*, **40**, 2170–2183.

Cole, L. C. (1954). The population consequences of life history phenomena. *Quarterly Review of Biology*, **19**, 103–137.

Derrickson, E. M. & Ricklefs, R. E. (1988). Taxon-dependent diversification of life-history traits and the perception of phylogenetic constraints. *Functional Ecology* **2**, 417–423.

Dingle, H. & Hegmann, J. P. (1982). *Evolution and Genetics of Life Histories*. Springer-Verlag, New York, Heidelberg, Berlin.

Doyle, R. W. & Hunte, W. (1981). Demography of an estuarine amphipod (*Gammarus lawrencianus*) experimentally selected for high '*r*': a model of the genetic effects of environmental change. *Canadian Journal of Fisheries and Aquatic Sciences*, **38**, 1120–1127.

Emlen, J. M. (1970). Age specificity and ecological theory . *Ecology*, **51**, 588–601.

Falconer, D. S. (1981). *Introduction to Quantitative Genetics.* Longman, London and New York.

Feldman, M. W. & Lewontin, R. C. (1975). The heritability hang-up. *Science*, **190**, 1163–1168.

Fisher, R. A. (1930). *The Genetical Theory of Natural Selection.* Oxford University Press, Oxford (Revised edition 1958, Dover Publications, New York).

Gause, G. F. (1942). The relation of adaptability to adaptation. *The Quarterly Review of Biology*, **17**, 99–114.

Giesel, J. T. (1986). Genetic correlation structure of life history variables in outbred, wild *Drosophila melanogaster*: effects of photoperiod regimen. *American Naturalist*, **128**, 593–603.

Giesel, J. T., Murphy, P. A. & Manlove, M. N. (1982). The influence of temperature on genetic interrelationships of life history traits in a population of *Drosophila melanogaster*: what tangled data sets we weave. *American Naturalist*, **119**, 464–479.

Gould, S. J. & Lewontin, R. C. (1979). The spandrels of San Marco and the Panglossian paradigm: a critique of the adaptationist programme. *Proceedings of the Royal Society, London B*, **205**, 581–598.

Istock, C. A. (1983). The extent and consequences of heritable variation for fitness characters. *Population Biology: Retrospect and Prospect* (Ed. by C. E. King & S. Dawson) pp. 61–96. Columbia University Press, New York.

Lacey, E. P., Real, L., Antonovics, J. & Heckel, D. G. (1983). Variance models in the study of life histories. *American Naturalist*, **122**, 114–131.

Lack, D. (1947). The significance of clutch size. *Ibis*, **89**, 302–352.

Lack, D. (1954). Evolution of reproductive rates. *Evolution as a Process* (Ed. by J. S. Huxley, A. C. Hardy, & E. B. Ford) pp. 143–156. George Allen & Unwin, London.

Lande, R. (1979). Quantitative genetic analysis of multivariate evolution, applied to brain: body size allometry. *Evolution*, **33**, 402–416.

Lande, R. (1982). A quantitative genetic theory of life history evolution. *Ecology*, **63**, 607–615.

Lande, R. & Arnold, S. J.(1983). The measurement of selection on correlated characters. *Evolution*, **37**, 1210–1226.

Levin, L. A., Caswell, H., DePatra, K. D. & Creed, E. L. (1987). Demographic consequences of larval development mode: planktotrophy vs. lecithotrophy in *Streblospio benedicti*. *Ecology*, **68**, 1877–1886.

Lewontin, R. C. & Cohen, D. (1969). On population growth in a randomly varying environment. *Proceedings of the National Academy of Sciences USA*, **62**, 1056–1060.

Luckinbill, L. S. & Clare, M. J. (1985). Selection for life span in *Drosophila melanogaster*. *Heredity*, **55**, 9–18.

MacArthur, R. H. (1962). Some generalized theorems of natural selection. *Proceedings of the National Academy of Sciences*, **48**, 1893–1897.

MacArthur, R. H. & Wilson, E. O. (1967). *The Theory of Island Biogeography.* Princeton University Press, Princeton, New Jersey.

May, R. M. & Rubenstein, D. I. (1984). In Reproductive strategies. In *Reproduction in Mammals:* **4** (Ed. by C. R. Austin & R. V. Short) pp. 1–23. Cambridge University Press, Cambridge.

Maynard Smith, J. (1982). *Evolution and the Theory of Games.* Cambridge University Press, Cambridge.

Medawar, P. B. (1946). Old age and natural death. *Modern Quarterly*, **1**, 30–56.
Medawar, P. B. (1952). *An Unsolved Problem of Biology*. H. K. Lewis, London.
Moss, R. & Watson, A. (1982). Heritability of egg size, hatch weight, body weight, and viability in red grouse (*Lagopus leagopus scoticus*). *The Auk*, **99**, 683–686.
Nguyen,H. T. & Sleper, D. A. (1983). Genetic variability of seed yield and reproductive characters in tall fescue. *Crop Science*, **23**, 621–626.
Norton, H. T. J. (1928). Natural selection and Mendelian variation. *Proceedings of the London Mathematical Society*, **28**, 1–45.
Orzack, S. H. (1985). Population dynamics in variable environments. V. The genetics of homeostasis revisited. *American Naturalist*, **125**, 550–572.
Partridge, L. & Harvey, P. H. (1988). The ecological context of life history evolution. *Science*, **241**, 1449–1455.
Pianka, E. R. (1970). On *r*- and *K*- selection. *American Naturalist,*, **104**, 592–597.
Price, G. R. (1970). Selection and covariance. *Nature*, **227**, 520–521.
Real, L. A. (1980). Fitness, uncertainty, and the role of diversification in evolution and behavior. *American Naturalist*, **115**, 623–638.
Robertson, A. (1968). The spectrum of genetic variation. *Population Biology and Evolution*, (Ed. by R. C. Lewontin), pp. 5–16. Syracuse University Press, Syracuse, New York.
Rose, M. R. (1982). Antagonistic pleiotropy, dominance, and genetic variation. *Heredity*, **48**, 63–78.
Rose, M. R. (1983). Theories of life-history evolution. *American Zoologist*, **23**, 15–23.
Rose, M. R. (1984a). Artificial selection on a fitness-component in *Drosophila melanogaster*. *Evolution*, **38**, 516–526.
Rose, M. R. (1984b). Genetic covariation in *Drosophila* life history: untangling the data. *American Naturalist*, **123**, 565–569.
Rose, M. R. & Charlesworth, B. (1981). Genetics of life-history in *Drosophila melanogaster*. II. Exploratory selection experiments. *Genetics*, **97**, 187–196.
Roughgarden, J. (1971). Density-dependent natural selection. *Ecology*, **52**, 453–468.
Roughgarden, J. (1979). *Theory of Population Genetics and Evolutionary Ecology: an Introduction*. Macmillan, New York.
Scheiner, S. M. & Goodnight, C. J. (1984). The comparison of phenotypic plasticity and genetic variation in populations of the grass *Danthonia spicata*. *Evolution*, **38**, 845–855.
Schlichting, C. D. (1986). The evolution of phenotypic plasticity in plants. *Annual Review of Ecology and Systematics*, **17**, 667–693.
Schmalhausen, I. I. (1949). *Factors of Evolution*. Blakiston, Philadelphia.
Service, P. M. & Rose, M. R. (1985). Genetic covariation among life-history components: the effect of novel environments. *Evolution*, **39**, 943–945.
Sibley, R. & Calow, P. (1983). An integrated approach to life cycle evolution using selective landscapes. *Journal of Theoretical Biology*, **102**, 527–547.
Sibley, R. & Calow, P. (1986). *Physiological Ecology of Animals*. Blackwell Scientific Publications, Oxford.
Southwood, T. R. E. (1988). Tactics, strategies, and templets. *Oikos*, **52**, 3–18.
Stearns, S. C. (1976). Life-history tactics: a review of the ideas. *Quarterly Review of Biology*, **51**, 3–47.
Stearns, S. C. (1977). The evolution of life-history traits: a critique of the theory and a review of the data. *Annual Review of Ecology and Systematics*, **8**, 145–171.
Stearns, S. C. (1980). A new view of life-history evolution. *Oikos*, **35**, 266–281.
Stearns, S. C. (1982). The role of development in the evolution of life histories. *Evolution and Development* (Ed. by J. T. Bonner) pp. 237–258. Springer-Verlag, Berlin, Heidelberg, New York.

Stearns, S. C. (1983). The influence of size and phylogeny on patterns of covariation among life-history traits in the mammals. *Oikos*, **41**, 173–187.

Stearns, S. C. (1984). The effects of size and phylogeny on patterns of covariation in the life history traits of lizards and snakes. *American Naturalist*, **123**, 56–72.

Stearns, S. C. & Crandall, R. E. (1981). Quantitative predictions of delayed maturity. *Evolution*, **35**, 455–463.

Stearns, S. C. & Koella, J. C. (1986). The evolution of phenotypic plasticity in life-history traits: predictions of reaction norms for age and size at maturity. *Evolution*, **40**, 893–913.

Tuljapurkar, S. D. (1982). Population dynamics in variable environments. III. Evolutionary dynamics of *r*-selection. *Theoretical Population Biology*, **21**, 141–165.

Tuljapurkar, S. D. & Orzack S. H. (1980). Population dynamics in variable environments. I. Long-run growth rates and extinction. *Theoretical Population Biology* **18**, 314–342.

Via, S. & Lande, R. (1985). Genotype-environment interaction and the evolution of phenotypic plasticity. *Evolution*, **39**, 505–522.

Vitt, L. J. & Seigel, R. A. (1985). Life history traits of lizards and snakes. *American Naturalist*, **125**, 480–484.

Wong, L. S. L. & Baker, R. J. (1986). Selection for time to maturity in spring wheat. *Crop Science*, **26**, 1171–1175.

Wyngaard, G. A. (1986). Heritable life history variation in widely separated populations of *Mesocyclops edax* (Crustacea: Copepoda). *Biological Bulletin*, **170**, 296–304.

Zimmermann, M. J. O., Rosielle, A. A. & Waines, J. G. (1984). Heritabilities of grain yield of common bean in sole crop and intercrop with maize. *Crop Science*, **24**, 641–644.

10 OPTIMIZATION IN ECOLOGY

J.R. KREBS AND A.I. HOUSTON
Edward Grey Institute of Field Ornithology,
South Parks Road, Oxford OX1 3PS, UK

INTRODUCTION

Most biologists accept the neo-Darwinian view that differential survival and reproduction of individuals is the major force of evolution. In the past however, ecological theories have often been formulated in ways that are explicitly or implicitly at odds with the theory of natural selection. Although some ecologists are clearly aware that their theories fly in the face of the widely accepted view that evolution occurs by natural selection (e.g. Wynne Edwards 1962, 1987), more often, as the following examples illustrate, ecologists simply have not thought in enough detail about how selection acts as a mechanism of evolution to recognize that their ideas may be difficult to reconcile with it.

Chitty (1960, 1967) proposed an hypothesis for the generation of cyclic fluctuations in small mammals which, although explicitly couched in terms of natural selection, has potential problems in relation to neo-Darwinian selection. Chitty's idea was that the cycles are maintained by oscillating, density-dependent selection acting on a genetic polymorphism: in low density populations, non-aggressive, highly fecund individuals are favoured, while at high densities individuals that are aggressive at the expense of reproductive output are at an advantage. When the low density genotype predominates, the population increases rapidly because of high fecundity and when the high density genotype comes to predominate, the population crashes because of the drop in fecundity.

Without going into the question of whether or not the evidence supports Chitty's view, one can see problems with it as an evolutionary hypothesis, two of which are as follows: (i) It is not obvious why selection should maintain an unstable polymorphism as opposed to favouring a single intermediate genotype which is neither too aggressive nor too fecund and which consequently does not produce cycles; (ii) the hypothesis as formulated by Chitty does not explain how the aggressive genotype benefits from behaving aggressively. Explanations of aggression in terms of natural selection interpret aggressive behaviour as a means by which individuals compete for limited resources (Davies & Houston 1984; Huntingford & Turner 1987), but Chitty's hypothesis makes no reference to the resources that are in limited supply at high population density. In fact the hypothesis was specifically developed as an alternative to the idea of resource (e.g. food) limitation as an explanation of cycles. One possibility would be to suggest that aggression is favoured because it allows individuals to exclude others from breeding. This is a form of spiteful behaviour, in which an individual pays a cost to inflict a greater cost on another, and the conditions under which spite is likely to evolve are quite narrow (Parker & Knowlton 1980).

An alternative, perhaps more plausible suggestion for how selection might favour changes in aggression is that of Charnov & Finerty (1980). If at low points of the cycle, small populations consist of isolated kin groups, kin selection might favour lowered aggression at this stage of the cycle, whilst at the high point of the cycle there might be less structuring of the large population into kin groups and hence lack of selection for being 'nice' to neighbours.

A second example comes from discussions of population, community and ecosystem properties such as stability, resilience and succession. The notion that communities should have properties such as stablity and resilience seems to imply evolution at the community level, although these properties could be just incidental by-products of individual adaptations, and at least some discussions of these phenomena are couched explicitly or implicitly in terms of stability and resilience being properties that would be favoured by evolution. Holling (1973) for example, writes 'A population responds to any environmental change by initiation of a series of physiological, behavioural and genetic changes that restore its ability to respond to subsequent environmental change.' This would however, only be true if selection favours adaptations that are 'good for the population.'

The 'community selection' view also appears to creep into discussions of interactions between populations: '. . . Thus in addition to making more food available, the larger grazers also create a safer habitat for the Tommies by clearing their field of view' (Ehrlich 1987). In fact there is no reason from the standpoint of natural selection why wildebeest should have evolved to promote Thompson's gazelle populations.

Community-level adaptations have traditionally been explicitly assumed in discussions of succession. Clements (1916), who developed the monoclimax theory of succession, viewed biological communities and ecosystems (or formations as he called them) as 'superorganisms'; he wrote '. . . the life history of a formation is a complex but definite process, comparable in its chief features with the life history of an individual plant'. As Simberloff (1980) has pointed out, the superorganism concept is by no means dead in ecology. Odum (1971) for example, thinks of an ecosystem as having evolved a developmental 'strategy' which results in '. . . control of, or homeostasis with, the physical environment in the sense of achieving the maximum protection from its perturbations' and Ehrlich (1987), whilst admitting that the superorganism analogy was overdrawn, suggests that 'notions based on it are still important in the study of succession'.

Why does it matter to ecologists whether they think in terms of selection acting on individuals or on communities? There are at least two reasons. First, ecologists should mind whether or not their theories are at odds with currently accepted views in other, related disciplines. Just as one would expect any evolutionary theory to be compatible with currently accepted views about the mechanisms of inheritance (e.g. that it is non-Lamarckian), so one should expect any ecological theory to be consistent with current views on the mechanism of evolution. Compatibility with natural selection should act as a kind of filter through which ecological theories must pass. Second, the recognition that selection acts on individuals and not communities leads automatically to a reductionist approach to analysing communities and ecosystems, in which an attempt is made to account for their properties in terms of the properties of more fundamental units of organization such as individuals. If you believe that community level properties can evolve, then you would have good grounds for trying to analyse community processes in terms of community level organization, but as long as you accept that community properties arise from evolutionary pressures acting on individuals, it is clear that a reductionist approach is necessary. Space does not permit a more detailed discussion of reductionism and its application to ecology, but the reader is referred to Schoener's (1986) excellent discussion. A brief word on the question of emergent properties is, however, included.

Some ecologists may argue that properties such as stability, resilience and succession are emergent properties; properties that are relevant only at the community level and therefore explicable only in community ecology terms. An individual cannot have stability or succession, so the evolution of these kinds of phenomena cannot be explained at the level of individuals. This kind of argument is based on a misunderstanding of the term 'emergent property'. Properties are emergent only when the theory that accounts for the property cannot be reduced to another more basic theory (Nagel 1961; Schoener 1986; Churchland 1987). Thus emergence (or its converse reducibility) is a concept that applies to theories not to the phenomena they explain. For example, the fact that water appears blue in large quantities is a property of large aggregates, but theories of the structure of water molecules allow the prediction that large aggregates will be blue, so although blueness is a property shown only by aggregates of water, it is not an emergent property.

SELECTIONIST THINKING AND OPTIMALITY MODELS

In this section we sketch some aspects of the history of selectionist thinking and in particular optimality modelling as an explicit way of thinking about features that arise through natural selection.

As Schoener (1987) has pointed out, the lack of selectionist thinking in ecology is clearly demonstrated by the fact that most ecology textbooks published before the 1960s (e.g. Elton 1927; Odum 1953 (and its 1959 & 1971 revisions); MacFadyen 1957) made little if any reference to the importance of individual adaptations in ecological processes, and to the extent that they did refer to them, their views of how selection works were usually incorrect. One of the key people to integrate evolutionary and ecological theory was David Lack. First, in his work on Darwin's finches (Lack 1947) he introduced the idea that to avoid competition, ecologically similar species might have evolved differences in their feeding niche and morphology by selection. Today this is a commonplace view but it was regarded with great suspicion in 1944. Second, and perhaps even more important, he proposed the first specific, experimentally testable, optimality model in ecological theory; this was his theory of clutch size in birds (Lack 1954). It stated that birds which feed their young lay a clutch that maximizes the number of young fledged. This, together with Williams' 1957 classic paper modifying Lack's theory by introducing the cost of reproduction, formed the basis for much experimental work on reproductive strategies. Another early paper on life histories, Cole's (1954) classic also used explicitly selectionist thinking (see also Caswell, this volume).

The major explosion of evolutionary ideas in ecology occurred in the late 1950s and early 1960s and was led more or less simultaneously by many workers including MacArthur, Orians, Brown, Crook and Lack. Between them, these workers applied evolutionary thinking to many ecological problems including niche separation, sex ratios, territoriality, group size and foraging strategies. The last of these topics, the theory of optimal foraging, has a remarkably well-defined beginning in 1966 with the simultaneous publication of two papers, one by MacArthur & Pianka (1966) and the other by Emlen (1966). Because optimal foraging theory has been at the centre of the stage in optimality models in ecology over the past 20 years, we take it as a representative case for more detailed consideration. Its history has been thoroughly documented, complete with amusing anecdotes, by Schoener (1987), so here we do no more than pick on a few key points.

OPTIMAL FORAGING THEORY

The justification for optimal foraging theory, as with any biological optimality models, is that natural selection influences many of the features of organisms that biologists study. In other words, that these features are adaptations. Natural selection, being a cumulative competitive and iterative process, tends to maximize the utility of adaptations in promoting survival and reproduction, so that organisms often have the appearance of having been designed to fit their environment. Optimality models are a tool used when asking what adaptations would look like if they had been favoured by selection to maximize survival and reproduction in a particular way. They provide a benchmark against which observed features can be compared. For example, one might ask in what group size would sparrows live if group size had been selected to achieve an optimal balance between benefits related to predator defence and the costs of competition for food arising from living in a group. Note that this argument includes the assertion that natural selection tends to maximize, not that it necessarily produces perfect outcomes.

An optimality model has three main components (Stephens & Krebs 1986).

1 Decision assumptions

An assumption or set of assumptions about the problem that the organism is faced with must be made. In foraging models two classic examples are decisions about which food items to eat when they are encountered (the contingency model of prey choice, Schoener 1971; Emlen 1973; Maynard Smith 1974; Pulliam 1974; Werner & Hall 1974; Charnov 1976a) and decisions about how long to spend foraging in a patch which is gradually being depleted (the marginal value model, Charnov 1976b; Parker & Stuart 1976).

2 Currency assumptions

The currency of an optimality model is the criterion by which alternative choices are evaluated, for example it might be 'net rate of energy intake', 'number of mates encountered per minute' or 'number of eggs laid per season'. In addition to specifying the currency itself, optimality models include an assumption about the choice principle. This is generally maximization, minimization, or in the case of problems with frequency dependent payoff, stablity. The currency may be chosen *a priori* because it is thought to relate to survival and reproduction, or *a posteriori,* because it gives a good description of what the organism does. Most biologists would perfer the former approach to currencies because the latter relegates optimality models, as in economics, to self-consistent descriptive accounts. In fact biologists usually compromise and choose currencies on the basis of both *a priori* intuition and descriptive success. In many of the first generation of optimal foraging theory models the currency was long-term average net rate of energy intake, and some ecologists have assumed that optimal foraging theory deals only with this currency (e.g. Crawley 1983). This is not so, as later models have considered, for example, other food-related currencies such as minimizing the probability of starvation (Caraco, Martindale & Whitham 1980), maximizing efficiency (Schmid-Hempel, Kacelnik & Houston 1985), and minimizing the combined probability of death due to starvation and predation (Lima, Valone & Caraco 1985; Houston *et al.* 1988, see below for more details). In other words, optimal foraging theory includes a diverse and continually changing set of models.

3 Constraint assumptions

It would be a very naive view of optimality to assume that evolution is free to produce any conceivable outcome. Clearly there are limits set by, among other things, accidents of history, developmental processes and genetic architecture. In a model these things are taken into account (whether or not they are taken into account satisfactorily is a matter of debate) under the general heading of constraints. Constraints define the limit of feasible choices and the limits of the payoff that can be obtained by the organism. Kacelnik & Cuthill (1987) distinguished between two kinds of constraint; those that are properties of the organism (e.g its ability to distinguish between sizes of prey, its speed of locomotion, its rate of ingestion) and those that arise from the environment, for example resource depletion. They call the former the 'strategy set' and the latter (following the terminology of Staddon 1983) the 'feedback function'. This distinction may be clear in some experimental set ups where the feedback function is deliberately chosen by the experimenter to be largely independent of the organism's behaviour, but in most natural situations the organism's own limitations will be closely intertwined with limitations imposed by the environment.

This breakdown of the formal structure of an optimality model shows that the optimality approach allows us to test assumptions about the decision, the currency and the constraints. When an optimality model produces predictions that agree with observations, confidence in these assumptions increases slightly, when a model fails to account for what happens in the world, confidence in the assumptions is diminished. The general background rationale of optimality modelling is itself not under direct test in any one study, although in the long term, repeated success or repeated failure of individual models would either sustain or undermine belief in the general assumptions surrounding individual models (Maynard Smith 1978; Kacelnik & Cuthill 1987; Schoener 1987). The extent to which concordance between observations and predictions are taken as support for an optimality model or the optimality approach itself, depends as with any model, on uniqueness of the predictions. Counter-intuitive or quantitative predictions are in this sense more valuable than intuitively obvious or qualitative predictions (Krebs & McCleery 1984; Gray 1987).

A final important point to bear in mind when considering tests of optimality models is that testing a model is not an end in itself, but a means to the end of gaining a greater understanding of the ecology and behaviour of living organisms. Thus the end result of testing many optimality models with positive results would not be 'to show that optimality works' but to draw general conclusions at the qualitative or quantitative level about the organization of behaviour and ecology.

It is an interesting fact that many ecologists and evolutionary biologists started to think in terms of optimality at about the same time as the first two papers on optimal foraging theory appeared in 1966. As Stephens & Krebs (1986) and Schoener (1987) point out, the sequential encounter model of prey selection was developed independently by at least six workers between 1966 and 1974, while two papers independently describing the marginal value theorem appeared in 1976 (Charnov 1976b; Parker & Stuart 1976). In the late 1960s and early 1970s and apparently more or less independently, the following key papers appeared. Royama (1970) used an optimal foraging model of patch choice to account for variation in the food items fed by parent great tits to their nestlings. Fretwell & Lucas (1970) wrote their seminal paper on the ideal free distribution as a model of how competitors should be distributed between habitats, and Orians (1969), following Verner & Willson (1966) developed his model of polygyny which is essentially the same as the ideal free model; similarly, Parker's (1970) model of the spatial distribution of male dungflies searching for females was another account of the ideal free distribution. Although the ideal free model was developed to describe a situation in which the payoffs for choosing an option are frequency dependent, the impetus to analyse a wide range of problems which include frequency dependence came only after Maynard Smith (1972) had introduced the terminology and methods of game theory into the literature on behavioural ecology.

Following this first flush of optimality models, the most significant new developments of the 1980s have been: (i) the introduction of the idea of risk sensitivity (Caraco, Martindale & Whitham 1980; Stephens 1981); (ii) the development of links between optimal foraging theory and psychological models of how choice and learning are related to differential reinforcement (Commons, Kacelnik & Shettleworth 1987); (iii) the application of dynamic optimization models both to long term decisions affecting life-history strategies (Gilliam 1982) and short term decisions affecting moment-to-moment behaviour (Houston *et al.* 1988).

Possibly the most characteristic distinguishing feature of optimal foraging as a body of ecological theory is the way in which application of the models to empirical data, especially experimental data, developed simultaneously with the models themselves (Schoener 1987). In the case of one of the models (the marginal value theorem), an experiment designed specifically to test the model was published 2 years before the model itself (Krebs, Ryan & Charnov 1974)! This contrasts with many of the theoretical developments in community and population ecology which developed in the 1960s, where the theory acquired a life of its own, independent of the real world. At least three factors contributed to the close link between theory and data. First, many ecologists were already collecting data on feeding ecology, either indirectly by analysing stomach contents or directly by observation, and foraging theory provided a convenient theoretical peg on which to hang essentially descriptive studies. Thus many empirical papers were not specifically designed to test foraging models; often they misunderstood the assumptions underlying the models and therefore applied them to data in inappropriate ways (Krebs, Stephens & Sutherland 1983, Stephens & Krebs 1986). Second, it happened that several of the early theoretical papers were developed either by theoreticians such as Charnov working in close collaboration with experimental ecologists, or by workers with good data (e.g. Royama 1970). Finally and perhaps most important, the kinds of problem represented in foraging models were especially amenable to experimental analysis in simple laboratory setups which nevertheless could capture the essence of the field problem.

Another striking aspect of the short history of optimality modelling, and particularly optimal foraging theory is that it acquired vociferous and severe critics almost as soon as it became established as a research method (e.g. Lewontin 1978; Gould & Lewontin 1979; Ollason 1980; Dennett 1983). The strength of feeling involved in these critiques is indicated by some of the titles, for example: 'Faith and foraging ...' (Gray 1987), 'Eight reasons why optimal foraging theory is a complete waste of time' (Pierce & Ollason 1987). The criticisms have been discussed at length by Stephens & Krebs (1986), Dupre (1987) and Stearns & Schmid-Hempel (1987) to which the reader is referred for further details.

The critics are by no means in agreement among themselves about optimization modelling. For example, in a recent edited volume on foraging theory (Kamil, Krebs & Pulliam 1987) one critic, Gray, sees optimization theory as '... the most rigorous form of functional explanation', where function is equivalent to the Darwinian concept of adaptation. Ollason, in contrast, thinks that '... optimal foraging theory has nothing to do with the theory of evolution'. Further, not all the critics are well-informed about the object of their criticism. Lewontin (1979) characterized optimal foraging as '... finding the most food with the least expenditure of energy', which as Schoener (1987) puts it, is '... a goal not characterizing any model of which I am aware and which would seem in principle impossible'. The criticisms themselves are diverse, ranging from general claims that optimization modelling in particular and the study of adaptation in general is tautological, to more specific claims about the success or failure of particular models in accounting for the data. Amongst these many criticisms, some are unlikely to be resolvable and so are probably not worth worrying about. For example, these include the problem of constraints (constraints of historical accident, of genetic interactions such as pleiotropy and heterozygote advantage, or of developmental pathways limiting the power of natural selection to optimize) and the problem of atomization. This latter refers to the fact that most optimization models involve treating one or at most a few aspects of an organism's total life history, structure and physiology (e.g. mating behaviour, territory size, foraging). However because all adaptations are inter-related (mating behaviour may depend on the way an animal feeds and vice versa), atomization is logically flawed. Further, by attempting to atomize, the optimality modeller is likely to end up asking trivial or pointless questions.

A potentially more damaging, and empirically resolvable, criticism is the one voiced by Lewontin (1987). As discussed earlier, an optimality model has two major classes of unknowns: the constraints and the currency. If on the one hand, all the constraints and currencies were known before starting, optimality modelling would be superfluous, while if none of the constraints or the currency is known, there are too many unknowns to cope with. The difficulty is whether or not optimality models can be useful in this situation. Practitioners of optimality modelling would claim that sensible guesses can often be made about constraints (for example by looking at the range of features found in organisms closely related to the one under study) or that constraints can be investigated independently, for example the nutrient requirements of an animal could be determined independently of any test of an optimal foraging model in which nutrient constraints were included. Similarly, a limited range of currencies is likely to be appropriate for any particular problem, and with intuition the number of choices can be kept within narrow limits. A related criticism is that optimality modellers are able to get out of any failure of their models by *ad hoc* modifications. This is however based on a misunderstanding of the nature of *ad hocism*. If a modification of a model produces a new testable prediction (for example, a diet model is modified to include a predation cost associated with handling prey items) this is not an *ad hoc* modification (Chalmers 1982). It is simply the normal practice of successively refining an hypothesis. A truly *ad hoc* modification would be one that is totally untestable in an independent experiment (e.g. 'the optimal diet model predicts choice accurately on March 18th 1988').

Finally, some authors think that optimality models do not add anything to other kinds of explanation. Ollason (1980) criticized the marginal value model by saying that a mechanistic model (a simple exponential decay model of memory) can produce the same results. Since there has to be a mechanism underlying any adaptive outcome, this criticism is pointless. Optimality models and mechanistic models are looking at different facets of the problem. Could one, however, do away with functional (including optimality) accounts altogether? No, on at least two counts. First, on intellectual grounds, a functionless biology would simply be a mass of descriptive detail with no organizing principles. Second, as a research strategy; mechanism questions are often easier to answer when placed in a functional context.

SOME EXAMPLES

Instead of attempting to review all the applications of optimality theory to ecology, we will concentrate on a few examples which show how optimization at the level of individual animals can have broader ecological implications.

Winter body mass of birds

It is well established that many small birds are heavier during the middle of winter than they are at either the start or the end of winter. There are also examples in which northern populations of a species have higher winter weights than southern populations. In some cases it has been shown that these differences in weight are the result of changes in the amount of fat that the birds carry on their bodies. Although fat may be of value in improving a bird's insulation, it is usually assumed that the main advantage of carrying fat arises from increased survival probability in an environment in which a bird is not always able to feed.

Most birds stop foraging at dusk and roost so their fat reserves at dusk must be sufficient to ensure that they do not starve during the night. In addition to this regular pattern of interruptions to feeding imposed by darkness, a bird may also be subject to irregular interruptions during daylight because of bad weather. This raises the possibility that the geographical and temporal patterns of body mass correspond to patterns of environmental harshness. Temperatures are usually lowest in the middle of winter, and Evans (1969) found that at any time the body mass of yellow buntings was strongly correlated with the minimum expected temperature. A problem with such a simple view is that the probability of starvation can always be decreased by increasing fat reserves. Why are not birds heavy throughout the winter? Furthermore, birds are often less heavy in winter than they are before migration, so why do they not carry the maximum amount of fat throughout the winter? King (1972) suggested that there will be a cost associated with high levels of fat reserves. One argument is that the foraging time required to maintain a high level of reserves both limits the time that can be devoted to activities other than foraging and increases the time for which the bird is exposed to predators. As it stands, this argument is inadequate because once the bird has done the extra foraging to get its reserves to a high level, it requires the same amount of time to *maintain* that level as to maintain a lower level. This objection can be overcome by assuming that metabolic expenditure increases as mass increases (perhaps because of the extra energy required for flight) so that the time spent foraging to maintain a given level of reserves does increase with the level of reserves (McNamara & Houston, in press).

An alternative argument is that as fat increases, a bird's ability to escape from a predator decreases. As a result the probability of being killed when a predator attacks increases as body mass increases. Lima

(1986) presented a model of optimal body mass that incorporates both these effects. By increasing its fat a bird can reduce its probability of starvation, but the associated increase in foraging time and mass result in an increase in the probability of being killed by a predator. The optimal body mass (OBM) is the mass at which total mortality (starvation plus predation) is minimized. Body mass changes throughout the day as the bird feeds and through the night as the bird rests. In order to talk about an optimal mass under given conditions, Lima assumed that the bird controls its foraging so as to maintain its mass at the start of the next day (starting mass) at some specified level. The OBM is thus defined as the starting mass that minimizes mortality over the winter.

Lima modelled the bird's environment in the following way. Days can either be 'good' or 'bad'. The bird can feed on good days but cannot feed on bad days. If a given day is good then there is a probability p that the next day is bad and probability $1-p$ that the next day is good. If a given day is bad there is a probability q that the next day is bad and a probability $1-q$ that the next day is good. On bad days the bird rests and it is safe from predation but its reserves decrease. If they reach zero then the bird starves. On good days the bird is not in danger of starving, but it can be killed by a predator.

The parameter q determines the length of the bad period; as q increases the mean number of consecutive bad days increases. This increase makes fat reserves more important, and so the OBM increases as q increases. The parameter p reflects the frequency of days on which the bird is unable to feed. As p increases, so does the OBM. Lima (1986) emphasizes that even though periods when food is unavailable may be rare, they can have a significant effect on body mass. The OBM decreases with an increase in the frequency of attacks by a predator and with an increase in the probability of capture as a function of mass, given that a bird is attacked.

All these results can be understood intuitively. The effect of the rate of energy intake r is perhaps not so obvious. The OBM decreases as this rate decreases. This effect arises because r has no effect on the starvation probability associated with a given level of reserves. On good days the bird gains energy deterministically at a rate r. Changing r changes the amount of time that must be devoted to foraging in order to obtain a given amount of energy, but it does not change the survival value of the energy, provided r is always big enough for the bird to gain any desired amount of energy during a day. What this means is that a change in r is equivalent to a change in predation risk, in that a decrease in r results in an increased time exposed to predation in order to achieve a given gain.

It is interesting to compare the results of Lima (1986) with those of McNamara & Houston (in press). Lima models the environmental stochasticity in terms of probabilistic transitions between good days (during which feeding is deterministic) and bad days (during which feeding is impossible). In the model developed by McNamara & Houston, all days are *a priori* the same but feeding is always stochastic. Days are divided into a series of equal intervals during which the animal is either uninterrupted and can forage, or is interrupted and cannot forage. The number of consecutive intervals that are uninterrupted has a geometric distribution with mean U. Similarly the number of consecutive interrupted intervals has a geometric distribution with mean I. By changing both U and I the mean time that is uninterrupted, $U/(U + I)$, can be kept constant while the scale of fluctuations in food availability is changed. When U and I are small, runs of good and bad intervals are short. Increasing U and I for constant $U/(U + I)$ increases the size of a run of interrupted intervals and results in birds carrying more reserves. This result has some resemblance to Lima's result based on changes in p and q, but these changes also affect the overall availability of food.

Lima (1986) found that a decrease in the rate of intake r resulted in a decrease in the optimal level of reserves. In contrast, McNamara & Houston (in press) found that a decrease in rate of intake increases the optimal level of reserves. This difference is the result of the different ways in which environmental stochasticity is modelled.

Lima (1986) used his model to make some general points. One is that although an event such as a severe snowstorm may be rare, it can nevertheless be an important influence on an animal's survival, and should influence the fat reserves that a bird carries. In a 'typical' winter this level of reserves may seem unnecessarily high. Another point is that attempts to isolate a single factor limiting the population size in winter are unlikely to be successful. In the models of Lima (1986) and McNamara & Houston (in press) it is not starvation or predation in isolation but the interaction between them that determines mortality. McNamara & Houston (1987) investigated this interaction. They assumed that an animal can determine its mean energetic gain μ for a day. The mean gain determines the probability $S(\mu)$ that the animal will starve. It also determines the probability $P(\mu)$ that the animal is killed by a predator. One possible way in which such a dependence can arise is that the animal chooses the habitat in which it feeds, and this in turn results in a given distribution of food items and danger of predation. McNamara & Houston (1987) considered an animal outside the breeding season for which fitness is maximized by the minimization of total mortaility $M(\mu)$, given by the equation

$$M(\mu) = S(\mu) + P(\mu)\,.$$

Thus, the optimal value of μ, denoted by μ^*, satisfies the equation

$$-\,S'(\mu^*) = P'(\mu^*)\,.$$

An important point emerges from this simple equation. There is no reason to expect natural selection to result in equal levels of starvation and predation. At the optimum, the slopes of $S(\mu)$ and $P(\mu)$ are equal and opposite, but their absolute levels may be very different [see Fig. 10.1(a).] McNamara & Houston illustrated this with a simple model in which the amount of food obtained by the animal during the day has a normal distribution with mean μ. In this context, μ can be thought of as the effort that the animal puts into foraging. If the amount of food that the animal obtains in a day is less than a critical amount G, then the animal dies of starvation. Thus an increase in G results in a deteriotation in the feeding environment. McNamara & Houston considered three ways in which $P'(\mu)$ depends on μ:

Case I: $P'(\mu)$ is constant.

Case II: $P'(\mu)$ is an increasing function of μ, i.e. $P(\mu)$ is an accelerating function of μ.

Case III: $P'(\mu)$ is a decreasing function of μ, i.e. $P(\mu)$ is a decelerating function of μ.

In each case, μ^* and hence $S(\mu^*)$ and $P(\mu^*)$ are found for a range of feeding environments. In all cases both μ^* and the total mortality increase as conditions deteriorate, but the changes in $S(\mu^*)$ and $P(\mu^*)$ depend on $P'(\mu)$. In case I, $S(\mu^*)$ is constant while $P(\mu^*)$ increases. This result is illustrated in Fig. 10.1(b). To keep $-\,S'(\mu) = P'(\mu)$ when the starvation curve moves to the right requires a higher value of μ^* but the same value of $S(\mu)$. In case II, both $S(\mu^*)$ and $P(\mu^*)$ increase, whereas in case III, $P(\mu^*)$ increases but $S(\mu^*)$ decreases [Fig. 10.1(c)]. What this simple model indicates is that under some circumstances the best response to a decrease in food availability can result in a decrease in the level of starvation. This sort of effect also occurs in a more complex model (McNamara & Houston, in press).

Diet choice in herbivores

Belovsky (1978, 1984a, 1986a, b) argued that the diet of many generalist herbivores can be understood in terms of maximization subject to various constraints. These constraints include general limitations (e.g.

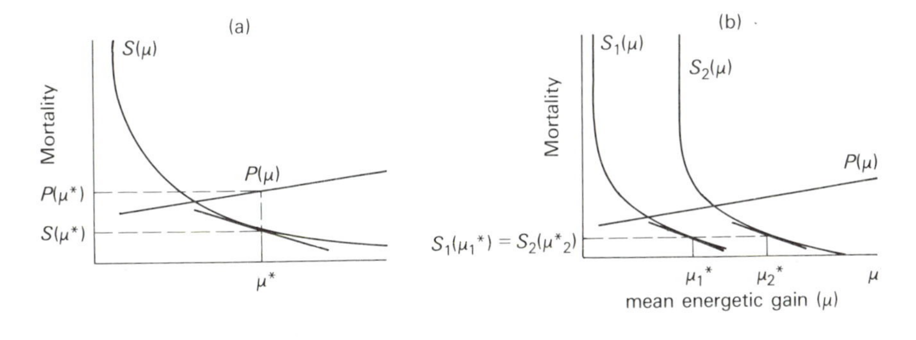

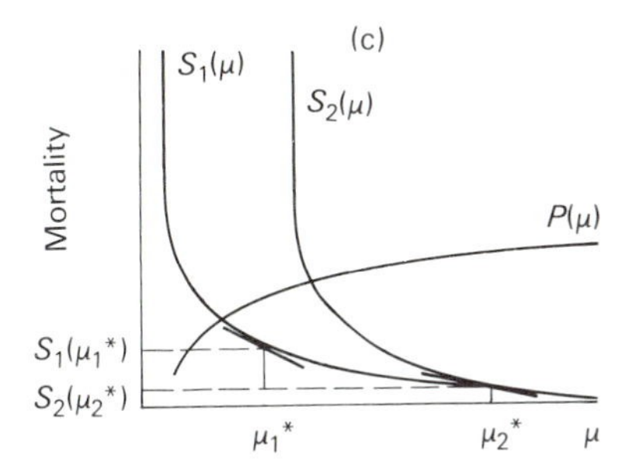

FIG. 10.1 Graphical representation of the Houston & McNamara (1987) model of the optimal tradeoff between starvation and predation. (a) Optimal tradeoff is where the slopes of the starvation and predation functions [$S(\mu)$ and $P(\mu)$, respectively] are equal and opposite (the slope of the $S(\mu)$ line is indicated by a tangent to the curve). This optimal point is indicated by the value of energetic gain $< S_1(\mu_1^*)$]. and the corresponding starvation and predation rates are $S(\mu^*)$ and $P(\mu^*)$. (b) Increased environmental harshness can be depicted by moving the $S(\mu)$ curve to the right [$S_1(\mu)$ to $S_2(\mu)$]. When $P'(\mu)$ is constant, μ^* increases ($\mu_2^* > \mu_1^*$) but the value of $S(\mu^*)$ does not change. (c) When $P(\mu)$ is a decelerating function, increased environmental harshness results in a decrease in $S(\mu)$ [$S_2(\mu_2^*) < S_1(\mu_1^*)$].

time constraints) and specific limitations that are associated with herbivory. We illustrate this approach using examples from Belovsky's work on herbivores in the National Bison Range, Montana (Belovsky 1986a). At this site during summer, plants can be divided into monocots, *m* (mainly graminoids) and dicots, *d* (forbs and deciduous leaves). Belovsky wished to understand why the diet of a given herbivore should contain a certain proportion of monocots as opposed to dicots. Belovsky considered four constraints that might limit diet composition. These constraints are linear, i.e. they are straight lines, in a space with the axes dicot intake and monocot intake.

1 Digestive capacity

The volume of a herbivore's digestive system, together with its digestive turnover will limit the amount of food that a herbivore can consume. For example, the bison cannot consume more than 26 310 g wet weight per day. Given that the wet wt/dry wt ratio of grasses is 1.64 and that of forbs is 2.67, then the constraint equation is

$$26\,310 \text{ g wet wt per day} \geq 1.64\, m + 2.67\, d,$$

where m is the intake of monocots (grasses) (g dry wt per day) and d is the intake of dicots (forbs) (g dry wt per day). This constraint is shown as line D in Fig. 10.2(a).

2 Feeding time

This constraint is based on the maximum possible amount of time during a day that a herbivore can devote to feeding. The upper limit depends on the herbivore's physiology, the climate that it experiences, and the time for which the herbivore's digestive organ is empty. For the bison the constraint equation is

$$178 \text{ min per day} \geq 0.1\, m + 0.045\, d.$$

This constraint is shown as line T in Fig. 10.2(a).

3 Energetic requirements

This constraint represents the minimum amount of energy that the herbivore requires in order to stay alive. The bison requires 74 250.4 kJ per day. From measurements of the energy content and digestibility of monocots and dicots it is found that monocots provide 12.40 $\text{kJ} \cdot \text{g}^{-1}$ dry wt per day and dicots provide 14.66 $\text{kJ} \cdot \text{g}^{-1}$ dry wt per day. Thus the energy constraint is

$$74\,250.4 \text{ kJ per day} \leq 12.40\, m + 14.66\, d.$$

This constraint is shown as line E in Fig. 10.2(a)

4 Nutrient requirements

It is often argued that the diet of herbivores is limited by the need to obtain at least a certain amount of sodium or protein. Such limitations can be incorporated in a linear programming model as additional constraints (e.g. Belovsky 1978, 1984b). Belovsky (1986a) argued that

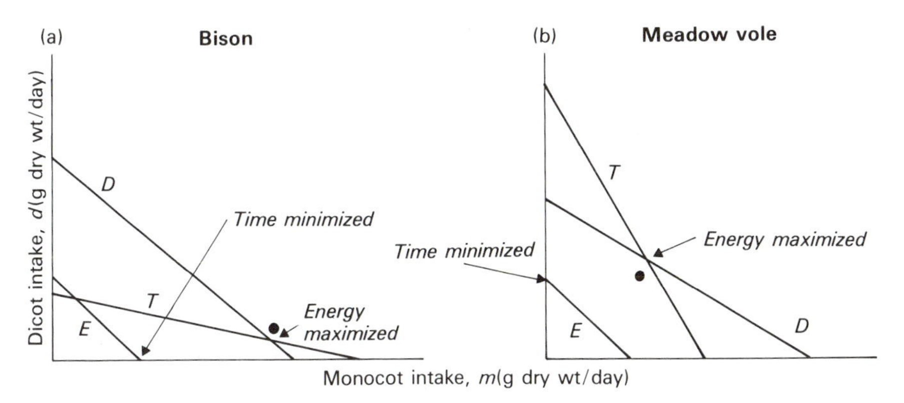

FIG. 10.2 Belovsky constraint model for herbivore diets applied to (a) bison, and (b) the meadow vole. In each graph the various constraints are indicated as follows. D = digestive capacity constraint, E = energetic constraint, T = time constraint. The actual diet is indicated by the filled circle. [After Belovsky (1986a).]

such constraints are not relevant for the herbivore species studied at Bison Range, and so they are not included in our example. Thus Fig. 10.2 (a) gives the three relevant constraints for the bison. The same constraints for the meadow vole (*Microtus pennsylvanicus*) are shown in Fig 10.2 (b).

With the constraints established, we can consider diets that maximize various currencies subject to these constraints. For any given currency, only two of the three possible constraints are relevant, because the third possible constraint is the currency. For example, when energy is being maximized, only the digestive capacity and feeding time constraints must be satisfied. A general feature of the two-dimensional linear programming problems that we are describing is that the optimal solution occurs at the intersection of two constraints or at the intersection of a constraint and an axis. In the case of energy maximization for bison and *Microtus*, the optimal diet is given by the intersection of the digestive capacity and time constraints, as is shown in Fig. 10. 2(a) and (b). In contrast, the time-minimization solutions occur where the energy constraint meets an axis. For *Microtus*, the intersection is on the dicot axis (i.e. no monocots are taken), whereas for bison the intersection is on the monocot axis (i.e. no dicots are taken). Belovsky (1986a) found that for the fourteen species of herbivores at the Bison Range, the minimization of digestive capacity is a very poor predictor of the observed diets, whereas the maximization of energy predicts the observed diets very well. There is a positive correlation between the diets predicted by time minimization and the observed diet, but it is much less impressive than the energy maximization correlation. Belovsky (1986b) reviewed a total of twenty herbivore species for which the linear programming approach has been applied and reported that the currency of energy maximization provides a much closer fit to the data than the currency of time minimization. Furthermore, in all but one of the species, the diet predicted by energy maximization involves a mixture of food types that is significantly different from the food available (i.e. from a 'random diet'). The exception is the elk, *Cervus elaphus*, at Bison Range. As the elk's diet is not significantly different from the energy maximization diet, Belovsky (1986a) argued that under these circumstances the elk's best policy is to eat food types as they are encountered.

The success of the linear programming model based on energy maximization suggests that a wide range of herbivores are limited by their digestive capacity and foraging time. Belovsky (1986b) (see also Belovsky 1981, 1984b) argued that, within a food type, this leads to two criteria that a food item must satify: a minimum digestibility/nutrient content and a minimum cropping rate. For a given herbivore, these minimum values can be used to specify the usable, as opposed to the actual, amount of food. Belovsky (1984c, 1986b) took the argument a step further by using the overlap between the usable food of two herbivores as an indication of competition between the species. Belovsky (1984c) carried out this analysis in the case of the moose (*Alces alces*) and the snowshoe hare (*Lepus americanus*), using sympatric and allopatric populations in the Isle Royale archipelago. He compared the fit of various competition models (Schoener 1974) to moose and hare population densities and the best fit was provided by a model involving shared and exclusive resources. The equations for population growth are

$$dN_1/dt = R_1N_1(I_{E1}/N_1 + I_{01}/(N_1 + a_2N_2) - C_1)$$

$$dN_2/dt = R_2N_2(I_{E2}/N_2 + I_{02}/(N_2 + a_1N_1) - C_2),$$

where N_i = number of individuals of species i, R_i = conversion of resources of species i into births minus deaths, I_{Ei} = quantity of resources that can be used exclusively by species i, I_{01} = quantity of resources for species 1 that is shared with species 2, I_{02} = quantity of resources for species 2 that is shared with species 1, a_i = competition parameter that converts resource use by an individual of species i into equivalent resource use by a member of the other species, C_i = minimum quantity of resources required by an individual of species i if it is to survive and reproduce.

Belovsky (1984c) found a close agreement between the estimates of shared and exclusive resources that emerged from a parameter estimation in the competition equations and from the minimum digestibility and cropping rate conditions. Thus, the foraging model provides independent support for the interpretation of the population equations in terms of competition. A similar result was obtained for grasshoppers at Bison Range, with the additional feature that population trajectories and competitive isoclines could be measured in population cages and compared to Schoener's equations (Belovsky 1986b).

It is worth emphasising that usable food, as determined by the minimum digestibility and cropping rate conditions, is not correlated with total plant biomass (e.g. Belovsky 1986b, figure 6.) An attempt to interpret the competition equations on the basis of actual biomass would be unsuccessful. Belovsky (1986b) suggested that the failure to distinguish between the actual amount of food and the usable amount of food may be a reason why some studies have not found that herbivores are food limited.

Choice of habitat and the ideal free distribution

A widely discussed model for how animals should be distributed over a range of habitats is the ideal free distribution (IFD). This model was proposed Fretwell & Lucas (1970), who assumed that all animals were free to go wherever they wanted to and were 'ideal' in that they had a perfect knowledge of the various habitats. When a group of individuals with the same abilities settle in an area, the first settlers should choose the best habitat, but as this habitat is settled, the density of animals will reduce the quality of the habitat below that of the next-best habitat. This habitat will then be chosen until in its turn its quality is reduced below the level of another unexploited habitat. The end result of this process is that in all the habitats that are used, all animals do equally well, and no animal can improve its payoff by moving to another habitat.

The IFD has been investigated experimentally in a simple procedure in which a group of animals can choose between two locations (patches) in which to feed. For example, Milinski (1979) gave six sticklebacks a choice of two patches, with the rate at which food was delivered being twice as high at one patch as at the other. If it is assumed that the probability that a given fish gets a food item at a given patch is the reciprocal of the total number of fish at a patch, then the feeding rates of all fish are equal when four fish are at the good patch and two fish are at the poor patch. This distribution can be characterized by the input matching rule

(Parker 1978) which says that the number of animals at a patch should match the input rate of resources at the patch. Milinski (1979) found that his fish divided their time between the two patches in this way.

Milinski's experiment is an example of what Parker & Sutherland (1986) called a 'continuous input' study. In these there is a continuous supply of resources at the various possible patches, and these resources are used as they arrive. Parker and Sutherland contrast these studies with 'interference' studies in which the distribution of resources is more or less constant. Patches differ in the density of resources that they contain, and this, together with the number of other animals in a patch, determines an animal's intake rate. Sutherland (1983) explored the effect of interference between animals, using the interference equation of Hassel & Varley (1969). He found that the distribution of predators as a function of resource density has the same form as the equation for aggregation used by Hassell & May (1973).

At first sight continuous input studies provide considerable support for the ideal free distribution (e.g. Parker & Sutherland 1986, table 1), but a closer inspection of the data shows that the assumption that all animals have the same competitive ability is often violated. Thus both Harper (1982) and Milinski (1984) found that the average intake rate of animals was the same in the two patches, but some animals did consistently better than others.

Sutherland & Parker (1985) developed a model in which animals differ in their ability to compete for food. Each animal has a competitive ability, and the probability that a given animal obtains food items at a given patch is equal to its competitive ability divided by the sum of the competitive abilities of all the animals at the patch. Working on the general principle that at equilibrium no animal should be able to improve its pay-off by moving to another patch, Sutherland & Parker (1985) and Parker & Sutherland (1986) found that there is now no longer a unique equilibrium distribution, but a range of possible distributions. These distributions all satisfy the condition that the sum of the competitive abilities at a patch matches the input rate at a patch. Parker's input matching rule thus becomes a special case of this general rule. Sutherland and Parker illustrate the possible equilibrium distributions with a simple example. There are two patches, a good patch providing twelve items per unit time and a poor patch providing six items per unit time. There are twelve animals, six good competitors and six poor competitors. The good competitors are twice as good as the poor competitors. One possible distribution (distribution A) is to have all the good competitors in the good patch and all the poor competitors in the poor patch. The complete set of possible distributions is given in Table 10.1.

TABLE 10.1. Possible distribution of twelve animals that result in the sum of competitive abilities at a patch matching the patch quality. The good patch is twice as profitable as the poor patch and the good competitors have twice the competitive ability of the poor ones. [After Houston & McNamara (1988).]

	Good Patch		Poor Patch		
	No. good Competitors	No. poor Competitors	No. good Competitors	No. poor Competitors	No. ways distributions can occur
A	6	0	0	6	1
B	5	2	1	4	90
C	4	4	2	2	225
D	3	6	3	0	20

Sutherland & Parker drew attention to distribution *C*, in which each ability class is divided in the ratio of the patch input rates, so that there are four good animals and four poor animals in the good patch and two good animals and two poor animals in the poor patch. This distribution mimics the IFD, in that the number of animals matches the patch input rates. Sutherland and Parker make the interesting suggestion that the tendency of animals to follow the IFD even when competitive abilities differ may be the result of chance. If we consider the four possible distributions shown in Table 10.1, there is only one way in which distribution *A* can be achieved. In distribution *D*, there is only one way in which the poor competitors can be distributed whilst the six good competitors can be divided into two groups of three in 6!/(3! 3!) = 20 ways. The number of ways for the four distributions are given in the table. It can be seen that distribution *C* can be achieved in the largest number of ways, and so might be expected to occur by chance if all possible number of ways that satisfy the equilibrium condition are equally likely.

While the distribution that mimics the IFD is the most likely when the number of animals is relatively small, this is no longer the case as the number of animals becomes large (Houston & McNamara 1988). Using an analysis based on statistical mechanics, Houston and McNamara showed that for large numbers of animals the most likely distribution is one in which the proportion of all animals on the good patch is less than the proportion of the resources that this patch provides.

So far, we have assumed that all possible distributions satisfying the equilibrium condition are equally likely, but this may not be so because of behavioural constraints (Houston & McNamara 1988). One form of constraint, discussed by Abrahams (1986), is the ability of animals to discriminate between patches. Another (Bernstein, Kacelnik & Krebs

1988) arises from the fact that competitors learn about the distributions of their prey and of other competitors. Extending previous work by Regelmann (1984) and Lester (1984), Bernstein, Kacelnik and Krebs examined by simulation the response of learning predators to a population of prey distributed among forty-nine patches. Predators were assumed to 'know' instantaneously their capture rate within a patch (determined by a disc equation with interference), but to learn about the average capture rate for the environment as a whole. The process of learning was represented by an exponentially decaying average of past capture rates, and animals moved at random between patches at the end of a simulation step if their estimated average for the environment was higher than the capture rate in the present patch.

In simulations with no depletion of prey, learning predators converged on the ideal free distribution, while with depletion of prey the fit was not so good. The lag resulting from learning meant that predators often failed to track changes in the distribution of prey. For a given 'speed of learning' the rates of depletion of prey had a dramatic effect on the spatial density dependence of predation: with slow depletion, density dependence was positive, with very rapid depletion it was negative. A somewhat unexpected result of the simulations was that predators with more knowledge of the environment as a whole, so that they could selectively migrate to better-than-average patches, produced a poorer fit to the ideal free distribution. In part this arose because all predators migrated synchronously, and many moved to the good patches, immediately reducing their profitability, so that at the next opportunity, migration to another site was advantageous.

Competition between species

Rosenzweig and his associates have applied the idea behind the IFD to interactions between species (e.g. Rosenzweig 1981, 1986; Pimm, Rosenzweig & Mitchell 1985). Consider two species, A and B, in an environment that contains two habitats (1 and 2). At very low densities, species A has a higher fitness in habitat 1 than in habitat 2, whereas species B has a higher fitness in habitat 2 than in habitat 1. Members of both species suffer a reduction in fitness as the density of either increases. How will the two species be distributed across the habitats as a function of species density? We can start by assuming that species B has zero density and see how the distribution of species A depends on its density (Fig. 10.3). When this density is very low, species A will choose habitat 1 (region S). As density increases, habitat 1 becomes less profitable and at

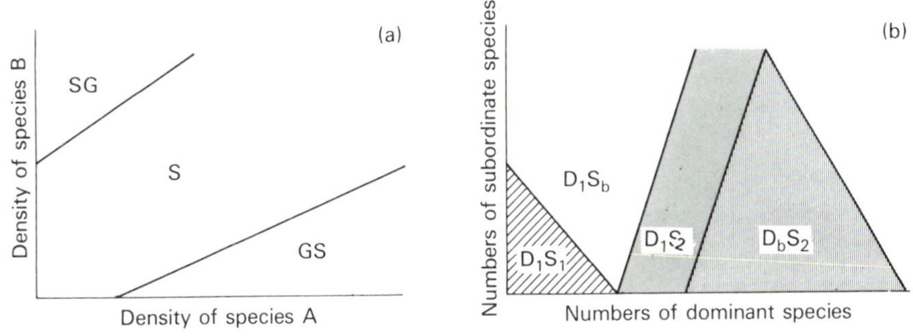

FIG. 10.3 Rosenzweig's model of habitat choice of two species as a function of their densities. (a) The two isolegs divide the figure into three regions. In region SG, species A specializes on its preferred habitat and species B is a generalist (i.e. uses both habitats). In region S, both species specialize and in region GS species A is a generalist and species B specializes. See text for further details. (b) Habitat selection isolegs for dominant and subordinate competing species. D = dominant; S = subordinate; 1 = uses habitat 1 (better habitat) exclusively; b = uses both; 2 = uses habitat 2. [After Rosenzweig (1986).]

some critical density the fitness in habitat 1 will equal the fitness in habitat 2. Above this critical level, both habitats will be used (region GS). Now assume that species B has a very low density. Species B will choose habitat 2 and thus devalue it. In turn this will mean that the density of species A will have to reach a higher critical level before its fitness in habitat 1 equals that in habitat 2. Repeating this argument for a range of densities of species B gives a line in species density space at which species A switches from choosing habitat 1 (G) to using both habitats (GS). Rosenzweig calls such a line an isoleg (from the Greek for equal choice). 'An isoleg is a line in a state space of animal densities such that some aspect of a species' habitat selection is constant at every point on this line' (Rosenzweig 1985). By considering various densities of species A, an isoleg for species B can be obtained, as is shown in Fig. 10.3(a). Species A specializes on habitat 1 to the left of its isoleg, and species B specializes on habitat 2 to the right of its isoleg. This gives us three regions. In region S both species specialize on their preferred habitat. In region SG, species A specializes and species B is a generalist, whereas in region GS species A is a generalist and species B specializes.

Pimm, Rosenzweig & Mitchell (1985) and Rosenzweig (1986) applied the isoleg model to three species of hummingbirds: blue throated (*Lamporius clemenicae*), Rivoli's (*Eugenes fulgens*) and black-chinned (*Archilochus alexandri*) competing at artificial feeders. In this study the isoleg model was modified to take into account the fact that one species (blue throated) is dominant to the other two. The birds were provided with two feeders, a 'good habitat' offering 1–2 molar sucrose and a 'poor habitat' with 0.35 molar sucrose. The data were analysed by looking at utilization of the two feeders by dominant–subordinate pairs of species as a function of their densities. Isoleg theory predicts the following [Fig. 10.3(b)]. First consider the subordinate species. At low densities of both species the subordinate should use the good habitat: the isoleg bordering this area has a negative slope, since an increase in density of either species reduces the suitability of the good habitat and the combined effects are assumed additive. Above the combinations of densities described by this first isoleg the subordinate species generalizes, and feeds on both habitats. As density of the dominant increases further, there comes a point at which it pays the subordinate species to feed only in the poor habitat, where competition is less severe. If members of the subordinate species compete amongst themselves, this second isoleg, in contrast to the first, has a positive slope. As the density of subordinates in the poor habitat increases and reduces its suitability, so a greater number of dominants is required to make it profitable for subordinates to feed exlusively in the poor habitat. Thus, the subordinate species should show three patterns of habitat distribution; feed only in the good site, feed in both, or feed only in the bad site, according to the density of the two species.

For the dominant species there are only two predicted patterns of habitat distribution. At low densities of the dominant, it should feed in the good habitat and at high densities, in both. The second isoleg has a positive slope if there is some negative effect of the subordinate on the suitability of the poor habitat for the dominant.

Rosenzweig estimated variations in density of the competitors by variations in the time spent by each species at the feeders. For two of the species, observed behaviour was in agreement with the isoleg model. The subordinate black-chinned showed three regions of habitat choice: exclusively using the good feeder at low densities of blue throats (dominant), using both at intermediate densities and using the poor feeder only at high densities. In the intermediate region there was a mixture of behaviours, sometimes being selective for poor sites and sometimes being non-selective. The dominant blue throat, as predicted, switched from exclusive use of the good site, to a generalist strategy at high blue throat density, but was unaffected by black-chin density.

Effects on population size

Holt (1985) considered the effect of habitat distribution between two patches on total population size. In particular, he examined how deviations from the ideal free distribution resulting from migration between patches alters population size. The assumption is that migration is symmetrical and that a fixed proportion of individuals migrates. The effect of migration depends on the relative slopes of the density-dependent effects on *per capita* growth rate within each patch. If the two slopes are equal the effect of migration from good to bad exactly compensates for reverse movement [Fig. 10.4(a)]. However, if the slopes are unequal, migration can have the effect of reducing total population size. In the example shown in Fig 10.4(b), the slope is steeper in the poor than in the good patch. Hence, the effect of migration is more severe when animals move to the bad patch than to the good, and since the number moving in the former direction is greater, the net effect is to reduce the total population size (the cross is to the left of the filled circle).

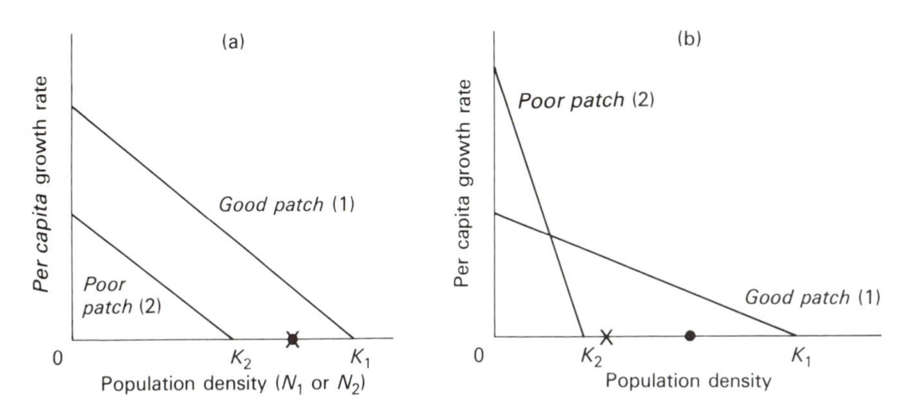

FIG. 10.4 Effects of migration between habitats on the total population size. There are two habitats, 1 & 2, in each of which *per capita* growth rate is negatively related to population density. *Per capita* growth is zero when $N_1 = K_1$, $N_2 = K_2$. With no migration, the equilibrium population per patch is $(N_1 + N_2)/2$ (filled circle). If individuals migrate symmetrically, and a fixed proportion of the population moves, the equilibrium population is given by the cross. In (a) the slope of the density-dependent function is equal in the two patches; in (b) it is steeper in the poor patch. [After Holt (1985).]

CONCLUSIONS

Our main points are as follows.

In the past ecologists have sometimes failed to analyse population and community processes in terms that are consistent with the neo-Darwinian view of selection.

One way of expressing selectionist hypotheses is in terms of optimization models.

Some examples of the use of optimization models in ecology show how processes at the individual, population and community level can be analysed and brought together in a unified approach.

ACKNOWLEDGMENT

We thank the SERC for financial support.

REFERENCES

Abrahams, M. V. (1986). Patch choice under perceptual constraints: a cause for departures from the ideal free distribution. *Behavioral Ecology and Sociobiology*, **19**, 409–415.

Belovsky, G. E. (1978). Diet optimization in a generalist herbivore: the moose. *Theoretical Population Biology*, **14**, 105–134.

Belovsky, G. E. (1981). Food plant selection by a generalist herbivore: the moose. *Ecology*, **62**, 1020–1030.

Belovsky, G. E. (1984a). Herbivore optimal foraging: a comparative test of three models. *American Naturalist*, **124**, 97–115.

Belovsky, G. E. (1984b). Snowshoe hare optimal foraging and its implications for population dynamics. *Theoretical Population Biology*, **25**, 235–264.

Belovsky, G. E. (1984c). Moose and snowshoe hare competition and a mechanistic explanation from foraging theory. *Oecologia*, **61**, 150–159.

Belovsky, G. E. (1986a). Optimal foraging and community structure: implications for a guild of generalist grassland herbivores. *Oecologia*, **70**, 35–52.

Belovsky, G. E. (1986b). Generalist herbivore foraging and its role in competitive interactions. *American Zoologist*, **26**, 51–69.

Bernstein, C., Kacelnik, A. & Krebs, J. R. (1988). Individual decisions and the distribution of predators in a patchy environment. *Journal of Animal Ecology*, **57**, 1007–1026.

Caraco, T., Martindale, S. & Whitham, T. S. (1980). An empirical demonstration of risk-sensitive foraging preferences. *Animal Behaviour*, **28**, 820–830.

Chalmers, A. F. (1982). *What is This Thing Called Science?*, 2nd edition. Open University Press, Milton Keynes.

Charnov E. L. (1976a). Optimal foraging: attack strategy of a mantid. *American Naturalist*, **110**, 141–151.

Charnov E. L. (1976b). Optimal foraging: the marginal value theorem. *Theoretical Population Biology*, **9**, 129–136.

Charnov E. L. & Finerty J. (1980). Vole population cycles: a case for kin selection. *Oecologia*, **45**, 1–2.

Chitty, D. (1960). Population processes in the vole and their relevance to general theory. *Canadian Journal of Zoology*, **38**, 99–113.

Chitty, D. (1967). The natural selection of self-regulatory behavior in animal populations. *Proceedings of the Ecological Society of Australia*, **2**, 51–78.

Churchland, P. S. (1987). *Neurophilosophy.* MIT Press, Cambridge, Massachusetts.

Clements, F. E. (1916). *Plant Succession: An Analysis of the Development of Vegetation.* Carnegie Institute Publications No. 242, Washington, DC.

Cole, L. C. (1954). The population consequences of life history phenomena. *Quarterly Review of Biology*, **29**, 103–137.

Commons, M. L., Kacelnik, A. & Shettleworth, S. J. (1987). *Quantitative Analyses of Behavior. Vol. VI. Foraging*, L. Erlbaum, New York.

Crawley, M. J. (1983). *Herbivory.* Blackwell Scientific Publications, Oxford.

Davies, N. B. & Houston, A. I. (1984). Territory economics. *Behavioural Ecology: an Evolutionary Approach*, 2nd edition (Ed. by J. R. Krebs & N. B. Davies), pp. 148–169. Blackwell Scientific Publications, Oxford.

Dennett, D. C. (1983). International systems in cognitive ethology. The Panglossian paradigm defended. *Behavioral Brain Science*, **6**, 343–390.

Dupre, J. (1987). *The Latest on the Best.* MIT Press, Cambridge, Massachusetts.

Ehrlich, P. R. (1987). *The Machinery of Nature.* Simon & Schuster, New York.

Elton, C. (1927). *Animal Ecology.* Sidgewick & Jackson, London.

Emlen, J. M. (1966). The role of time and energy in food preference. *American Naturalist*, **100**, 611–617.

Emlen, J. M. (1973). *Ecology: an Evolutionary Approach.* Addison Wesley, New York.

Evans, P. R. (1969). Winter fat deposition and overnight survival of yellow buntings (*Emberiza citrinella L.*). *Journal of Animal Ecology*, **38**, 415–423.

Fretwell, S. D. & Lucas, H. L. Jr. (1970). On territorial behavior and other factors influencing habitat distribution in birds. *Acta biotheoretica*, **19**, 16–36.

Gilliam, J. F. (1982). *Foraging under Mortality Risk in Size-structured Populations.* PhD Thesis. Michigan State University.

Gould, S. J. & Lewontin, R. C. (1979). The spandrels of San Marco and the Panglossian paradigm: a critique of the adaptationist programme. *Proceedings of the Royal Society of London B*, **205**, 581–598.

Gray, R. D. (1987). Faith and foraging: a critique of the 'paradigm argument from design'. *Foraging Behavior* (Ed. by A. C. Kamil, J. R. Krebs & H. R. Pulliam), pp. 69–140. Plenum Press, New York.

Harper, D. G. C. (1982). Competitive foraging in mallards: 'ideal free' ducks. *Animal Behaviour*, **30**, 575–584.

Hassell, M. P. & May, R. M. (1973). Stability in insect host-parasite models. *Journal of Animal Ecology*, **42**, 693–726.

Hassell, M. P. & Varley, C. G. (1969). New inductive population model for insect parasites and its bearing on biological control. *Nature*, **223**, 1133–1136.

Holling, C. S. (1973). Resilience and stability of ecological systems. *Annual Review Ecology and Systematics*, **4**, 1–23.

Holt, R. D. (1985). Population dynamics in two-patch environments. Some anomalous consequences of an optimal habitat distribution. *Theoretical Population Biology*, **28**, 181–208.

Houston, A. I., Clark, C. W., McNamara, J. M. & Mangel, M. (1988). Dynamic models in behavioural and evolutionary ecology. *Nature*, **332**, 29–34.

Houston, A. I. & McNamara, J. M. & (1988). The ideal free distribution when competitive abilities differ: an approach based on statistical mechanics. *Animal Behaviour*, **36**, 166–174

Huntingford, F. A. & Turner, A. K. (1987). *Animal Conflict*, Chapman & Hall, London.

Kacelnik, A. & Cuthill, I. C (1987). Starlings and optimal foraging theory: modelling in a fractal world. *Foraging Behavior* (Ed. by A. C. Kamil, J. R. Krebs & H. R. Pulliam), pp. 303–334. Plenum Press, New York.

Kamil, A. C. Krebs, J. R. & Pulliam, H. R. (eds) (1987). *Foraging Behavior*, Plenum Press, New York.

King, J. R. (1972). Adaptive periodic fat storage by birds. *Proceedings of the XVth International Ornithological Congress* (Ed. by K. H. Voous), pp. 200–217 Brill, Leiden.

Krebs, J. R. & McCleery, R. H. (1984). Optimisation in behavioural ecology. *Behavioural Ecology: An Evolutionary Approach*, 2nd edition (Ed. by J. R. Krebs & N. B. Davies) pp.

91–121. Blackwell Scientific Publications, Oxford.

Krebs, J. R. Ryan, J. C. & Charnov, E. L, (1974). Hunting by expectation or optimal foraging? A study of patch use by chickadees. *Animal Behaviour*, **22**, 953–964.

Krebs, J. R., Stephens, D. W. & Sutherland W. (1988). Perspectives in optimal foraging. *Perspectives in Ornithology* (Ed. A. H. Brush & G. A. Clark), pp. 165–216. Cambridge University Press. New York.

Lack, D. (1947). *Darwin's Finches.* Cambridge University Press, Cambridge.

Lack, D. (1954). *The Natural Regulation of Animal Numbers.* Oxford University Press, London.

Lester, N. P. (1984). The feed-feed decision: how goldfish solve the patch depletion problem. *Behaviour*, **85**, 175–199.

Lewontin, R. C. (1978). Adaptation. *Scientific American*, **239**, 156–169.

Lewontin, R. C. (1979). Fitness, survival and optimality. *Analysis of Ecological Systems* (Ed. by R. D. Horn, E. R. Stairs & R. D. Mitchell), pp. 3–21. Ohio State University Press. Columbus, Ohio.

Lewontin, R. C. (1987). The shape of optimality. *The Latest on the Best* (Ed. by J. Dupre), pp. 151–159. MIT Press, Cambridge, Massachusetts.

Lima S. L. (1986). Predation risk and unpredictable feeding conditions: determinants of body mass in birds. *Ecology*, **67** 377–385.

Lima S. L., Valone T. J. & Caraco, T. (1985). Foraging efficiency – predation risk tradeoff in the grey squirrel. *Animal Behaviour*, **33**, 155–165.

MacArthur, R. H. & Pianka, E. R. (1966). On the optimal use of a patchy environment. *American Naturalist*, **100**, 603–609.

MacFadyen, A. (1957). *Animal Ecology: Aims and Methods.* Pitman, London.

Maynard Smith, J. (1972). *On Evolution.* Edinburgh University Press, Edinburgh.

Maynard Smith, J. (1974). *Models in Ecology.* Cambridge University Press, Cambridge.

Maynard Smith, J. (1978). Optimization theory in evolution. *Annual Review of Ecology and Systematics*, **9**, 31–56.

McNamara, J. M. & Houston, A. I. (1987). Starvation and predation as factors limiting population size. *Ecology*, **68**, 1515–1519.

McNamara, J. M. & Houston, A. I. (in press). Starvation and predation in a patchy environment. *Living in a patchy environment* (Ed. by I. Swingland, N. C. Stenseth & B. Shorrocks). Oxford University Press, Oxford.

Milinski, M. (1979). An evolutionarily stable feeding strategy in sticklebacks. *Zeitschrift fur Tierpsychologie*, **57**, 36–40.

Milinski, M. (1984). Competitive resource sharing: an experimental test of a learning rule for ESS's. *Animal Behaviour*, **32**, 233–242.

Nagel, E. (1961). *The Structure of Science: Problems in the Logic of Scientific Explanation.* Harcourt, Brace & World, New York.

Odum, E. P. (1953). *Fundamentals of Ecology*, 1st edition. W. B. Saunders, New York.

Odum, E. P. (1959). *Fundamentals of Ecology*, 2nd edition. W. B. Saunders, New York.

Odum, E. P. (1971). *Fundamentals of Ecology*, 3rd edition. W. B. Saunders, New York.

Ollason, J. G. (1980). Learning to forage — optimally? *Theoretical Population Biology*, **18**, 44–56.

Orians, G. H. (1969). On the evolution of mating systems in birds and mammals. *American Naturalist*, **103**, 589–603.

Parker, G. A. (1970). The reproductive behaviour and the nature of sexual selection in *Scatophaga stercoraria*, L. (Diptera: Scatophagidae). VII, The origin and evolution of the passive phase. *Evolution*, *24*, 774–788.

Parker, G. A. (1978). Searching for mates. *Behavioural Ecology: an Evolutionary Approach*,

1st edition (Ed. by J. R. Krebs & N. B. Davies), pp. 214–244. Blackwell Scientific Publications, Oxford.

Parker, G. A. & Knowlton, N. (1980). The evolution of territory size – some ESS models. *Journal of Theoretical Biology*, **84**, 445–476.

Parker, G. A. & Stuart, R. A. (1976). Animal behavior as a strategy optimizer: evolution of resource assessment strategies and optimal emigration thresholds. *American Naturalist*, **110**, 1055–1076.

Parker, G. A. & Sutherland W. J. (1986). Ideal free distributions when individuals differ in competitive ability: phenotype-limited ideal free models. *Animal Behaviour*, **34**, 1222–1242.

Pierce, G. J. & Ollason, J. G. (1987). Eight reasons why optimal foraging theory is a complete waste of time. *Oikos*, **49**, 111–118.

Pimm, S. L., Rosenzweig, M. L. & Mitchell, W. A. (1985). Competition and food selection: Field tests of a theory. *Ecology*, **66**, 798–807.

Pulliam, H. R. (1974). On the theory of optimal diets. *American Naturalist*, **108**, 59–75.

Regelmann, K. (1984). Competitive resource sharing: a simulation model. *Animal Behaviour*, **32**, 226–332.

Rosenzweig, M. L. (1981). A theory of habitat selection. *Ecology*, **62**, 327–335.

Rosenzweig, M. L. (1985). Some theoretical aspects of habitat selection. *Habitat Selection in Birds* (Ed. by M. L. Cody), pp. 517–540. Academic Press, New York.

Rosenzweig, M. L. (1986). Hummingbird isolegs in an experimental system. *Behavioral Ecology and Sociobiology*, **19**, 313–322.

Royama, T. (1970). Factors governing the hunting behaviour and selection of food by the great tit (*Parus major L*). *Journal of Animal Ecology*, **39**, 619–668.

Schmid-Hempel, P., Kacelnik, A. & Houston, A. I. (1985). Honeybees maximise efficiency by not filling their crop. *Behavioral Ecology Sociobiology*, **17**, 61–66.

Schoener, T. W. (1971). The theory of feeding strategies. *Annual Review of Ecology and Systematics*, **2**, 369–404.

Schoener, T. W. (1974). Competition and the form of the habitat shift. *Theoretical Population Biology*, **6**, 265–307.

Schoener, T. W. (1986). Mechanistic approaches to community ecology: a new reductionism. *American Zoologist*, **26**, 81–106.

Schoener, T. W. (1987). A brief history of optimal foraging theory. *Foraging Behavior* (Ed. by A. C. Kamil, J. R. Krebs & H. R. Pulliam), pp. 5–67. Plenum Press, New York.

Simberloff, D. S. (1980). A succession of paradigms in ecology: essentialism to materialism and probabilism. *Synthese*, **43**, 3–39.

Staddon, J. E .R. (1983). *Adaptive Behavior and Learning.* Cambridge University Press, New York.

Stearns, S. C. & Schmid-Hempel, P. (1987). Evolutionary insights should not be wasted. *Oikos*, **49**, 118–125.

Stephens, D. W. (1981). The logic of risk sensitive foraging preferences. *Animal Behaviour*, **33**, 667–669.

Stephens, D. W & Krebs, J. R. (1986). *Foraging Theory.* Princeton University Press, Princeton, New Jersey.

Sutherland, W. J. (1983). Aggregation and the 'ideal free' distribution. *Journal of Animal Ecology*, **52**, 821–828.

Sutherland, W. J. & Parker, S. A. (1985). Distribution of unequal competitors. *Behavioural Ecology: Ecological Consequences of Adaptive Behaviour* (Ed. by R. M. Sibly & R. H. Smith). pp. 255–274. Blackwell Scientific Publications, Oxford.

Verner, J. & Willson, M. F. (1966). The influence of habitats on mating systems of North

American passerine birds. *Ecology*, **47**, 143–147.

Werner, E. E. & Hall D-J. (1974). Optimal foraging and the size selection of prey by the bluegill sunfish (*Lepomis macrochirus*). *Ecology*, **55**, 1042–1052.

Williams, G. C. (1957). Pleiotropy, natural selection and the evolution of senescence. *Evolution*, **11**, 398–411.

Wynne Edwards, V. C. (1962). *Animal Dispersion in Relation to Social Behaviour*, Oliver & Boyd, Edinburgh.

Wynne Edwards, V. C. (1987). *Evolution through Group Selection*. Blackwell Scientific Publications, Oxford.

11 LEVELS OF ORGANIZATION IN ECOLOGY

R. M. MAY
Department of Biology, Princeton University, Princeton, New Jersey 08544, USA

INTRODUCTION

Recent advances in ecology and evolution reach down to the molecular biology of the gene and up to the interplay between biological and physical processes on a global scale. More than ever, ecological questions must be studied at many different levels and on many different spatial and temporal scales, from the way that evolutionary accident, environment and biomechanical constraints shape the life histories of individual organisms at particular times and places, to the way plate tectonics and global climate changes have helped shape biogeographical realms.

The pursuit of understanding on many different levels and scales creates a variety of problems. It makes it difficult to organize the study of ecology and evolution within universities, where departmental boundaries are usually themselves the result of evolutionary accident. It makes difficulties of selection and organization in ecology textbooks (and these difficulties are not really resolved by North American texts that encyclopaedically include everything, but which then have to be trundled around in a wheelbarrow). Beyond this, the practical problems that ecological science must help us solve arise at every level of organization and on a variety of spatial and temporal scales.

Consider, for example, the problem of releasing genetically-engineered micro-organisms that are designed to control an insect pest or to enhance the productivity of a crop. Such a project will necessarily be based on a molecular-level understanding of one particular aspect of the functioning of the micro-organism in question. But it is essential to think more broadly about other possible effects that may follow from the release of a new living thing (and new genetic material) into the natural environment. Much of the early debate on these issues has been sadly polarized. At one extreme, some molecular biologists seem to be under the misapprehension that 'evolution has tried everything', so that there are no worries. At the other extreme, some Ludditіes seem to be urging that we never pick any flower lest we disturb some star. There is in my mind no doubt that, my over the next few decades, the biotechnological revolution in general — and genetically engineered micro-organisms and crops in particular — will lead to economic and social changes even more marked than those caused over the past few decades by the computer revolution. Before many of these potential gains can be safely realized, however, we must have protocols for testing genetically-engineered organisms, to guard against unforeseen effects upon natural ecosystems. But on what spatial and temporal scale must tests be carried out?

For producing genetically-engineered organisms, the motivating concern is at the level of the molecular biology of the gene, but the emerging problems require us to lift our sights to deal with population interactions and ecosystem properties. Conversely, the answers to many of the problems loosely grouped under the rubric 'global change' ultimately lie at lower levels, in the interaction of populations or even individual organisms with a changing environment. Uncertainties about the long-term effects of the input of CO_2 into the atmosphere by burning wood and fossil fuels, for instance, derive essentially from uncertainties about rates of photosynthetic fixation of carbon by particular categories of plants in specific marine and terrestrial regions.

This essay offers a scattered selection of remarks, all dealing with one or other aspect of the interplay between different levels of approach to ecological questions. I begin with a review of past and present thoughts about the ecology of individuals, populations, communities and ecosystems, and of the relationships between them. Some consideration is given to ways in which dynamic processes at one level can give rise to patterns at a higher level. I then briefly discuss some recent insights and speculations that emerge from explicitly recognizing that evolution has, over the history of life on Earth, acted at different levels, beginning with self-replicating molecules, going on to complex molecular aggregates and cells, and then to individual multicellular organisms (Buss 1987). Next I review the different spatial and temporal scales on which different biological and physical processes take place on land and, rather differently, in the sea. The essay concludes by re-emphasizing the importance of all these considerations to any systematic analysis of the long-term ecological impact of human activities on various scales.

LEVELS OF ORGANIZATION

It is conventional to begin a catalogue of 'levels of organization' with the ecology of individuals. In the earlier years of this century, autecological studies tended to emphasize biomechanical design, seeing organisms as living machines interacting with their physical environment. Characteristic of this approach are the laws — Allen's Law, Bergmann's Law, Leibig's Law and many others — that summarize apparent trends in growth and form within taxa, over geographical gradients (see e.g. Krebs 1983). There is, of course, much excellent work in this vein today (for example, the 3/2 thinning law for plants, or the biomechanical adaptations of organisms to the stresses of intertidal environments). Over the past few decades, however, there has been an enormous growth in studies of the behavioural ecology of individuals and groups, with explicit emphasis on how evolutionary forces have shaped this behaviour. The aim of all this is to understand the life history and behaviour of individual organisms in evolutionary terms. That is, the aim is to reach down from what is usually thought of as the lowest level of organization of ecological studies — the opening chapters of the average textbook, the ecology of individuals — to the underlying theory of evolution.

Before turning my sights in the opposite direction, and discussing the relationships between the dynamics of populations and the behaviour of individuals, I emphasize that there are many interesting and relevant questions at the sub-organismal level, which are essentially ecological ones. The immune system in vertebrates provides an example, and one measure of its importance is that it comprises approximately 3% of the body weight of any average human. In general, the immune response mounted by a vertebrate host to a viral or bacterial infection is characterized by phenomenological parameters, such as the average latent period, the duration of infectiousness and the duration of acquired immunity (often lifelong). In their chapter, Anderson and Hassell (pp. 147–196) show how these individual-level parameters provide a basis for understanding the overall dynamics of associations between populations of hosts and pathogens or parasites, and thence for understanding 'herd immunity' and for designing immunization programmes. But the immune response of an individual organism involves interactions among populations of different kinds of cells, so that the dynamics of an individual's immune system (and its characteristic parameters) can, in principle, be understood on a cellular, or even molecular, basis.

For one thing, antibody diversity in any one individual is largely generated by somatic processes in the first few years of life , at rates that depend somewhat on nutrition. What eventually halts this generation of diversity, setting some limits to the number of different types of antibodies? Jerne (1974) has suggested the process may be akin to those ill-understood processes that determine species diversity in ecosystems, with an increasingly complex web of interactions among idiotypes (or species) ultimately attaining some rough balance where distinct types are, on average, removed at the same rate as others appear. In formulating these ideas, Jerne borrowed explicitly from ideas about stability and complexity in model ecosystems, and he emphasized the evolutionary analogy: each individual is an island, generating its own diverse array of antibody types. The subsequent elaboration of Jerne's notions has been rather abstract (see for example, Farmer *et al.* 1987), and I am not aware of any effort, for example, to calculate the rough number of antibody types in an average individual (around 10^7 to 10^8) from these ideas. But they do indicate that many properties conventionally regarded as pertaining to the fundamental level of individuals can, from a different perspective, be viewed as the properties of 'ecosystems' of interacting populations of cells or molecules.

For another thing, it is possible that many basic features of the immune system may be understood in terms of the dynamic behaviour of populations of cells which stimulate or inhibit each others' reproduction (Hoffmann 1980; Kaufman, Urbain & Thomas 1985). Figure 11.1 illustrates one such dynamical landscape, derived from a deliberately oversimplified model in which the interactions among an infectious agent and different populations of lymphocytes and antibodies are based on empirical observations (Anderson & May 1988). The figure exhibits the essential features of the real immune system. Originally, antibodies to the particular infectious agent are at low levels, but, after being challenged, the system moves permanently to an alternative stable state with an elevated level of the relevent antibody, corresponding to acquired immunity. If, however, the challenge is too weak, the system returns to the 'susceptible' state or 'low zone tolerance'. Note that here acquired immunity is a dynamical phenomenon, resulting from a disturbance (infection) which moves the system to an alternative stable state. More elaborate models can capture further aspects of the immune system. The essential point is that ecological models for the dynamics of the sub-organismal populations that correspond to the immune system might provide insights that go beyond a bewildering taxonomic catalogue of cell types (including putative 'memory cells' to explain lasting immunity). Further steps in such a programme of ecological research require data that are not commonly available, such as the average birth and death rates of different kinds of lymphocytes and antibodies.

After this extended digression on sub-organismal levels of organization that might benefit from a more explicitly ecological or population biological approach, I return to the conventional progression and move up from individuals to populations.

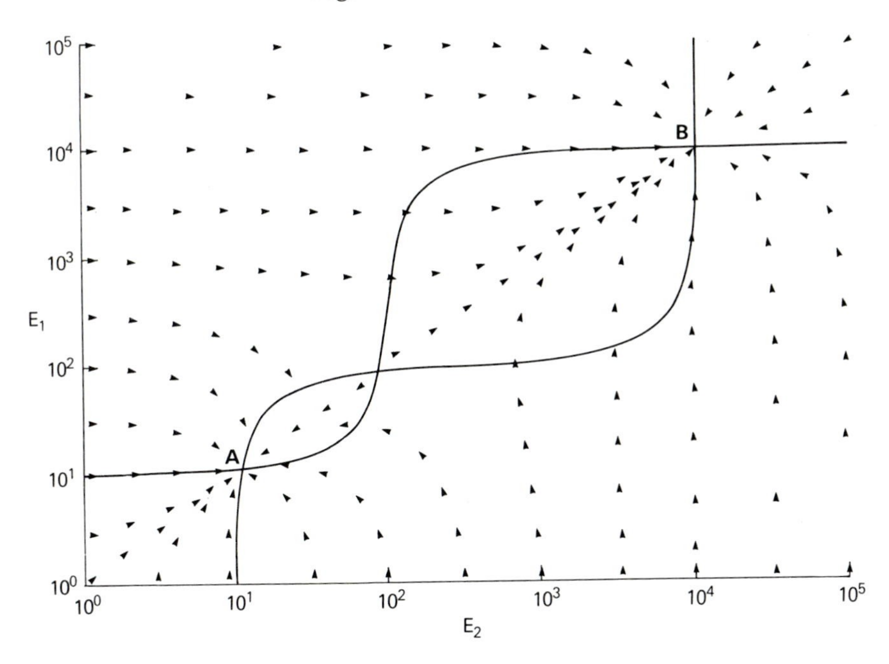

FIG. 11.1 Phase plane, depicting the changes in concentrations of two idiotypes (E_1 and E_2 lymphocytes or antibodies), in a very simple model for the immune system; for details, see Anderson & May (1988). Before being challenged by infection, the system is in the 'virgin state' (idiotypes at low concentrations) labelled A. Significant disturbance by an infectious agent can leave the system in an alternative stable state, labelled B, with antibody levels permanently raised (corresponding to acquired immunity against reinfection). If the infection is sufficiently weak, the system may not cross the 'breakpoint' to the immune state, but may instead return to the point A ('low zone tolerance').

In some simple situations, the properties of populations are indeed derived from those of their constituent individuals. Thus, in the absence of any density dependent effects, overall rates of population growth or decline, r, can be deduced from the age-specific fertility and survival probabilities of individuals, m(a) and l(a), via the Euler equation:

$$1 = \int m(a)\, \mathrm{l}(a) \exp(-\mathrm{r}a)\, \mathrm{d}a$$

(see e.g. Krebs 1983). More commonly, however, we need to deal with aggregations of individuals for which birth, death and movement are partly dependent on population density. The population is then usually treated as if it were some meta-level organism, with the effects of population density being described by phenomenological parameters that are divorced from the underlying behaviour of individuals. Two among many practical examples are the parameter b in the Haldane–Morris relation, $N_{t+1} = \lambda (N_t)^{1-b}$, that has been fitted to so many data on insect populations (Haldane 1953; Southwood *et al.* 1974) and the parameter z in the International Whaling Commission's formula, $R(N) = N[a - b(N/K)^z]$, for the effects of density on recruitment in specific whale stocks (see e.g. May 1980). Such a phenomenological approach undoubtedly has its uses, especially when management decisions are needed immediately while basic understanding of the regulatory mechanisms is still lacking. But if we are ever to understand the dynamics and spatial distribution of populations in evolutionary terms, it is likely to be through an understanding of how the population dynamics derive explicitly from the behaviour of individuals (itself forged by evolution). In 1984, the British Ecological Society devoted a well-attended symposium to this topic of the relation between individuals and populations, but few of the papers actually integrated the two levels of organization (Sibley & Smith 1985).

Some intriguing suggestions do arise from the relatively few studies that have sought systematically to deduce the dynamics of a population from explicit assumptions about the life-history strategy and behavioural ecology of its constituent individuals (e.g. Hassell & May 1985; Lomnicki 1988). For instance, some of this theoretical and empirical work raises doubts about whether we can always elucidate regulatory factors simply by collecting long time-series of data about average population densities (May 1986, Hassell *et al.* 1987). Specifically, suppose we have a population which redistributes itself over many patches in each generation, and where the movement patterns and reproductive behaviour of individuals are sufficiently non-linear that cyclic or chaotic dynamics can arise in any one patch. In the presence of some kinds of environmental noise, it may be impossible to tease out the regulatory signals (derived from the density dependencies in reproductive or movement behaviour) by conventional 'k-factor' or other analysis. These tentative ideas come from the analysis of mathematical models that relate overall population dynamics to underlying individual behaviour in patchy and noisy worlds, but they appear to be borne out by empirical data since 1962 for whiteflies on the leaves of viburnum bushes at Silwood Park (Hassell *et al.* 1987).

Moving onward and upward, there are many empirical and theoretical studies of the interaction among two, three or more populations. Such work seeks to make a beginning in understanding communities as aggregations of interacting populations. Surprises soon appear. Systems of two prey and one predator, or one prey and two predators (or a pathogen and a predator, or two pathogens) can easily produce chaotic dynamics, even without any environmental noise (Gilpin 1979; Anderson & May 1986). Systems where a prey is attacked by a specialist and a generalist predator can exhibit alternative stable states, either or both of which may be cyclic or chaotic. Systems of three competitors, even when described by the excessively simple Lotka–Volterra equations, can give rise to strange cyclic dynamics if one has the intransitive 'scissors/paper/stone' configuration in which A beats B beats C beats A (Gilpin 1975; May & Leonard 1975; Buss & Jackson 1979). Of the many interesting consequences that I must hurry past, one is that intransitive competition may play an important part in maintaining diversity, especially when levels of disturbance are relatively low. In simulations, high levels of disturbance facilitate diversity and thus diminish the relative importance of intransitive competition (Karlson & Buss 1984).

There are many ways in which communities of interacting organisms can be viewed. Paine (1980) has given an incisive review of the strengths and weaknesses of some of the main approaches. One approach, developed at greater length in Lawton's chapter (pp. 43–78), is to attempt to identify all the explicit 'links' between species (or groups of species) in a 'food web'. The goal is then to understand the statics and dynamics of the food web in terms of the number of constituent species/populations and the patterns of connection among them (May 1974; Cohen, 1978, 1988; Pimm 1982; Briand & Cohen 1987). A disadvantage is that all interactions are given equal weight in such a scheme. A second approach is to identify broad patterns of energy flow, hoping to understand the structure and function of the community in terms of the thermodynamic properties of constituent species or groups of species. Here the disadvantage is that some paths may carry little of the system's energy, yet be crucial for its functioning (nectar-consuming pollinators, for example). A third approach is to catalogue patterns in the relative abundance of different species in a community or, more generally, patterns in the overall commonness or rarity of species. The hope is to identify patterns that are characteristic of different ecological circumstances (earlier succession versus climax; undisturbed versus highly disturbed environments), and ultimately to understand how these patterns derive from population properties. It seems to me that the first of these approaches is currently enjoying more success than the other two.

Schaffer & Kot (1985, 1986) and Sugihara *et al.* (1988) offer a variation on this explaining-communities-in-terms-of-populations theme. Consider a time series of data on the abundance of some population, such as a species of plankton in the ocean. This population is likely to be embedded in, and affected by, a community of other biological populations. That is, our time series represents but one variable in a non-linear system possessing many independent variables. It may nevertheless be possible to extract, by phenomenological means, a description of the dynamical behaviour of the population in question without fully understanding the n-dimensional system. This description may be capable of short-term predictions. The technique for doing this has been developed for physical systems by Packard *et al.* (1980), Takens (1981), and others. Even if some multi-dimensional 'strange attractor' underlies the observed time series, this dynamical entity may be reconstructed without any understanding of the fundamental mechanism that generates it, by choosing some fixed time lag, T, and plotting values of the variables $x(t)$, $x(t + T)$, $x(t + 2T)$, . . ., $x(t + [m - 1]T)$, in m-dimensional space; the value chosen for T is not critical, while the value of m is selected so that increasing its value by unity does not apparently result in any additional structure.

Schaffer & Kot (1985, 1986) have applied these techniques to recorded numbers of cases of chickenpox, mumps and measles per month, $N(t)$, in New York and Baltimore before mass vaccination. They construct three-dimensional phase plots of $N(t)$, $N(t + T)$, $N(t + 2T)$, with T fixed around 2–3 months. For all three phase plots, Poincare cross-sections suggest the flows are indeed confined to a nearly two-dimensional conical surface, corresponding to some nearly one-dimensional relationship. Olsen & Degn (1985) have reviewed this work, and given a parallel but independent analysis of measles data from Copenhagen which yields a one-dimensional humped relationship almost identical with that found for measles by Schaffer & Kot. Schaffer (1984) has also provided a similar analysis of the Canadian data on apparent cycles in lynx abundance, arguing that this system also is chaotic and governed by a nearly one-dimensional map. This phenomenological approach is clearly different in spirit from conventional approaches which seek to understand dynamical behaviour in terms of specific models based on underlying biological mechanisms; it is a kind of 'top-down' approach rather than the usual, synthetic 'bottom-up' approach. This work holds the promise of providing new insights. The main problem is that the approach needs longer runs of data than are commonly available to population biologists.

The ecosystem level of organization is concerned with flows of nutrients or energy from primary producers via consumers to decomposers, seeking to understand how such systems work and how they respond to specific perturbations. The ecosystem may be a particular community (e.g. Harper's grassfield at Aber in North Wales) viewed in a different light; it may be some environmental type (e.g. the 'mangrove ecosystem'), or a geographic region (e.g. the Amazonian ecosystem), or even the whole world (the global ecosystem). In this last and grandest interpretation, the entire biosphere may be seen as a superorganism that influences and is influenced by its physical environment of atmosphere and ocean. It is clearly a poetic extravagance to see the film of life clinging to our globe as a fragile Goddess (Gaia: Lovelock 1979), but there may be some truth in this metaphor for the complex web of biological and physical interactions that have made our planet habitable. Although much has been learned in recent years (e.g. Kasting, Toon & Pollack 1988), there are still major uncertainties about how specific disturbances (such as input of CO_2 or of chlorofluorocarbons) may produce global changes, and how abrupt these non-linear processes may be.

HIERARCHICAL STRUCTURE AND EMERGENT PROPERTIES

There have been various attempts to codify formally the interplay among the kinds of understanding that may be attained at various levels of ecological organization, from individuals to ecosystems.

There are, for example, recurring attempts to borrow ideas from statistical mechanics and thermodynamics, where physicists can, by appropriate averaging, deduce many of the properties of macroscopic substances from knowledge of the basic dynamical properties of their constituent molecules. The idea of similarly deducing patterns of relative abundance or of overall fluctuations in biomass within a community, from the vital rates of constituent organisms, is clearly beguiling (e.g. Kerner 1959). The problem is that the techniques of statistical mechanics rest on exact conservation laws (or, equivalently, on exact symmetries or Hamiltonians). Thus ecological analogies can only be found by working, for instance, with populations obeying Lokta–Volterra equations in which the coefficients are exactly antisymmetric. I think this is silly. Other studies seem to me to involve drawing dubious analogies between some biological quantity, such as diversity (defined as

$$\sum p_i \ln p_i ,$$

where p_i is the relative abundance of the ith species) and thermodynamic entropy, and then playing semantic games.

Hierarchy theory draws its inspiration more from the social sciences, particularly the work of Simon (1962). It offers a systematic way of thinking about nested levels of organization, like those discussed in the last section. These ideas appear to give concrete insights in some scientific contexts, as in the hierarchical structure of polymeric materials (Baer, Hiltner & Keith 1987). O'Neill *et al.* (1986) and Sugihara *et al.* (1984) have given lucid accounts of the potential use of hierarchy theory in ecology. Although I am not persuaded that there have been any substantial applications as yet, I do recognize that formal codification of ways of thinking about things can be useful in itself and Houston *et al.* (1988) gives several telling examples where systematic application of the methodology of dynamic optimization theory resolved specific problems in behavioural ecology. In retrospect, the solutions did not demand the methodology of dynamic optimization, but this methodology led routinely to insights that were not so easily found by more existential means.

One major problem that hierarchy theory does highlight, although it offers no systematic solution, is how to aggregate variables and how to define the boundary of the system (e.g. O'Neill & Rust 1979; Sugihara *et al.* 1984). For example, in the synoptic catalogue of 113 food webs compiled to date by Briand, some of the webs comprise individual species as their nodes, others comprise 'collections of organisms that feed on a common set of organisms and are fed by a common set of organisms' (e.g. zooplankton), while yet others mix these different kinds of categorization [Briand & Cohen (1987) and references therein]. While I doubt it, the possibility exists that some of the patterns found in these diversely-assembled food webs may derive in part from the psychology of how people go about organizing collections, and here it is not irrelevent to recall Miller's (1956) classic finding that people tend consistently to recall 7 ± 2 objects from lists of arbitrary length. The problem of putting some sensible bound on the ambit of a study is also often difficult: in a study of an intertidal community, does one include the owl that eats the odd mouse that eats the odd mollusc? And how do we deal with Southwood's 'tourist' species, that wander through from time to time, sometimes having non-trival impact? All these questions await theoretical, much less practical, resolution.

There is one generalization that I do find useful about the formation of pattern at one level of organization (or on a large scale) by mechanisms operating at a lower level (or on a more local scale). This body of work is often called 'reaction-diffusion theory'. In essence, it shows how patterns can be formed without any detailed blueprints, provided the underlying pattern-generating mechanisms have two basic features: short-range activation (that stimulates production or enhances local differences) and long-range inhibition (that reduces production or retards the spread of disturbances). For reviews, consult Murray & Oster (1984), Levin &

Segel (1984) and Murray (1988). The patterns in question may be developmental processes (resulting from the dynamics of cell division and movement), markings on shells or hides (resulting from the dynamics of biochemical agents), distributions of prey and predator populations (resulting from the dynamics of their interaction, with prey as activators and predators as inhibitors), geographical distributions of gene frequencies (resulting from the interplay between selection and migration: see Fig 11.2), or any other of a host of phenomena. As Meinhardt (1982) has emphasized in a developmental context, a great variety of different mechanisms may account for the short-range activation, and for the long-range inhibition; what is essential is that both these basic features be present. Levin (1988) has stressed the corollary '. . . the fact that these two characteristics are all that are needed to produce a wide range of patterns makes clear the impossibility of discovering processes from pattern; quite distinct underlying processes can give rise to identical sets of patterns'.

Most of the applications of these thoughts about patterns generated by underlying mechanisms of reaction and diffusion are to problems involving spatial scale, a topic which is discussed below. The basic idea has been outlined here, however, because it does show how some ecological patterns in complex systems may be derived from generic kinds of mechanisms at a lower level of organization.

LEVELS OF ORGANIZATION IN EVOLUTION

It is commonly thought that no matter what level of ecological organization we are dealing with, any evolutionary explanation must ultimately be developed at the level of individuals, because it is on individuals that selective forces act. Buss (1987) has, however, recently argued that things are more complicated, and that the history of life on Earth is itself a history of selection operating at successive levels of organization. In hierarchical fashion, new self-replicating units have appeared as elaborations or aggregations of the self-replicating sub-units contained within them. Self-replicating molecules created self-replicating complexes which gave rise to simple cells. Cells became more complicated by acquiring organelles and other internal structures, and finally multicellular individuals appeared. Buss's focus is on developmental phenomena, which he places squarely within an evolutionary framework. This sets the stage for specific evolutionary predictions about the workings of development, cell structure, the organization of the genome and (most interesting for our present purposes) the observed diversity of life-history strategies.

Buss begins by elaborating a theory of developmental processes in

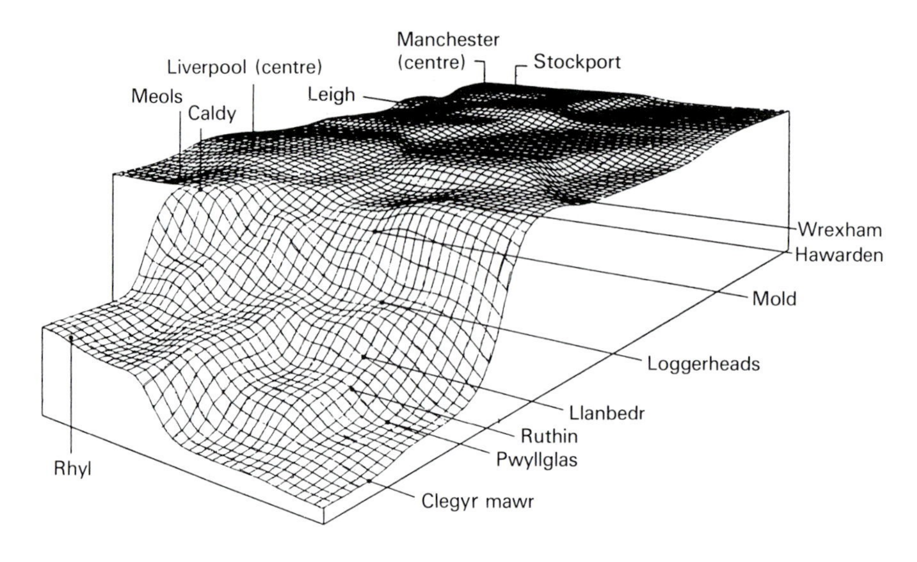
Manchester (centre)
Stockport
Liverpool (centre)
Meols
Caldy
Leigh
Wrexham
Hawarden
Mold
Loggerheads
Llanbedr
Ruthin
Pwyllglas
Rhyl
Clegyr mawr

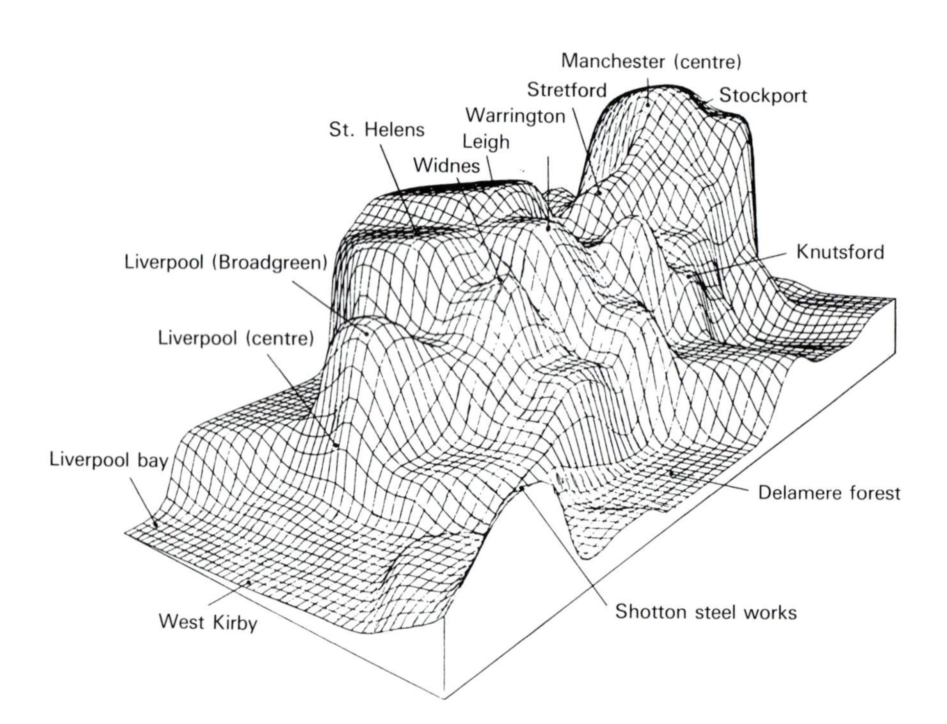
Manchester (centre)
Stretford
Stockport
Warrington
St. Helens
Leigh
Widnes
Liverpool (Broadgreen)
Knutsford
Liverpool (centre)
Liverpool bay
Delamere forest
West Kirby
Shotton steel works

metazoans. On the one hand, he argues that the conservatism of early ontogeny (as evidenced by patterns of cleavage, gastrulation and mosaicism) is the outcome of conflicts and comprises between selection at the level of the individual and selection at the level of the cell lineage, all subject to the ancestral constraint that the protist groups which gave rise to metazoans bore only a single, uncommitted microtubule-organizing centre per cell. Turning on the other hand to the diversity of late ontogeny, Buss notes that (following a variable period of maternal control of the fate of embryonic cells) metazoan embryos are organized into one of a discrete number of 'bauplans', as a result of interactions among embryonic cell lineages. In all these interactions, one cell lineage acts to restrict the replication of another, while enhancing its own replication. Buss thus sees the establishment of metazoan 'bauplans' as mechanisms that resolve potential conflicts. Subsequent evolutionary changes must then come primarily from changes in the relative timing of developmental events (called 'heterochrony').

By systematically applying his ideas about developmental processes as the evolutionary outcome of an ancient transition from the cell as a unit of selection to the multicellular organism as a unit of selection, Buss provides a compelling explanation for the different developmental pathways found in the animal, plant and fungal Kingdoms. He shows how his ideas about ontogeny can provide a basis for restricting the number of qualitatively different kinds of life cycles found in nature. As an example, he lists various categories of options among which life-history choices may be made. Cell architecture may be coenocytic, or cellular with rigid or non-rigid walls; reproduction may be sexual, primitively asexual or secondarily asexual; the division of labour may be unicellular, or multicellular with or without cellular differentiation; germ-line sequestration may be present or absent; a histo-recognition system may be absent, or present in various forms; and various choices of ploidy are possible, including alternation. This catalogue generates some 1000 or so distinct life cycles. The rules Buss deduces from the interplay

---

FIG. 11.2 Gene frequency of the melanic allele as a function of geographical location in the Industrial Midlands in Britain, for two different species of peppered moth exhibiting industrial melanism. Both patterns are the outcome of selection for the melanic allele near cities (and against it on cleaner trees in the countryside), opposed by gene flow associated with moth movement; that is, both are 'reaction–diffusion' patterns. In (a) the moths of this species characteristically move relatively large distances, tending to produce a smooth distribution; the moth species in (b) tends to exhibit significantly less movement, producing a more textured pattern. [For details, see Bishop & Cook (1975).]

among units of selection and from the role of development in limiting elaboration, however, successively reduce the number of possible life cycles. For example, sex is a necessary precondition to cellular differentiation; germ-line sequestration precludes primitive asexuality; non-rigid cell walls or coenocytic conditions require histo-recognition; vegetative diploidy is decreed by non-rigid cell walls; and so on. In this conceptual example, Buss thus ends up with only twenty-seven possible life cycles (for a slightly different number see Adams 1982), which is well within the range of life-cycle diversity seen in reality. In short, Buss's work on the consequences of evolution acting at different levels may tell ecologists important things about the limits to diversity of life histories.

Buss focuses on the transition from selection at the level of the cell to selection at the level of the individual. This is, however, only one of the grand transitions in levels of selection that mark the evolutionary record. Other transitions must have preceded it, and some would argue that selection at the superorganismal level has in some situations succeeded it. Gould (1982), for example, has argued that natural selection should be viewed as operating simultaneously on levels above and below the individual, from gene through clade. Wilson & Sober (1988) consider a population which is structured into groups, such that selection operates both within groups and between groups. They then ask what will be the fate of a mutant trait that reduces selection within groups but increases the variation among groups. Such a trait may spread, not by increasing its frequency within groups (as is usual), but rather by increasing the productivity of the groups that contain it, relative to other groups. In this way, we can have selection for group or 'higher level' properties, without any mysticism or violation of evolutionary rules. Wilson and Sober discuss the evolution of some such 'superorganisms', as exemplified by colonies of social insects or phoretic associations (in which mites, nematodes, fungi and microbes that specialize in patchy and ephemeral habitats ride around in multispecies associations with insects that lay eggs on carrion or dung). The extent to which these ideas about evolution at levels above the individual are relevant to an understanding of the dynamics and organizational structure of populations, communities and ecosystems is, to my mind, not yet clear. But, after reading Buss's book, I am less sceptical than I was.

SCALING IN SPACE AND TIME

Cutting across any discussion of ecological processes, whether at the level of individuals, populations or ecosystems, are questions of the spatial and temporal scales on which processes operate.

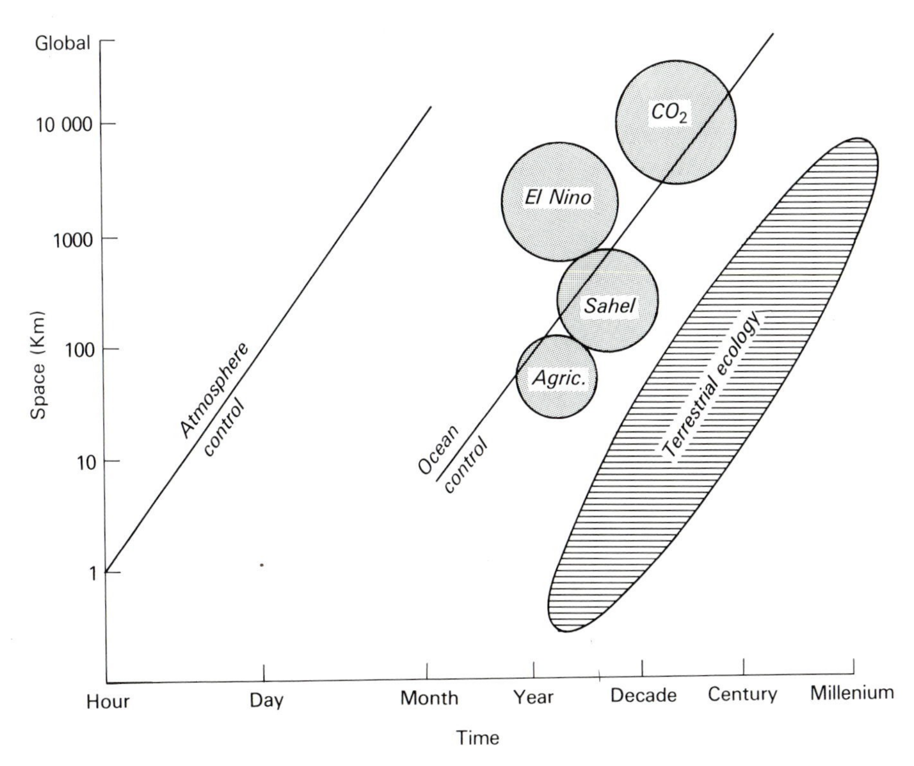

FIG. 11.3 Schematic representation of characteristic spatial and temporal scales for physical and biological processes on land and in the sea. Events with large spatial scales generally have associated with them long time scales: physical processes in the atmosphere exert control over time scales of weeks or less; physical processes in the ocean exert control over years or more. As discussed more fully in the text, space–time scales for biological processes in the ocean tend to coincide with those for physical processes; this is not the case for terrestrial ecosystems, where physical processes in the atmosphere tend to represent external noise. For further dicussion, see the text. [This figure is after Clark (1985), Steele (1985) and Powell (1988), where much more detailed accounts are given.]

Figure 11.3 attempts to give a schematic impression of the relationship between scales on which physical and biological processes operate in space and time. The patterns are clearly different on the land and in the sea.

For the ocean, the relationship between the spatial and the temporal scales for typical physical processes were first sketched by Stommel (1963). Haury, McGowan & Weibe (1978) portrayed the corresponding scaling laws for variability in biomass of ocean plankton, and Steele (1978), Powell (1988) and others have elaborated this 'Stommel Diagram' to include other biological processes in marine food chains. As indicated by the line labelled 'ocean control' in Fig. 11.3, there is coincidence between the way variability in physical and in biological processes scale with space and time in the sea. This is almost surely because organisms in the marine environment have evolved adaptations to the ever-present dynamics of the ocean; populations must accept and ultilize the patterns of variability dictated by the physics of the ocean (Steele 1985).

Conversely, for terrestrial environments, the scaling for physical (atmospheric) processes and for natural ecological processes are very different. As shown in Fig. 11.3, the physical environment fluctuates on much shorter time scales than do physical processes in the ocean; weather, as an atmospheric phenomenon, ranges from local events (at scales of km per hour), such as thunderstorms, to cyclonic systems (at scales of 1000 km per day). But in natural ecosystems on the land, the processes governing rates of change, such as soil development and the consequent growth of forest and grassland systems, take place on slower time scales and at smaller space scales than do ecological processes in the ocean (see the region labelled 'terrestrial systems' in Fig. 11.3). For such terrestrial ecosystems, it is obviously appropriate to regard the atmospheric variability as short-term noise, which is very different from the typical situation in the ocean. For a more detailed account of these ideas, see Clark (1985).

As Steele (1985) has emphasized, physical processes operating in the ocean, with their characteristically longer spatial and temporal scales, are essentially the flywheel for a larger system encompassing land, sea and atmosphere. As the 'atmosphere control' line in Fig. 11.3 suggests,our ability to predict the weather has an upper limit of about 10 days. Beyond this timescale, we necessarily become involved in the interactions between atmosphere and ocean, which are essentially represented by the 'ocean control' line in Fig. 11.3, and which are exemplified by El Niño.

In the past, terrestrial ecosystems (with their characteristically slow timescales and small space scales) could be regarded as reacting to climatic changes rather than driving, or interacting with them. Forest composition and human activities passively tracked the ice ages. More recently, however, the capacity for human activity to influence global systems on relatively short timescales may have shifted us from passive reaction, into the truly interactive domain. As indicated in Fig. 11.3, changes in the CO_2 cycle, or events such as the Sahel drought, may be contemporary examples where the scope of human activities has moved us into a new domain of scaling relations between biological and physical processes.

On longer timescales, the flow of nutrients from the land to freshwater, thence to estuaries and coastal waters, and finally to the deep ocean are increasingly being seen as a driving force in climate change. We are currently accelerating such flows. We do not yet have a clear understanding of the consequences of such changes in the characteristic coupling among land, sea and atmosphere, but in Steele's (1985) words it depends, among other things, on '... an understanding of ocean dynamics, particularly at time scales of decades. This, in turn, requires a knowledge of events in the wider spectrum for days to centuries; from near-surface eddies to the general deep circulation; from the basic physics to the biological processes'.

Implications for experimental design and environmental management

As indicated in Fig. 11.3, and explained in detail by Haury, McGowan & Weibe (1978), Steele (1978, 1985), Clark (1985) and others, we usually have a good idea of the appropriate spatial and temporal dimensions for studying large-scale ecological processes in the sea. But as we move down to the level of individuals and populations, especially on land, the appropriate scales of space and time are often uncertain.

Earlier theoretical models for ecological processes tended to assume spatial homogeneity. It is now widely recognized that spatial heterogeneity is often crucial for understanding the behaviour of individuals, populations or communities. In plant ecology and forestry, ideas about gap dynamics help explain how communities or ecosystems form intricate mosaics over large areas (e.g. Pickett & White 1985; Hubbell & Foster 1987; Horn, Shugart & Urban 1988). Levin & Paine (1974) have independently developed similar ideas for intertidal communities. The dynamics of shifting patches are increasingly seen as essential in understanding the persistence of associations between plants and phytophagous insects or between prey and predators or parasitoids (Hassell & May 1974; Chesson 1986; Kareiva & Odell 1987). Indeed, it may be that associations between plants and large vertebrate herbivores (such as elephants) are a shifting mosaic of recovering vegetation, vegetation currently being exploited and thus degraded, and fully-recovered vegetation awaiting the next cycle of depredation, all on spatial and temporal scales vastly larger than might be expected intuitively from the 'trivial movements' (*sensu* Moran & Southwood 1982) of elephants (May & Beddington 1981). A laboratory metaphor for much of this is Huffaker's (1958) classical experiment in which predatory mites searched for prey in a patchy archipelago of oranges separated by greasy seas. At the orange level the system is wildly unpredictable, but as the universe is made larger and larger the overall prey–predator association persists longer. In this metaphor, as in many of the plant–herbivore associations which are of concern to conservationists, the scale (the number of oranges, as it were) necessary for long-term persistence is simply unknown; we do not know how to relate the population-level processes to the time and size scales of individual behaviour.

Questions about the appropriate scales for studying or managing a population, community or ecosystem have many practical implications. In the Introduction, I outlined the obvious problems involved in designing research protocols addressed to the safe release of genetically-engineered organisms. Many practical problems in conservation biology, whether at the level of individuals, populations, communities or ecosystems, are essentially problems of determining the appropriate scale of a reserve in order for it to serve its designated purpose (Noss 1985; Bourgeron 1988).

Well-designed experiments of a manipulative kind are, of course, most easily done on a relatively small spatial scale, and over relatively short times. Figure 11.4 shows Kareiva & Anderson's (1988) survey of manipulative field experiments published in *Ecology* between 1980 and 1986, as a function of the maximum linear dimension involved in the study and the number of replicates. This figure includes only studies that addressed population- or community-level questions (excluding autecological, demographic or evolutionary studies), and worked with 'real organisms' (i.e. insects or larger, and excluding plankton or micro-organisms). The results in Fig. 11.4 are striking. Half the studies have a maximum linear dimension of 1 metre or less. Furthermore, the larger scale experiments, especially those over a characteristic linear size of 10 metres, are poorly replicated.

Figure 11.4, as such, should not disturb us. It is understandable that well-designed and well-replicated manipulations in the field should be more easily accomplished on small spatial (and temporal) scales, and there are plenty of interesting questions on these scales. However it is possible that one of the factors producing the trends manifested in Fig. 11.4 is a dogmatic belief, held in some circles, that 'real science' necessarily means manipulative experiments, of the kind fondly imagined to pervade physics. These are actually found in introductory physics courses, though not always in the real and messy world of astrophysics or elementary particle physics. Insofar as such factors play a part in shaping the patterns in Fig. 11.4, we might worry. We should resist the temptation to find some problem that can be studied on a

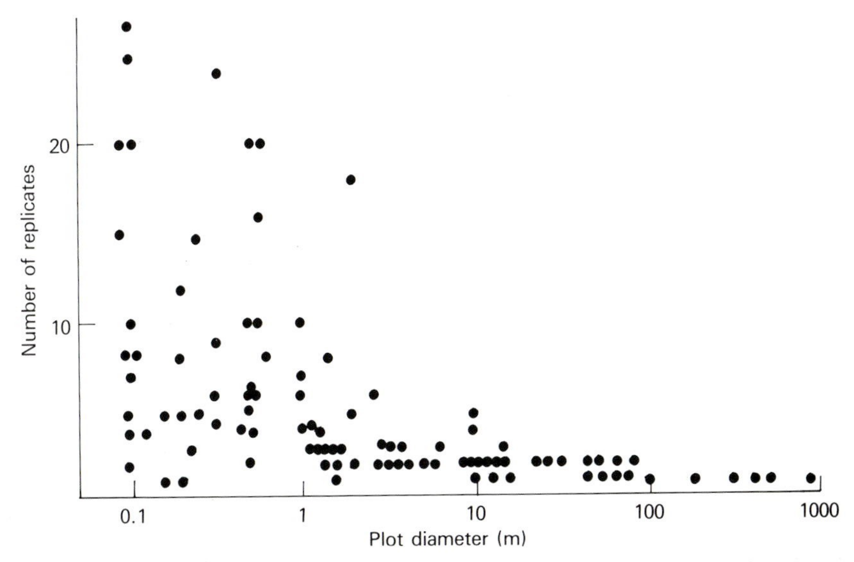

FIG. 11.4 Relative abundance of manipulative field experiments in community ecology, as functions of the characteristic physical dimensions of the study and of the number of replicates (see text for full discussion). Each point is taken from a different published paper in *Ecology*, between January 1980 and August 1986. [After Kareiva & Anderson (1988).]

convenient scale, but rather should strive to identify the important problems and then ask what is the proper scale on which to study them, and how such studies might be carried out if the scale is large.

Fractal geometry of nature

Until recently, discussions of scale in relation to ecological processes rested on conventional Euclidean assumptions about the geometry of the natural world. Coastlines had a length, and areas scaled as $(\text{length})^2$. Mandelbrot (1977) (see also Loehle 1983; Sugihara 1988) has, however, shown us the complexities of asking 'how long is the coastline of Britain?' The answer is not simple. If we measure on a 10 km scale using an atlas, we get one answer. On a 1 m scale walking around it we get another, larger answer. Thus the coastline of Britain is not one-dimensional, but has a 'fractal dimension', D, such that the perceived length, $L(\lambda)$, depends on the step-length of measurement, λ, according to

$$L(\lambda) = c\,\lambda^{1-D}. \tag{1}$$

Here c is a constant. If $D = 1$, we have the familiar Euclidean situation where L is independent of the measuring scale. If $D = 1.5$ then a ten-fold reduction in the measurement scale (from 10 to 1m) will result in the apparent length increasing by a factor of $10^{0.5} \simeq 3$.

We can now proceed to measure the fractal dimension of the coastline of Britain and other countries, or of a landscape pattern, or of foliage on a plant. One operational scheme is to cover the object under investigation with successively finer meshes (effectively traversing the coastline on different step lengths), and essentially determine D from equation (1). An alternative procedure involves determining the relation between perimeter P and area A (for areas of different sizes), and deducing D from the scaling relation $P = cA^{D/2}$. In this way it is found that most coastlines, including that of Britain, have fractal dimension around 1.3.

The fractal dimension itself, however, may vary, having different values at different scales. This is likely to happen, for example, if one kind of ecological process underlies the landscape pattern up to some limiting scale, and a different process determines the landscape pattern at larger scales. Under these circumstances, such a change in the fractal dimension may tell us something about the underlying physical and/or biological processes. Even without such understanding, a change in fractal dimension of this kind may at least help us decide the appropriate scale of experiment or management.

Three examples may help make these ideas more concrete.

Hastings *et al.* (1982) have examined the cumulative frequency distribution of various size patches of two kinds of vegetation (one characteristic of a later stage of succession than the other) in the Okefenokee Swamp in the southeastern USA. From these data, they deduce the fractal dimensions for the patch patterns of these two different kinds of vegetation, discovering that it is larger for the earlier successional stage. The empirical finding is consistent with simulation studies carried out by these authors.

Krummel *et al.* (1987) evaluate the fractal dimension D of deciduous forest patterns in the Natchez Quadrangle in Mississippi (USGS 1973), using the 'perimeter/area' method. This region has experienced conversion of forest cover to crop land relatively recently. They found a marked ($P < 0.001$) discontinuity in D, with $D \simeq 1.20 \pm 0.02$ at small scales and $D \simeq 1.52 \pm 0.02$ at large scales; the discontinuity occurred at areas around 60–70 hectares. Krummel *et al.* suggest their results stem from human processes predominating at smaller scales (making for smoother geometry, and D closer to unity), while natural processes continue to predominate at larger scales. These results encourage the hope that such analyses may help us to evaluate changes on the Earth's surface using remote sensing data.

For the third example, I turn to a much smaller ecological scale. As part of a broader study of how numbers of individual phytophagous insects scale with their physical size, Morse *et al.* (1985) measured the fractal dimension of various kinds of vegetation at different scales, finding that D was typically around 1.5, but that it tended to have larger values at smaller scales (e.g. for vine leaves, the figure is $D \simeq 1.55$ at small scales and $D \simeq 1.30$ at large scales). This is in the opposite direction from the landscape example just given, but such a trend can be seen to follow generally from the branching structure of vegetation (Horn, personal communication).

A different and less familiar implication of ideas about fractal geometry is that many actual distribution patterns, in space or time, may not be described well by smooth distributions derived essentially from Gaussian and like distributions, which are well characterized by means and variances. Rather, many real distributions are likely to be, as it were, jagged on every scale. As these ideas are developed further, they may well affect the way ecologists think about, and collect data on, the spatial and temporal distribution of populations.

CODA

The main theme of this paper, which has kept recurring amidst the many digressions, is that different spatial and temporal scales are revelant to different ecological processes. There is some rough correlation between levels of ecological organization and such scales, with the behavioural ecology of individuals typically understandable on smaller scales than are required in studies of communities or ecosystems. But there are no universal rules, and the overall persistence of many populations may require vastly larger areas that would be guessed from the characteristic scale of individual movement.

The interactions between the ecological drama, at levels ranging from individuals to ecosystems, and the physical dimensions of the stage on which it is played out, are important in understanding how the world works. They are also important for practical problems such as the release of genetically engineered organisms, the management of forests, fisheries and other resources, the conservation of endangered species and communities and the overall patterns of global change in response to inputs of CO_2 and other materials. I hope this paper makes it plain that while ecologists have a lot to learn about the natural world, they have much to contribute to the understanding of environmental problems that are currently on accelerating trajectories.

ACKNOWLEDGMENTS

I am indebted to R. M. Anderson, H. Hastings, H. S. Horn, P. Kareiva, S. A. Levin and particularly to G. Sugihara for their help. This work was supported in part by the NSF, under grant DMS87–03503.

REFERENCES

Adams, D. (1982). *Life,the Universe and Everything.* Pan Books, London.

Anderson, R. M. & May, R. M. (1986). The invasion, persistence and spread of infectious diseases within animal and plant communities. *Philosophical Transaction of the Royal Society,* **B 314,** 533–570.

Anderson, R. M. & May, R. M. (1988). *Infectious Diseases of Humans: Dynamics and Control.* Oxford University Press, Oxford.

Baer, E., Hiltner, A. & Keith, H. D. (1987). Hierarchical structure in polymeric materials. *Science,* **235,** 1015–1022.

Bishop, J. A. & Cook, L. M. (1975). Moths, melanism and clean air. *Scientific American,* **232,** 90–99.

Bourgeron, P. S. (1988). Advantages and limitations of ecological classifications for the protection of ecosystems. *Conservation Biology,* **2,** 218–219.

Briand, F. & Cohen, J. E. (1987). Environmental correlates of food chain length. *Science,* **238,** 956–960.

Buss, L. W. (1987). *The Evolution of Individuality.* Princeton University Press, Princeton.

Buss, L. W. & Jackson, J. B. C. (1979). Competitive networks: nontransitive competitive relationships in cryptic coral reef environments. *American Naturalist,* **113,** 223–234.

Chesson, P. L. (1986). Environmental variation and the coexistence of species. *Community Ecology* (Ed. by J. M. Diamond & T. J. Case), pp. 240–256. Harper & Row, New York.

Clark, W. C. (1985). Scales of climate impacts. *Climatic Change,* **7,** 5–27.

Cohen, J. E. (1978). *Food Webs and Niche Space.* Princeton University Press, Princeton.

Cohen, J. E. (1988). Food webs and community structure. *Perspectives in Ecological Theory* (Ed. by J. Roughgarden, R. M. May & S. A. Levin), pp. 181–202. Princeton University Press, Princeton.

Farmer, J. D., Kauffman, S. A., Parkard, N. H. & Perelson, A.S. (1987). Adaptive dynamic networks as models for the immune system and autocatalytic sets. *Annals of New York Academy of Sciences,* **504,** 118–131.

Gilpin, M. E. (1975). Limit cycles in competition communities. *American Naturalist,* **109,** 51–60.

Gilpin, M. E. (1979). Spiral chaos in a predator–prey model. *American Naturalist,* **113,** 306–308.

Gould, S. J. (1982). Darwinism and the expansion of evolutionary theory. *Science,* **216,** 380–387.

Haldane, J. B. S. (1953). Animal populations and their regulation. *New Biology,* **15,** 9–24.

Hassell, M. P. & May, R. M. (1974). Aggregation of predators and insect parasites and its effect on stability. *Journal of Animal Ecology,* **43,** 567–594.

Hassell, M. P. & May, R. M. (1985). From individual behaviour to population dynamics. *Behavioural Ecology* (Ed. by R. Sibly & R. Smith), pp. 3–32. Blackwell Scientific Publications, Oxford.

Hassell, M. P., Southwood, T. R. E. & Reader, P. M. (1987). The dynamics of the viburnum whitefly (*Aleurotrachelus jelinekii*): a case study on population regulation. *Journal of Animal Ecology*, **56**, 283–300.

Hastings, H. M., Pekelney, R., Monticciolo, R., Vun Kannon, D. & Del Monte, D. (1982). Time scales, persistence and patchiness. *BioSystems*, **15**, 281–289.

Haury, L.R., McGowan, J. A. & Weibe, P. H. (1978). Patterns and processes in the time–space scales of plankton distribution. *Spatial Pattern in Plankton Communities* (Ed. by J. H. Steele), pp. 277–328. Plenum, New York.

Hoffmann, G. W. (1980). On network theory and H–2 restriction. *Contemporary Topics in Immunobiology*, Vol. **11** (Ed. by N. Warner), pp. 185–226.

Horn, H. S., Shugart, H. H. & Urban. D. L. (1988). Simulators as models of forest dynamics. *Perspectives in Ecological Theory* (Ed. by J. Roughgarden, R. M. May & S. A. Levin), pp.256–267. Princeton University Press, Princeton.

Houston, A., Clark, C., McNamara, J. & Mangel, M. (1988). Dynamic models in behavioural and evolutionary ecology. *Nature*, **332**, 29–34.

Hubbell, S. P. & Foster, R. B. (1987). The spatial context of regeneration in a neotropical forest. *Colonization, Succession and Stability* (Ed. by A.J. Gray, M. J. Crawley & P. J. Edwards), pp. 395–410. Blackwell Scientific Publications, Oxford.

Huffaker, C. B. (1958). Experimental studies on predation: dispersion factors and predator–prey oscillations. *Hilgardia*, **27**, 343–383.

Jerne, N. K. (1974). Toward a network theory of the immune system. *Annales d' Immunologie (Institute Pasteur)*, **125C**, 373–393.

Kareiva, P. & Anderson, M. (1988). Spatial aspects of species interactions: the wedding of models and experiments. *Theoretical Community Ecology* (Ed. by A. Hastings). American Mathematical Society, Providence, Rhode Island (in press).

Kareiva, P. & Odell, G. M. (1987). Swarms of predators exhibit 'preytaxis' if individual predators use area restricted search. *American Naturalist* **130**, 233–270.

Karlson, R. H. & Buss, L. W. (1984). Competition, disturbance and local diversity patterns of substratum-bound clonal organisms: a stimulation. *Ecological Modelling*, **23**, 243–255.

Kasting, J. F., Toon, O. B. & Pollack, J. B. (1988). How climate evolved on the terrestrial planets. *Scientific American*, **258(2)**, 90–97.

Kaufman, M., Urbain, J. & Thomas, R. (1985). Towards a logical analysis of the immune response. *Journal Theoretical Biology*, **114**, 527–561.

Kerner, E. H. (1959). Further considerations on the statistical mechanics of biological associations. *Bulletin of Mathematical Biophysics*, **21**, 217–255.

Krebs, C. J. (1983). *Ecology: The Experimental Analysis of Distribution and Abundance*, 3rd edition. Harper & Row, New York.

Krummel, J. R., Gardner, R. H., Sugihara, G. & O'Neill, R. V. (1987). Landscape patterns in a disturbed environment. *Oikos*, **48**, 321–324.

Levin, S. A. (1988). Pattern, scale, and variability: an ecological perspective. *Perspectives in Ecological Theory* (Ed. By J. Roughgarden, R. M. May, & S. A. Levin). pp. 242–255. Princeton University Press, Princeton.

Levin, S. A. & Paine, R. T. (1974). Disturbance, patch formation and community structure. *Proceedings of National Academy of Sciences*, **71**, 2744–2747.

Levin, S. A. & Segel, L. A. (1984). Pattern generation in space and aspect. *SIAM Review*, **27**, 45–67.

Loehle, C. (1983). The fractal dimension and ecology. *Speculations in Science and Technology*, **6**, 131–142.

Lomnicki, A. (1988). *Population Ecology of Individualists.* Princeton University Press, Princeton.

Lovelock, J. E. (1979). *Gaia: A New Look at Life on Earth.* Oxford University Press, Oxford.

Mandelbrot, B. B. (1977). *Fractals: Form, Chance, and Dimension.* W. H. Freeman, San Francisco.

May, R. M. (1974). *Stability and Complexity in Model Ecosystems,* 2nd edition. Princeton University Press, Princeton.

May, R. M. (1980). Mathematical models in whaling and fisheries management. *Some Mathematical Questions in Biology,* Vol. **13** (Ed. by G. F. Oster), pp. 1–64. American Mathematical Society, Providence, Rhode Island.

May, R. M. (1986). When two and two do not make four: nonlinear phenomena in ecology (the Croonian Lecture). *Proceeding of Royal Society,* **B 228**, 241–266.

May, R. M. & Beddington, J. R. (1981). Notes on some topics in theoretical ecology, in relation to the management of locally abundant populations of mammals. In *Problems in Management of Locally Abundant Wild Mammals* (Ed. by P. A. Jewell & S. J. Holt), pp. 205–216. Academic Press, New York.

May, R. M. & Leonard, W. J. (1975). Nonlinear aspects of competition between three species. *Society for Industrial and Applied Mathematics. Journal on Applied Mathematics,* **29**, 243–253.

Meinhardt, H. (1982). *Models of Biological Pattern Formation.* Academic Press, New York.

Miller, G. A. (1956). The magical number seven plus or minus two: Some limits on our capacity for processing information. *Psychology Reviews,* **63**, 81–97.

Moran, V. C. & Southwood, T. R. E. (1982). The guild composition of arthropod communities in trees. *Journal of Animal Ecology,* **51**, 289–306.

Morse, D. R., Lawton, J. H., Dodson, M. M. & Williamson, M. H. (1985). Fractal dimension of vegetation and the distribution of arthropod body lengths. *Nature,* **314**, 731–732.

Murray, J. D. (1988). How the leopard gets its spots. *Scientific American,* **258**, 80–87.

Murray, J. D. & Oster, G. F. (1984). Cell traction models for generating pattern and form in morphogenesis. *Journal of Mathematical Biology,* **19**, 265–279.

Noss, R. F. (1985). On characterizing presettlement vegetation: how and why. *Nat. Areas J.,* **5**, 5–19.

Olsen, L. F. & Degn, H. (1985). Chaos in biological systems. *Quarterly Reviews of Biophysics,* **18**, 165–225.

O'Neill, R. V., DeAngelis, D. L., Waide, J. B. & Allen, T. F. H. (1986). *A Hierarchical Concept of Ecosystems.* Princeton University Press, Princeton.

O'Neill, R. V. and Rust, B. (1979). Aggregation error in ecological models. *Ecological Modelling,* **7**, 91–105.

Packard, N. H., Crutchfield, J. P., Farmer, J. D. & Shaw, R. S.(1980). Geometry from a time series. *Physical Review Letters ,* **45**, *712–716.*

Paine, R. T. (1980). Food webs: linkage, interaction strength and community infrastructure. *Journal of Animal Ecology,* **49**, 667–685.

Pickett, S. T. A. & White, P. S. (1985). *The Ecology of Natural Disturbance and Patch Dynamics.* Academic Press, New York.

Pimm, S. L. (1982). *Food Webs.* Chapman & Hall, London.

Powell, T. M. (1988). Physical and biological scales of variability in lakes, estuaries, and the coastal ocean. *Perspectives in Ecological Theory* (Ed. by J. Roughgarden, R. M. May & S. A. Levin) pp. 157–176. Princeton University Press, Princeton.

Schaffer, W. M. (1984). Stretching and folding in lynx fur returns: evidence for a strange attractor in nature? *American Naturalist,* **124**, 798–820.

Schaffer, W. M. & Kot, M. (1985). Nearly one dimensional dynamics in an epidemic. *Journal of Theoretical Biology,* **112**, 403–427.

Schaffer, W. M. & Kot, M. M. (1986). Differential systems in ecology and epidemiology. *Chaos* (Ed. by A. V. Holden), pp. 158–178. Princeton University Press, Princeton.

Sibley, R. M. & Smith, R. H. (1985). *Behavioural Ecology,* Blackwell Scientific Publications, Oxford.

Simon, H. A. (1962). The architecture of complexity. *Proceedings of American Philosophical Society,* **106**, 467–482.

Steele, J. H. (1978). *Spatial Pattern in Plankton Communities.* Plenum, New York.

Steele, J. H. (1985). A comparison of terrestrial and marine ecological systems. *Nature,* **313**, 355–358.

Stommel, H. (1963). Varieties of oceanographic experience. *Science,* **139**, 572–576.

Southwood, T. R. E., May, R. M., Hassell, M. P. & Conway, G. R. (1974). Ecological strategies and population parameters. *American Naturalist,* **108**, 791–804.

Sugihara, G. (1988). Applications of fractals in ecology. *Trends in Ecology and Evolution* (in press).

Sugihara, G. (rapporteur) *et al.* (1984). Ecosystems dynamics. *Exploitation of Marine Communities* (Ed. by R. M. May), pp. 131–153. Springer–Verlag, New York.

Sugihara, G., Palka, D., Tont, S., Trombola, A. & Tynan, C. (1988). Hierarchical decomposition of strange attractors in phytoplankton populations. (under review)

Takens, F. (1981). Detecting strange attractors in turbulence. *Lecture Notes in Mathematics,* **898**, 366–381.

USGS (1973). Land use and land cover digital data tape for the 1:250,000 Natchez, Mississippi Quadrangle. US Geological Survey, Reston, Virginia.

Wilson, D. S. & Sober, E. (1988). Reviving the superorganism. *Journal of Theoretical Biology* (in press).

AUTHOR INDEX

Page numbers in italics indicate the main entries in the reference lists.

SUBJECT INDEX

Page numbers in italics indicate figures and/or tables.